PSYCHOLOGY OF

DECEPTION

ANALYSIS OF SEXUALLY PSYCHOPATHIC SERIAL CRIME

Don JACOBS

AUTHOR OF THE FORS RUBIC & FORENSIC PSYCHOLOGY
AMONG TEXAS COLLEGES

Kendall Hunt
publishing company

Cover images:
Forest image © Vicki France, 2009. Used under license from Shutterstock, Inc.
Woman image © Ostill, 2009. Used under license from Shutterstock, Inc.
Skull image © Pokaz, 2009. Used under license from Shutterstock, Inc.

Kendall Hunt
publishing company

www.kendallhunt.com
Send all inquiries to:
4050 Westmark Drive
Dubuque, IA 52004-1840

ISBN 978-0-7575-6515-1

Printed in the United States of America
10 9 8 7 6 5 4 3 2 1

Contents

About the Author

Don Jacobs is the architect of the FORS academic rubric allowing students to sample forensic science courses in the first two years of college or university studies. From 2004 to 2008, his forensic science labs were offered on a trial basis at Weatherford College as unique needs courses. Soon, other campuses followed and forensic science began to dot the academic landscape from Amarillo to North central Texas.

On August 8th, 2008 (8–08–2008), the Texas Coordinating Board gave final approval for both courses—FORS 2440 *Introduction to Forensic Science* and FORS 2450 *Introduction to Forensic Psychology*—as ACGM (Academic Course Guide Manual) transferable courses eligible for state funding and articulation to a variety of university programs.

Psychology of Deception: Analysis of Sexually Psychopathic Serial Crime (The POD) is written to accompany the lab manual *Brainmarks: How the Brain Marks Pathways in Life featuring Headquarters for Things that Go Bump in the Night.* A test bank and instructor manual accompanies both text and lab manual, available upon adoption.

Don Jacobs has been a full faculty member and/or behavioral science department chair for the past twenty-five years. He has written numerous textbooks reflecting his passion for neuropsychology. He offers nationally known speakers connected to Forensic Science Conferences. Please send all comments or questions to djacobs@wc.edu.

Note to Colleagues

One of my more assertive psychology majors once asked "what I made?" He rubbed his fingers together in a common gesture meaning money. After I told him the relatively modest amount WE ALL MAKE as educators I retreated to my office to prepare a keynote speech I had been paid a handsome stipend to present. "What do I make?" still resonated within my inner most thoughts. Like a flash, it occurred to me those of us who chose education as our career care less about 'what we make' and more about 'what we do.' While a carefully constructed budget will take care of *what we make,* a gift far greater than accounting makes all the difference in the world in *what we do.*

What we do in education is to help mold a *thriving mind out of a surviving brain.* We nurture the two most important organs in the body to excel in life—the brain and the heart. The more I have discovered about the adolescent brain, the more I am sure of the importance of *what we do.* The more we know about the workings of our marvelous brain, the more we can do for the heart—courage—to untangle the pretzel of living an adventurous and fulfilling life. The most important years of formal schooling is highlighted by the rapid *growth and pruning back* of the dynamic brain. Do we lose what we don't use? Yes, absolutely.

So, "What Do We Make?" With knowledge of the developing brain and its immense potential, *we make minds ready to achieve far more than the owner ever imagined. By that standard, I say we make a fortune!*

Don Jacobs (2009)

Prologue

Netherworld

"Netherworld—my term for a culture of toxic, often predatory, parenting—has always existed. Usually, it existed behind closed doors and is an example of deception characterized by severely *dysfunctional family milieus.* This type of parenting produces children who become deceptive themselves when out in public. They may become violent human predators—what society calls serial killers—and what forensic psychologists label sexual psychopaths. Camouflaged by societal persona *calculated for deception,* sexual psychopaths are prowlers intent upon *rapacious behavior*—stalking live prey for sexual thrills. As an end result, few, if any, safe neighborhoods exist today as they restlessly troll upscale neighbors as well as lonely communities along dirt roads. Where can children run and play with innocent abandonment? Without parental vigil, can such a *utopia* (meaning—*no place*) exist?

Roaming freely in society, killing a string of strangers sequentially, serial killers have made citizens uptight and almost panicky. For good reason: they target children and young women. Yet, they move through society hidden in plain sight. *They don't appear to be what they really are.*

Toxic parenting aside, netherworld has created a class of desensitized citizens. An increasing number of individuals (many of them parents) in the age group most populated by serial killers (early twenties to almost forty years of age), put their heads in the sand and choose to ignore the reality that not all parents are competent, and nurturing caregivers.

Sexual psychopaths stalk and kill young girls and women, often in multiple states across America before capture. They stalk children in department stores waiting to see if adult chaperons look elsewhere, even for a few seconds. They watch checkout lines for a quick steal and rapid exit though sliding doors to an accomplice waiting at curbside with the motor running.

Sexually psychopathic serial killers never stop. They have to be found and extracted from society like a bad tooth.

Psychology of Deception: Analysis of Sexually Psychopathic Serial Crime is an academic text with *POD Readings—a* collection of brilliant essays by top students showing the diversity of interdisciplinary topics available in modern forensic psychology with insight into the crimes of society's most elusive predators. In a section called *Predator Profiles, selected serial killers and their crimes are presented and analyzed.*

Netherworld, toxic family milieus where children are conditioned *to be fearful of parents,* has become a permanent social milieu that can no longer be ignored. We must address what makes some families so dysfunctional. In the meantime, we must become more vigilant if we want our children to survive and thrive. The reality of toxic, netherworld parenting may move society into futuristic legal benchmarks such as *granting of a parental license to have children.* Are parents mentally healthy enough to raise children? Do they have enough financial resources to support them? Legally, we cannot drive a car without a license. Perhaps in the near future we may not be able to have children without one either.

Don Jacobs (2009)

Fact

Nature's Deceptive Nurturing

Nature has nurtured some of her selected species with three seemingly underappreciated abilities, unless of course, the species in question is self-aware of the survival agenda of camouflage, regeneration, and metamorphosis. *Camouflage* allows predators to hide in plain sight. They blend into the surrounding environment so well—rocks, trees, or vegetation. Predators are permitted to stalk, sneak around, and blindside unsuspecting prey. Simultaneously, camouflage also allows prey to hide in plain sight from predators. Nature is full of curve balls. She stands as the *original pattern for deception.*

Does camouflage ability suggest self-awareness? Do animals know they are virtually hidden? Considering what we know today about animal intelligence, it is most likely true. The splendid white polar bear hides his black nose with its great paw with razor-sharp black claws clutched beneath. Unsuspecting prey see pure white snow ahead until, of course, they see red.

Due to color and markings, a big cat—the lynx—blends in with surrounding tree bark; prey have no idea what's happening until fang and claw tear them to shreds. Meanwhile, a species of butterfly looks like a green leaf as it sits upon leafy twigs.

Want further proof of nature's power? Look at regeneration. A bird dives into the path of a crustacean—a crab, lobster, or crayfish. It grabs one of the crabs' ten legs in its strong beak and takes to the sky, only to lose its prey seconds later, left "holding the leg," so to speak. The crab frees itself by losing the trapped leg, only to *grow it back in about two weeks.* Fish, salamanders, and some mammals show regenerative ability. Humans have regenerated fingertips, ribs, and entire livers with as little as twenty-five percent of the original organ left to sprout another.

The most spectacular example of *nature's changing room* is biological metamorphosis where species conspicuous and abruptly morph—*changing their entire physical form* to something else. What once was a repulsive worm becomes a beautiful butterfly.

Do we ever really know what stands before us?

POD Forecast

Must We Be Perpetually Afraid & Paranoid?

Privately, does everyone have a Mr. Hyde, yet in public personify the essence of respectability as Dr. Jekyll? How shocked are we when we see how easily we can be duped? As citizens, must we be perpetually afraid and paranoid of everyone we meet?

The literature is replete with examples of the deception that surrounds sexually psychopathic serial killers. Those close to Ted Bundy, for instance, were absolutely amazed when Bundy was first arrested as the prime suspected in the murders of several college coeds. "They have the wrong guy" remarked an acquaintance who was also an attorney. Bundy would finally admit to over thirty of the most sexually sadistic crimes in the history of sexual psychopathy.

A friend of John Wayne Gacy offered to extend to Gacy employment upon his parole for a sexual indecency charge he felt was wrongly visited upon his friend; the same Gacy who would eventually have over thirty bodies discovered buried in the crawlspace of his home.

Ed Gein, ostensibly a gentle Wisconsin farmer, gained the trust of neighbors who often asked him to watch their children. This was the same man who would fashion furniture coverings and trinkets out of the skin and bone he ripped from victims. Gein became the postmodern *archetypical boogeyman* inspiring such film classics as *Psycho* and characters such as Leatherface in *Texas Chainsaw Massacre.* But, who knew at the time a monster resided behind his gentle and slow-witted persona.

The psychology of deception is fully engaged in sexually psychopathic serial crime as predators appear to be one thing yet turn out to be another.

Deception is not new to forensic psychology. For example, it is important to understand *malingering* as a deceptive strategy. In the DSM (the Diagnostic and Statistical Manual of Mental Disorders—the manual used by clinical psychologists) malingering is defined *as the intentional production of false or grossly exaggerated physical or psychological symptoms, motivated by external incentives such as avoiding military service, avoiding work, obtaining financial compensation, evading criminal prosecution, or obtaining drugs* (2002). A defendant may intentionally fake a mental illness or may be exaggerating the degree of symptomatology.

What about *competency to stand trial?* In such cases, a forensic psychologist is appointed by the court to examine and assess the individual who may be in custody or recently released on bail. Based on the forensic assessment, the psychologist recommends whether or not the defendant is competent or incompetent to stand trial.

If the defendant is considered incompetent to proceed, a report will include recommendations for the interim period during which an attempt at restoring the individual's competency to understand the court and legal proceedings, as well as participate appropriately in their defense, will be made. If the individual is deemed incompetent to stand trial and if competence is not regained after a suitable period of time, he or she may be involuntarily committed, on the recommendation of the forensic psychologist, to a psychiatric treatment center until such time as a panel of clinical forensic psychologists and psychiatrists agree competency has returned.

What about the perpetrator's state of mind at the time of the offense? Was he or she impaired? Should he plead not guilty by reason of insanity? In these cases, the presiding judge may appoint forensic psychologists or psychiatrists while defense attorneys are busy locating their own "hired guns" to tip the verdict in their favor.

Once found guilty, can the person's *state of mind, past history of abuse, or mental illness* be used to mitigate the sentence? This may be absolutely true relative to his medical history, family history of abuse, violence, or trauma that may have triggered a *violent brainmark*—soon to be addressed in the lab manual *Brainmarks* accompanying this text. Might the person re-offend upon release? Would the person be a good candidate for rehabilitation? Is rehabilitation possible for a person who has never been habilitated? Forensic psychology has evolved into an important branch of neuropsychology—psychology at the tissue level—sure to become one of the most important disciplines in history as it attempts to peer deep within the brain to tease away the pretzel of sexually psychopathic serial crime.

UNIT I

Deception & Sexuality

CHAPTER ONE

Psychology of Deception

There Is No Such Thing as a Freak Accident

Author Nick Jacobs co-wrote a novel of suspense—*Freak Accident* (Jacobs & Jacobs, 2002)—in which he exposed an entire community of shape-shifter werewolves living as humans in the far west Texas community of Mt. Kyle, near Lajitas, Texas. (Mt. Kyle is fictional, Lajitas is not). Although literary fiction, how true is it that individuals may not always be who or what they claim to be? Might the clandestine side—Mr. Hyde—be violent, predatory, and serial? The novel delves deep into the *psychology of deception*. As we will document, *deception is all around us hidden in plain sight.*

Peter Marx, the rich developer of the mountain top community of Mt. Kyle turned out to be the Alpha Male werewolf—a fierce enforcer and recruiter for his shape-shifter trainees. He had worldwide ties to similar groups. To see him in person—rich, handsome, and impeccably dressed—the only hint of his *beastly side* was an undeniable *animal magnetism.* Even in his transformation, Marx was far from the stinky wolf-man with matted, oily hair, and yellow teeth.

Julie, the Brewster County deputy sheriff, turned out to be the Alpha Female—chief nurturer of the clandestine society and recruiter of adolescent trainees. Again, the fierce she-wolf lived beneath the flesh of a beautiful, intelligent, and articulate woman. Who could have guessed her true identity?

The entire wolfen society was growing steadily according to a carefully constructed strategic plan going back centuries with the eventual goal of replacing the human population with shape-shifter wolves. After all, the shifters—also known as "Lycans" or "Loopers"—had been, secretly, Homo sapiens' chief rivals for centuries, but due to lack of adaptation fused into the evolution of their brain's prefrontal cortexes, the shifters had taken a back seat to humans until, through meticulous effort, they perfected shape-shifting. Homo sapiens had no idea they existed and would prove to be their chief rivals.

The one event that uncovered the entire society of deception in Mt. Kyle was a freak accident of nature—an un-forecasted blizzard that caused a recent recruit—"Glass Eye" Eddie Hawkins—to morph into a fierce wolf. Without the insight and courage of a single lodger—Dr. Shadow Darwin—the remaining lodgers would have perished. Today, how many *sexually psychopathic serial predators—serial killers* in FBI lingo—walk around freely in society only to be exposed upon capture as a violent *predatory psychopath* completely absent of remorse and conscience—a cold-blooded killer—equivalent to a wolfen predator.

Psychology of Deception: Analysis of Sexually Psychopathic Serial Crime presents violent and sexual serial predators as *masters of deception.* Forensic neuropsychologists armed with neuroimaging studies of the brain (neuroscans) have become a new breed of scientist showing how the brain is "marked" for violence in sexually psychopathic crimes. What they are showing in courtroom presentations across the nation suggest that typical citizens should develop a *healthy skepticism of others* until they have moved well beyond being a mere acquaintance.

What is a psychopath?

> **Psychopathic personality:** an emotionally and behaviorally disordered state characterized by *clear perception of reality* except for the (1) *individual's social and moral obligations,* and often by the (2) *pursuit of immediate personal gratification in* (3) *criminal acts,* (4) *drug addiction, or* (5) *sexual perversion.*
>
> —Webster's Collegiate Dictionary (10 Ed.)

View to a Kill

In 1998, after leaving his halfway house to spend the night on the street, twenty-one year old Donta Page followed a young woman to her home. When the woman, twenty-four year old Peyton Tuthill, left her apartment for a job interview, Page broke into her apartment to steal something to sell. On her return, she surprised him. A fierce physical battle ensued. After he subdued her, he tied her hands and arms with an electrical cord. Then he left. Peyton was safe. But, Page changed his mind. He returned.

When she saw him, she cried and begged for her life. He slit her throat to stop her screams. He cut her hands and wrists to stop her from fighting. Then, he stabbed her repeatedly until he killed her. Upon his arrest after the murder, he confessed to the horrific crime.

According to forensic psychiatrist Dr. Dorothy O. Lewis *(Guilty By Reason of Insanity),* Donta Page was a *walking textbook* in the clinical evaluation of severe child abuse. Repeatedly, his mother physically abused her son. As early as first grade he appeared markedly withdrawn. By second grade, his inactivity in the classroom called for a mental health examination. By age ten, medical records showed repeated sodomy. Warnings—red flags—of his growing *psychopathology and psychopathy* showed he desperately needed intervention and treatment by a caring adult.

According to neurologist Jonathan Pincus, *violent criminals* may have had horrific home experiences (and maybe they didn't), but all share observable *extensive prefrontal lobe damage* to the brain courtesy of high-resolution brain scans. Neuroscientist Adrian Raine of USC performed a brain scan of Donta Page's prefrontal cortex, and the results showed minimal functioning of the vital area noted for cognitive and emotional restraint—the prefrontal cortex within the frontal lobes of Page's brain—where it was remarkable by its blue color, or "cool-coded," a term suggested by Jacobs (2003), to describe the blue color of PET Scans due to scarcity of blood flow, showing *decreased prefrontal cerebral activity*.

Can severe damage to regions of the prefrontal cortex within the human frontal lobes—the seat of restraint and strategic thinking—unleash violent behavior? Neuroscientist Adrian Raine found frontal lobe damage in forty-one out of forty-one murderers in his landmark study. The intrusion of *forensic neuroscience* into criminal trials has significantly changed courtrooms. Now, jurors can see evidence of the damaged brains of the accused.

Pat Tuthill, the mother of Peyton Tuthill, believes Donta Page murdered her daughter because "he is evil."

With neuroscience firmly in the middle of the courtroom and on the minds of the jurors, would Donta Page's documented brain damage save him from lethal injection? At the end of the trial, jurors returned a verdict of guilty on all counts. Subsequently, the trial judge sentenced Page to life without possibility of parole. In this case, the defense of prefrontal neurological damage mitigated the judge's ruling of first-degree murder.

Innocence Destroyed

In 1995, twenty-three year old Jeremy Skocz (pronounced "Scotch") committed a vicious and inhuman crime. He murdered a four year old child. In his own childhood, Skocz spent five horrific years with parents who were heroin-addicted and convicted felons. They never attempted to hide their antisocial lifestyle from their son.

Every day he lived amid hateful, unloving, and drug-addled parents. When Skocz reached seven years old, a healthy family adopted him. The family, in his words, seemed almost "too perfect." Everything appeared to turn around for the young boy, but something went tragically wrong. When walking home through a park at age fifteen, an assailant violently attacked him with a baseball bat in the face and head. With an already damaged brain (from the emotional backlash of harsh toxic parenting), his head trauma permanently and radically *changed his personality*. In the following weeks, he became increasingly withdrawn, depressed, and brooding.

Without warning, Skocz left the safety of his adopted home to find his birth father. He found him in his house trailer smoking a joint. Nothing had changed. Nevertheless, Skocz remained in the trailer with his birth father and soon began using hard drugs. Although he looked normal, he barely kept the lid on his growing rage, frustration, and disappointment about feeling different.

The Cox family lived next door to Jeremy's father. One evening near dusk, four year old Shelby Cox disappeared from her front porch. Worried, her parents quickly formed a search party. Jeremy volunteered in the search. In a shocking surprise, after being questioned by police, Skocz led authorities directly to the shed were he had hidden her body. Criminal investigation showed Skocz bound her with duct tape to prevent her from running away and raped her while suffocating her to death with duct tape. He stuffed her small body into a garbage bag and covered it with a gym bag. He then stuffed her body into a small cabinet.

When asked by an investigator in prison, "Why did you kill her?" Skocz replied with a distant look and with an expressionless face (known as *blunt affect*), "I don't know." After a long pause he continued, "I can't answer that question." As we will see, neurological damage can be spotted by blunted affect.

After the murder, a series of neurological scans of Skocz showed the same pattern of "cool-coded" prefrontal lobe damage found by Raine in his study of the forty-one murderers. Neuroscientists concluded that, due to the (1) emotional abyss of antisocial, toxic parenting (our term is *predatory parenting*), (2) the attack on his head from the baseball bat, and (3) the use of hard drugs in attempts to self-medicate, Skocz's brain had become severely and irreparably damaged.

A forensic psychologist testified, "Skocz lived in a broken body."

The jury accepted evidence of brain damage mitigating cognitive *premeditation* required for first-degree murder. Jeremy Skocz will spend the rest of his life in prison, signaling the presence of neuroscience in the courtroom. Apparently, juries are no longer blinded by science; rather, they are embracing its technology by incarceration of violent offenders and throwing away the key.

Although not serial killers by definition, would Donta Page and Jeremy Skocz have become serial predators if not for their capture and subsequent incarceration? The same question will be posed in an upcoming chapter for adolescent killers who kill classmates and adults in home and school place violence. Before we look closer into the mind of predatory psychopaths, let's explore a short history lesson in *folklore* by addressing the world's first serial killers.

Folklore—The First Serial Killers

Harking back to the novel *Freak Accident,* obscured by centuries, displaced by time and *zeitgeist,* and far from the modern practice of criminal profiling—exist instances of society's first serial killers (sexual psychopaths). In rural areas of France and Germany during the sixteenth and seventeenth centuries, vicious, apparently motiveless murders occurred along country roads and pastureland. Sometimes weekly, farmers discovered bodies of mutilated children and adults. As the body count steadily rose, fear turned to outrage. Hunting parties tried to track the vile predators, but without modern profiling what did the crime scene disclose? What could they find? How could they analyze it to suggest the identity of the UNSUB (*unknown subject* in FBI lingo)? What about the physical characteristics of the predator? How could they know his motivation? What did villagers have to go on, other than bloodhounds and anger?

Remarkably, deranged human males, the world's first serial killers, became known as snarling, devil-spawned werewolves. The word **lycanthropy** refers to a modern psychiatric condition characterized by patients who imagine themselves wolves. They may howl at the moon, run around on all fours, and imagine a heightened sense of smell. The *lycanthropes,* documented by historical records, savagely attacked victims, tearing them to shreds, and repeating their crimes in *sequential fashion with a period of inactivity* (the "cooling off period" observed in *serial rape* and *serial homicide.*)

Apparently, some lycanthropes could alter their appearances with wolfen characteristics (such as growing facial hair and filing their teeth to a sharp, canine point). A French hermit, Gilles Garnier, stalked and killed numerous children and then cannibalized their corpses. Stalked like an animal himself by angry villagers, Garnier died at the stake as a werewolf in 1573, thus preventing further instances of serial savagery.

Over a twenty-five year period another self-proclaimed werewolf, Peter Stubbe, terrorized adults and children around Cologne, Germany in the sixteenth century. According to records, Stubbe could change his shape (hence the term *shape shifting* in werewolf literature) displaying . . . "eyes great and large, which in the night sparkled like brands of fire; mouth great and wide, with sharp and cruel teeth, a huge body and mighty claws" (Everitt, 1993).

Like Garnier, Stubbe stalked his prey and staged a bloody, savage attack just like a real wolf attacking prey. In another practice reflecting his deranged mind, Stubbe repeatedly committed incest with his own daughter and recruited her to help him lure unsuspecting victims, an instance of becoming a *compliant co-offender.*

Hunted and eventually captured, Stubbe's execution took place in the following way:

> ". . . strapped to a wheel, ten hot pincers were used to tear away his flesh in ten places, then his arms and legs were crushed with the head of an axe, his head decapitated and his body burned. His decapitated head was put on the top of a pole bearing the picture of a wolf" (Everitt, 1993).

In our enlightened society, convicted serial rapists and murderers—modern werewolves—sit on death row for approximately ten years on average, eat three hot meals a day, and receive free medical and dental attention. Victim advocacy groups ask "why?"

The fear of wolves killing children and livestock in rural areas may have influenced the perception of wolfen characteristics to the unkempt appearance of both Garnier and Stubbe. The vile killers suffered from *psychotic delusions* imagining themselves as demonic wolves. It is true that wolves, in fact all predatory wild animals, are *serial predators*—they kill sequentially as a mere act of survival.

Only human predators derive emotional and *sexual thrills* from stalking, dominating, and controlling other persons, then from deciding when they will die. Analyzed by modern psychiatric criteria, Garnier and Stubbe would be likely diagnosed with *psychosis* driven by sexual obsessions and/or delusions of murderous fantasy. An alternative diagnosis of paranoid schizophrenia might apply. Today, experts know that *most serial killers are not psychotic nor are they schizophrenic*—they *know exactly what they're doing,* and that's the real scary part.

The study of *modern serial homicide began in the nineteenth century* with the murders of prostitutes by Jack the Ripper in Whitechapel, London in 1888. Both the appearance of serial killers and the method of punishment (electrocution, lethal gas, lethal injection, or incarceration) have changed drastically.

Beyond Folklore-Today's Existential Reality

Today, serial killers (technically: sexual psychopaths) remain partially human, mostly animal. They rape and murder mostly strangers, not for survival, but for macabre fascination to temporarily satisfy their sexual perversion and the compulsive desire to control life and death in unsuspecting victims. Controlling, manipulating, and dominating unsuspecting victims characterize their pronounced compulsivity, since normal pursuits have left them unsatisfied.

Rapacious behavior describes their actions—they seek to *prey upon live and unsuspecting victims.*

The psychological payoff is the thrilling high of perverted sexual self-gratification often marked by *sexual experimentation.* To the world, the sexual psychopath is known as a serial rapist or serial killer. Upon capture, he must be incarcerated for life without the possibility of parole, or be put to death. Because of brain rewiring similar to what transpires in addiction, there exists no treatment or therapy.

He lives in a scary, inner world where every aspect of rape and/or killing is *sexualized*—to the last detail. Plain and simple, *he is a sexual deviant and a lifelong sexual predator.* In addition, he is a sexual addict who tries to convince others that he is deeply religious or a model citizen. Actually, sexual psychopaths are predatory **nihilists**—they do not value life. In a small number of cases, they are diagnosed as paranoid schizophrenic (such as David Berkowitz, "The Son of Sam"); seldom are they **psychotic.**

As studies continue to show, he never becomes anything more than a cowardly, pathetic loser, who stalks and preys upon unsuspecting victims for sexual gratification—a gratification ultimately unfulfilling, unlike the sense of impassioned lovers who fulfill each other's emotional and sexual needs. From the neuropsych perspective, the *midbrain limbic system* exacerbated by regions of the brainstem merge to produce *compelled behavior* as obsessive-compulsive behavior emblazoned by unrelenting *brutal urges that are not suppressed by regions of the prefrontal cortex.*

Just as no one wakes up one morning and decides to become a serial killer, **antecedent** events from horrific childhoods rewire his brain and thus diminish the freedom in his life. According to Jacobs (2006) and others, sexual predators' *abusive histories* come from several sources, namely

- hateful, antisocial, toxic parenting (or predatory parenting), where parents prey upon their own children in abusive ways
- addiction to pornography and/or deviant Web sites
- perverted neurocognitive mapping—powerful "thinking maps" of behavior, (in some cases, such as Ed Gein, sexual repressions stemming from a hateful and domineering mother contribute to predatory brain mapping)
- addiction or poly-addiction (two or more drugs), most notably alcohol
- a pronounced *biological predisposition* toward violence soon to be addressed and amplified in the *Brainmarks Lab Manual*

How many times have we witnessed children who come from horrific home milieus with the most horrific conditions imaginable (or who have a history of drug abuse and other social problems), but they turn out to be relatively normal with perhaps a little therapy? So, *bad treatment alone*—even physical or sexual abuse—simply cannot create the violent monsters we observe today. There's always more.

The Neuro School (or neuropsychology) of which this text and accompanying lab manual are based upon, has accumulated considerable evidence that show *biological predispositions* to violence when coupled with *severe parental mistreatment* producing a disconnect with society and peers can lead to severe personality maladjustment that "mark" the brain for violence.

Precision *psychometric* instruments such as the Hare Psychopathy Checklist (HPC) document what brain scans show in real time that a violent psychopath's brain is wired differently.

Predatory Behavior

At night, he prowls under any moon, and he may strike in broad daylight on a busy street. He's cunning, deliberate, and becoming increasingly arrogant. In his **narcissism,** he feels much smarter than those who pursue him. His victims don't know him, but according to notorious serial killer Ted Bundy, the stalker "knows his subjects very well." Often, he visits them in dreams. Or he brushes up against them in a crowded mall in a moment of *frottage.*

He chooses the time, and he chooses the place. He bets his victims are naïve, unsuspecting, and trusting. He hopes their parents either have ignored them or hovered around them (as "helicopter" parents) for so long that they ignore signs of fear. He wagers they never thought it would happen to them. He bets he will catch them off guard, completely unaware of his plans.

He stalks his victims from afar and near. With the mind of a compulsive drug addict and the face of a human masking a reptile, he's the modern version of the werewolf profiled as a serial rapist or serial killer. Perhaps as many as fifty or more at any given time prowl in North America alone, and his numbers are growing.

Now. . . and Then

The young musical artist raises the microphone to his lips in the semi-dark studio and sings the first words of his song as his band accompanies the words with music. Barely visible, audience members sing along with the well-known words. Band members move around the stage as only musicians can move to the cadence of music. The studio comes alive with movement and rhythm. The singer is Brandon Boyd. The song is "Warning" by the alternative rock group *Incubus.* The year is 2003.

In another place and another time, the singer in the studio could have been Charles Manson. Like Boyd, he loved music. He loved to sing. He loved to play the guitar. He associated with other musicians, such as Dennis Wilson of the *Beach Boys.* In 1969, he envisioned rock star status. For certain, Brandon Boyd, creative force behind his group, and Charles Manson, incarcerated mastermind of his fanatical "family," comprised mostly of teenage girls, are galaxies apart. Yet, interesting similarities exist.

Charismatic and engaging, Boyd writes his own music. Journalists who have interviewed Manson comment that he remains charismatic to a new generation of followers. He still writes his own music. More than any other person, Manson brought worldwide attention to his *brainwashed,* largely teenage family judged ultimately responsible for the Tate-LaBianca murders in Los Angeles in the late 1960s. The Manson Family will forever live in infamy.

Crushed by failure, the son of a prostitute who spent most of his youth in prison, Manson became bitter when his dream of a recording contract never materialized. Boyd portrays an accomplished artist, Manson represents the failed manipulator. Wealthy and free to move about society, Boyd recently recorded "The Morning View Sessions" in a California mansion near the Pacific.

Manson remains incarcerated in a cramped cell for the rest of his ruined life. He, too, resides in California. Boyd has tattoos. Manson has a swastika carved on his forehead right between his eyes. Boyd respects and nurtures life through his music; Manson disrespected life and, in his paranoid, nihilistic view of the world, predicted *Helter Skelter*—an out-of-control race war. Manson preyed upon others and took everything from them; in some instances, he took even their lives. The victims never saw it coming in the hot August nights of Southern California.

Had Manson persisted in his musical dreams, he might have succeeded. "The Manson Family" might have referred to his backup group instead of his mind-controlled murderers, who will spend the remaining years of their lives in prison. Boyd—a creator; Manson—a destroyer. Each an artist, yet each took a different path. Why? This text attempts to collect pieces of the puzzle of violent crime. What does *forensic neuropsychology* know about the development of sexual psychopaths? In the process, how does the mousetrap—the criminal profile—aid law enforcement in catching this perverted and elusive predator?

Sexual Psychopathy

The *Diagnostic and Statistical Manual of Mental Disorders* (DSM) (2002) Cluster B Personality Disorders (PD) diagnose violent offenders with antisocial personality disorder (APD). Yet, crime scene investigators (CSIs) encounter at crime scenes the *profundity of pathology* suggesting *sexual perversity* of the perpetrators simply does not adequately describe the deviance of this disgusting monster hiding behind a human face. By any stretch of the imagination, we are dealing with a sexual psychopath, not simply an antisocial criminal.

The condition known as psychopathy has experienced a rather interesting transformation.

According to Professor Robert D. Hare of the University of Vancouver, the world's acknowledged scholar on psychopathy, "although *psychopathy* is closely associated with antisocial and criminal behavior, psychopaths are *qualitatively different* from others who routinely engage in criminal behavior."

Words used to describe disorders continue to evolve, according to Jacobs (2006)—

> "It is predictable that the highly political *American Psychiatric Association (APA)* and the *World Health Organization (WHO)* would leave academics, clinicians, and educators completely confused over *diagnostic criteria* used to define, categorize, and diagnose potentially *violent individuals* by a charade of clinical taxonomies—(1) psychopath, (2) sociopath, (3) antisocial, and (4) dissocial PD. Experimental neuroscience (cutting-edge *clinical research* utilizing high resolution brain imaging, valid and historically reliable *psychometrics* (such as Hare PCL) is not so confounding, conflicted, or politically obvious. Hard evidence from research is not to be confused with clinicians who must abide by the inaccurate criteria in both DSM and ICD-10 manuals. Research clearly demonstrates time and again that *psychopathy* is *qualitatively different* from (1) *antisociality* or (2) sociopathy and (3) dissociality (both blatantly useless terms.) It's time for the so-called experts to wake up, and smell the species! Colloquially, the psychopath is more like a monkey's uncle (biologically predisposed) and the antisocial a societal thug."

According to the **DSM** (Diagnostic and Statistical Manual of Mental Disorders) (2002), the clinical reference book used by clinical psychologists and psychiatrists, the diagnostic feature of the **Antisocial Personality Disorder** reveals a pervasive pattern of disregard for, and violation of, the rights of others that begins in childhood or early adolescence and continues into adulthood. The DSM version of a sexually perverted, criminal personality appears *far too antiseptic and clinical* to portray the macabre perversion of sexual psychopathy.

While serial killers certainly fit antisocial criminality, the true nature of their heinous acts—*behaviorally and emotionally*—reaches far down into the *crawlspace of human depravity.* Deviance beyond human rationality—the sanctum of *sexual psychopathy*—succinctly describes the depraved mind and sexualized crimes of serial killers.

Plain and simple, serial killers are sexual psychopaths—sexual deviants with no remorse and no conscience, who are completely incorrigible.

Sexual psychopaths continue throughout life as sexual deviants.

Adolescent Psychopathy

Some children not yet old enough to drive make up a growing list of kids too young to qualify as serial felons but nonetheless kids who kill kids and adults in home place and school place violence. *They often target the faces of victims.* Due to the current *zeitgeist,* in 1998 the FBI felt compelled to publish the first ever *school place violence profile* presented in Chapter Six.

Criminal neuropsychology is suggesting that biological templates of tendencies toward violence are becoming known in violent criminality—especially the psychopathic version of more cold blooded, remorseless crimes and how psychosocial, familial, and developmental influences exacerbate an existing *underlying neurological condition.*

The Atlanta Murders

First Test of Profiling

An infamous string of murders in Atlanta, Georgia put the *criminal profile* in the crosshairs of media scrutiny. A viable investigative tool or mere hocus pocus? Criminal profiling first came under the harsh glare of media and public scrutiny when it claimed center stage in the high profile murders of many black youths in Atlanta in the summer of 1981. Due to pleas from some of the victim's parents and appeals from local law enforcement, the FBI Academy at Quantico, Virginia sent FBI profilers John Douglas and Roy Hazelwood to Atlanta to test the controversial tool.

Criminal profiling utilizes data from the crime scene and victimology—why this specific victim and not another—along with principles of criminal or forensic psychology, common sense, good luck, and timing, to narrow down the search to one suspect.

Known Offender Characteristics

Unknown to the general public, in the Atlanta Child Murders, the FBI had an ace up their sleeve—a large database of *known offender characteristics of serial killers* from the pioneering work of FBI special agents John Douglas and Robert Ressler. Still a large amount of skepticism troubled police officers working the case, as well as many agents within the FBI. How could "a profile" be worth the paper it was written on? How could *knowledge of psychology disclose probable behavior and personality* of violent UNSUBs (unknown subjects)?

Douglas and Hazelwood created the criminal profile that surprised and shocked everyone. The profile fit Wayne Williams with such uncanny accuracy that all doubt ended. The document suggested the offender was a young black male and a police buff who impersonated police officers. This **ruse** allowed the UNSUB ready access to young, unsuspecting victims. The profile's great accuracy caused observers to comment, "It was as though the agents were watching the killer's every move even when he committed the murders and dumped the bodies."

During the trial, Douglas predicted Williams would fake a heart attack in court as the tide of evidence turned against him. He did. He suggested a strategy in interrogating Williams on the witness stand that might cue the rage Williams hid under his carefully crafted persona. During questioning, Douglas instructed the prosecuting attorney to invade Williams' personal space, by grabbing his hands and in a low, barely audible voice, inches from William's face say: "What was it like to wrap your fingers around your victims' throats? Were you frightened, Wayne?" Shocked and caught off guard, Williams replied "No." Realizing he had implicated himself in a murder, Williams angrily jumped from his seat blasting the attorney with "You're not going to implicate me with your profile!" When jurors actually experienced the outburst of rage in the middle of the proceedings (along with forensic evidence of hair and fiber samples gathered at the crime scene matching samples found in his home), they became convinced. They convicted Wayne Williams of the Atlanta child murders and sent him to prison for life. Williams maintains his innocence to this day.

Psychology and Criminal Profiling

Before analyzing the mind of a *special category of killers* with psychopathic personality emblazoned with sexual psychopathy—*serial killers*—we first address an important question. Could there be accurate profiling without underlying psychological principles? The answer is clearly "No." Admittedly, borrowing from criminology, sociology, and criminal justice might offer a crude sketch of a perpetrator based on crime trends, statistics, and criminal typologies. But this rudimentary criminal *overview* would lack significant and important pieces of the *psychological pretzel of violence* such as *personality proclivities, emotional* and *sexual* components, and the *behavioral habits* and *patterns* characteristic of violent offenders.

It appears from the research that emotional factors—**affective** (feeling) components of potentially violent regions of the brain—the brainstem and midbrain limbic system—bully prefrontal regions known to be the final brake on inappropriate or violent behavior.

In the early 1990s, the key ingredients of profiling—psychological, emotional, and behavioral—comprised the profile's first media moniker, known as the *psycho-behavioral profile.* By the same moniker, *The Silence of the Lambs,* the Academy Award Best Picture of 1992, mentioned it. Psychological principles enable an in-depth analysis of the mind of a violent criminal capable of *rapacious behavior—preying upon live prey in sexualized ways.*

The POD text does not provide a primer or a tutorial on the ways and means of writing a criminal profile. At the end of the day, readers cannot hang out their shingles and offer services to law enforcement agencies as independent profilers. Nor should they. According to the ex-FBI special agents and modern founders of criminal profiling, Robert Ressler, John Douglas, Roy Hazelwood, and others, it takes years of training, expertise, research, maturity, and experience to author effective profiles.

PREDATOR PROFILE 1-1

UNSUB
"Jack the Ripper"

Ashleigh Portales
Edited by Don Jacobs

Alternative Media Moniker
Saucy Jack
Time Span Crimes Committed
Mid-to-late August 1888 to November 9, 1888

Physical Description of Offender
Unknown
Offenses Prior to Serial Killing
Unknown

MO & Victimology

The Ripper killed prostitutes and then "ripped" them open. Each crime occurred in an isolated area (all but the last outdoors); each crime scene provided means of easy escape. The Ripper and his victims stood facing each other, the victims distracted momentarily by voluntarily lifting their skirts. The Ripper strangled the victims until unconscious.

Signature

The Ripper lowered victims to the ground after strangling and positioned them with their heads tilted to their right sides. He slashed victims' throats with a straight razor blade (from left to right), with each victim positioned as above. Evidence of blood spatter at the scene indicates most of the victims died by strangulation prior to having the throat cut. Violent post-mortem mutilation, usually including graphic positioning of eviscerated intestines suggested hatred for women or the profession of prostitution. Collection of trophies varied, but included: visceral tissue, kidney, uterus, heart, or ears. Bodies remained at the crime scene. No apparent antemortem or postmortem sexual activity appeared with victims, suggesting Jack's impotence or possible penis malformation.

Current Status

Unsolved

Victimology (Preferred Prey)

Jack chose prostitutes who frequented the Whitechapel area of London. While the identity of the Ripper remains unknown, one cannot say for sure why he specifically chose prostitutes. Contemporary knowledge of serial predators, however, allows the postulation that he harbored a deep hatred of women, which probably began early in his childhood.

Jack's Playground—London, 1888

Conjuring up images of Victorian London, one imagines horse-drawn carriages carrying gentlemen in top hats and coattails and ladies in bustling petticoats and tightly cinched corsets. At night, the scene appears through the fog-hazed light of the gas-lit cobblestone streets. The cries of newsboys echo off the buildings and into the alleyways where flashy prostitutes in heavy makeup and flamboyant dress proposition those courageous enough to approach them.

Yet, there existed in London a darker, more feral, side which was the day-to-day reality for most of the city's poor inhabitants. Arthur Morrison best described this aspect of the legendary city in his book, *Tales of the Mean Streets*. Referring to the infamous East End (Whitechapel), he writes "it is a vast city . . . a shocking place . . . an evil plexus of slums that hide human creeping things; where filthy men and women live on Gin, where collars and clean shirts are decencies unknown, where every citizen wears a black eye, and none ever combs his hair."

The East End contained the outcasts of English society: the working poor, the sporadically employed, the unemployed, and the criminals. Approximately nine hundred thousand inhabitants drudged out their lives within this slum with streets stained with blood and excrement; the stench of garbage and liquid sewage pervaded the streets of Whitechapel. More than half of the children born died before the age of five, and those that survived the horrific conditions had a high rate of mental and physical handicaps.

The residents of the East End crowded together in tenement houses, often two families to one room. Sanitation records report finding a father, mother, three children, and four pigs living within the same four walls. Missionaries' reports reveal even greater horrors. In one room lay a man ill with small pox; his wife lay in a weak condition nearby, as she had just returned home from her eighth confinement, while their children ran around dirty and practically nude. In another account, seven people lived in an underground kitchen while their deceased infant lay in the room with them. Another report disclosed an impoverished widow and her three children living in the same room with her fourth child, dead for thirteen days.

Prostitution offered one of the most reliable sources of income for a single woman or a widow. Police estimate the number of prostitutes at around twelve hundred, a number that did not include those who supplemented their meager incomes by "occasionally" applying the sexual trade. Many of these women lived one day at a time, earning money in hope of being lucky enough to pay for a bed in one of two hundred lodging houses, little more than rooms containing row upon row of single cots in which tenants had to fight the rats and insects for space. Had a woman not earned enough to pay for a bed, she had either to rely on the generosity of one of the sympathetic customers or sleep on the streets, a potential death sentence. To live in the East End meant living in squalor, and here the phantom known as Jack the Ripper found the inspiration for his demented crimes.

In 1888, the techniques so familiar in modern forensics did not exist. Fingerprinting and blood typing still had to develop, and common practice did not include photographing victims' bodies and the surrounding crime scenes. Scotland Yard had no crime laboratory until the 1930s. Unprepared for his level of deviancy, London (and the world at large) had never seen anything like the crimes of Jack the Ripper. Neither the police force nor the bawdy society of 1888 London expected the Ripper. In spite of several brutal murders, too many people in the East End depended on the meager cash flow created by daily prostitution and other evening enterprises. A few killings could not provide motivation enough to curtail the sexual trade—people's preoccupation with their own festering poverty limited worry over another's misfortune and, though they denigrated the police force for its blatant incompetence in apprehending the killer, they carried on with their lives in much the same manner as they had before. The resilient atmosphere provided precisely what a predator like the Ripper needed to conceal his identity and continue his crimes. And thus Jack emerged as the *first modern serial killer*, born even before the coining of the term serial killer, and faded forever into lore and legend as the ultimate UNSUB.

Jack Be Nimble, Jack Be Quick

In fact, Jack acted nimbly, and Jack moved quickly. He set the bar high for sexual psychopaths who followed. While most eventually get caught, Jack became one of the few exceptions—the ultimate escape artist.

Uncertainty exists whether the murder of the first prostitute in 1888 in Whitechapel came from the hand of the Ripper. However, the similarities of her murder to the subsequent five justify her inclusion in the chronicles of The Ripper. August 6, 1888 began as just another despicable day in the East End of London, no different than all the ones that had come before—until the appearance of the body of thirty-nine year old prostitute Mary Tabram in George Yard. The TOD—time of death—approximated 2:30 A.M. According to the postmortem examination report by Dr. Timothy Killeen, over thirty-nine stab wounds appeared on "body, neck, and private parts with a knife or dagger." One wound came from a large, long-bladed knife, such as a dagger or bayonet, and the rest appeared inflicted with a penknife. No slashing of the throat or overzealous mutilation of the abdominal area existed, as noticed in later Ripper victims. However, possibly as the fledgling killer's first murder, it provided a learning experience based upon personal discovery in the building of *modus operandi.*

Before the month ended, The Ripper claimed another victim. Just before four o'clock on the morning of August 31, the body of a woman with her skirts raised to her waist appeared on Buck's Row. When Constable John Neil took a closer look at the body he found her nearly decapitated, with her throat slashed deeply and her eyes frozen open in a chilling death stare. Dr. Rees Llewellyn concluded from her cold hands and wrists and warm arms that the killing took place no more than half-an-hour earlier. The two slashes to her neck had severed both her trachea and esophagus. Upon removing the woman's clothing at the morgue, Llwellyn discovered not only wounds in her abdomen but mutilation created by many long, deep, jagged knife slashes in a downward direction. After the postmortem exam, the next step in identifying the victim became a daunting task, for only a comb, a broken mirror, and a handkerchief existed on her person. Her petticoats bore the mark of The Lambeth House, but her other garments, inexpensive and ragged, provided no further clues. Physically, the woman stood at five-feet-two-inches tall, had brown hair turning gray, brown eyes, and several missing front teeth. Aid to the police came from the locals who informed them of their prostitute friend, forty-two year old Mary Ann "Polly" Nichols. Later the deceased's father and estranged husband identified her. The couple had split after producing five children because of Polly's excessive drinking.

In the wake of two prostitute slayings in half as many months, everyone speculated about possible motives of the murder. An edition of the *East End Londoner* said of the Tabram and Nichols murders:

> "The two murders which have so startled London within the last month are singular for the reason that the victims have been of the poorest of the poor, and no adequate motive in the shape of plunder can be traced. The excess of effort that has been apparent in each murder suggests the idea that both crimes are the work of a demented being, as the extraordinary violence used is the peculiar feature in each instance."

In the days that followed, a suspect emerged in the story of a man known as "Leather Apron." Supposedly the man was a Jewish shoemaker who threatened prostitutes with physical violence if they did not pay him the amount he demanded. *The Star* published the following description of the man:

> "From all accounts he is five-feet-four or five-inches in height and wears a dark, close-fitting cap. He is thickset and has an unusually thick neck. His hair is black, and closely clipped, his age being about 38 or 40. He has a small, black moustache. The distinguishing feature of his costume is a leather apron, which he always wears . . . His expression is sinister, and seems to be full of terror for the women who describe it. His eyes are small and glittering. His lips are usually parted in a grin which is not only reassuring, but excessively repellant."

The resulting public awareness caused "Leather Apron" to go into hiding.

Unsure of what to expect or when to expect it, Londoners waited in fearful anticipation until September 8, when the mutilated body of "Dark Annie" Chapman appeared in the backyard of 29 Hanbury Street across from the Spitalfields market.

A homeless prostitute, the forty-seven year old Chapman began life in a much different way. In 1869 she had married coachman John Chapman and bore him three children, one of which became crippled. Another died of meningitis. The marriage soon broke up under the stress of their children's illnesses and the strain from the heavy drinking of both Annie and John. Not long after the separation, John died and Annie lost the financial security from his income. She turned to selling flowers and crocheting to make ends meet, but when that did not suffice, she turned to prostitution in spite of her plain features, plump figure, and many missing teeth. Though unaware of it, her rough lifestyle had resulted in tuberculosis. The day before her death, she told her friend Amelia that she felt ill and "must pull myself together and get some money or I shall have no lodgings." The next morning at around 2:00 A.M., she lost her lodging house for lack of funds. Four hours later, a resident of 29 Hanbury Street found her body, her skirts lifted to her pelvis and her throat slit. Dr. George Bagster Phillips, the police surgeon who examined the body, described it as such for the subsequent inquiry:

> "I found the body of the deceased lying in the yard on her back . . . The left arm was across the left breast, and the legs were drawn up, the feet resting on the ground, and the knees turned outwards. The face was swollen and turned on the right side, and the tongue protruded between the front teeth, but not beyond the lips; it was much swollen. The small intestines and other portions were lying on the right side of the body on the ground above the right shoulder, but attached. There was a large quantity of blood, with a part of the stomach above the left shoulder . . . The body was cold, except that there was a certain remaining heat, under the intestines, in the body. Stiffness of the limbs was not marked, but it was commencing. The throat was severed deeply. I noticed that the incision of the skin was jagged, and reached right around the neck."

He later described the extent of the abdominal mutilation in the September 29 issue of the *Lancet*.

> "The abdomen had been entirely laid open; that the intestines, severed from their mesenteric attachments, had been lifted out of the body, and placed by the shoulder of the corpse; whilst from the pelvis the uterus and its appendages, with the upper portion of the vagina and the posterior two thirds of the bladder, had been entirely removed. No trace of these parts could be found, and the incisions were cleanly cut, avoiding the rectum, and dividing the vagina low enough to avoid injury to the cervix uteri. Obviously the work was that of an expert, at least, who had such knowledge of anatomical or pathological examinations as to be enabled to secure the pelvic organs with one sweep of the knife."

He could not believe that the murderer could so quickly and skillfully do what he felt he would have taken close to an hour to do. The coroner later concurred. Both felt that the murderer probably used a thin, narrow knife with a blade between six and eight inches long, the type used by surgeons for amputations. In the doctor's estimate, she had died approximately two hours earlier than the six o'clock hour when found, yet sunrise arrived shortly after five that day, and the killer would have wanted to avoid daylight. No one near the scene of the murder heard a woman cry out, but the evidence showed that she had first been strangled into unconsciousness and then mutilated. Dr. Phillips also noticed a strange occurrence at the feet of the dead woman. In her pocket a small piece of cloth and two different combs all appeared, purposefully arranged by the killer. Investigators noticed that her fingers had the cheap brass rings she commonly wore forced off them, possibly mistaken for gold. They also found a leather apron among some other trash around the yard.

John Richardson, the son of Chapman's friend Amelia, had visited the cellar at 29 Hanbury to secure the locks. Satisfied of the cellar's security, he sat down to cut a piece of leather from his boot between 4:45 A.M. and 4:50 A.M. on the morning of the murder. He neither saw nor heard any disturbance. Additionally, the man living next door to 29 Hanbury told police he had heard voices coming from the neighboring yard just after 5:20 A.M. He heard the word, "No," and then a few minutes later, at 5:30 A.M., he heard the sound of something falling against the fence.

The final witness said she heard the clock at the nearby brewery strike 5:30 A.M. as she headed to the Spitalfields market by way of Hanbury Street. She testified that she saw a man and a woman talking "close against the shutters of No. 29." She later identified Annie Chapman's body as that of the woman she had seen, but she couldn't provide much of a description of the man as he had his back turned to her.

A few days later, on September 11, the police arrested the man known as "Leather Apron" and identified him as John Pizer who had served six months hard labor previously for stabbing a person. He admitted to going into hiding after the death of Annie Chapman under the advice of his brother for fear of public retaliation. However, he had solid alibis for the times of Chapman's and Nichols' murders. Cleared of the murders, they released him. The accusations that he had threatened and extorted area prostitutes remained unproven.

Although fresh out of suspects, unfortunately the police did not lack bodies. On September 30 at around 1:00 A.M., a man driving his pony cart to a nearby town stumbled upon the body of a woman just off Berner Street in Whitechapel. Dr. Frederick Blackwell and his assistant arrived on the scene at 1:16 A.M. and began to examine the body. The assistant's notes chronicled the following:

> "The deceased was lying on her left side obliquely across the passage, her face looking towards the right wall. Her legs were drawn up, her feet close against the wall of the right side of the passage. Her head was resting beyond the carriage-wheel rug, the neck lying over the rut . . . The neck and chest were quite warm, as were also the legs, and the face was slightly warm. The hands were cold. The right hand was open on and on the chest, and was smeared with blood. The left hand, lying on the ground, was partially closed, and contained a small pocket of cachous (breath sweeteners) wrapped in tissue paper . . . The appearance of the face was quite placid. The mouth was slightly opened. In the neck there was a long incision . . . commenced on the left side, two inches below the angle of the jaw, and almost in a direct line with it, nearly severing the vessels on that side, cutting the windpipe completely in two, and terminating on the opposite side."

The time of death approximated between 12:36 A.M. and 12:56 A.M. A witness walking down the street at around 12:40 A.M. had seen nothing, nor had the cart driver when he first pulled in at 1:00 A.M.

Before the Berner Street victim could be identified, another body surfaced one-fourth of a mile away in Mitre Square by Police Constable Edward Watkins of the City Police, who walked his beat in that area. All appeared quiet and normal when he passed through the square at 1:30 A.M., but when he passed by again at 1:44 A.M., he discovered the body of a woman he later described as "lying on her back with her feet facing the square, her clothes up above her waist . . . her throat was cut and her bowels protruding. The stomach was ripped up. She was lying in a pool of blood."

A physician that arrived on the scene at approximately 2:18 A.M. placed the TOD (time of death) "most likely within the half hour." Then, slightly before three in the morning, a constable patrolling on Goulston Street, also in Whitechapel, found a bloody piece of apron in the doorway of a building. Above the entrance, written on the brick in white chalk, appeared the words: *"The Juwes are The men That Will not be Blamed For nothing."*

The bloody apron turned out to belong to the woman murdered in Mitre Square, leading police to believe the writing belonged to the killer. Police guarded the wall while they prepared to have it photographed. Meanwhile, Sir Charles Warren, the Commissioner of the Metropolitan Police, gave a highly controversial order to have the writing destroyed, which occurred before taking the photographs. Though critics attacked Warren for over one hundred years for his actions, he defended himself at the time by saying,

> "The writing was on the jamb of the open archway . . . visible to anybody in the street and could not be covered up . . . I do not hesitate to say that if the writing had been left there would have been an onslaught upon the Jews, property would have been wrecked, and lives would probably have been lost."

Whatever the reasoning, a chance to document the writing of the probable killer disappeared forever, and investigators had no way of knowing its usefulness.

Their first real clue obliterated, police turned to the identification of the two bodies in the morgue. The first one, found off Berner Street, stood approximately five feet, two inches tall. Dressed in black with a single red rose decorating her jacket, she had a light complexion and dark brown, curly hair. Soon police identified her as Elizabeth Stride, born in Sweden in 1843. Upon arriving as a domestic worker in England, she concocted a story that she had survived the Princess Alice boating disaster of 1878, in which her husband and two children had drowned, in order to obtain charity from the Swedish Church in London. Actually, her husband, John Stride, had survived the Thames River tragedy but later he had died in the poorhouse. She had spent the last three years prior to her death living with a laborer by the name of Michael Kidney, supplementing the wages she earned from sewing and cleaning with occasional prostitution, which she worked under her common nickname, "Long Liz." On the evening of her death, she had left her lodging house early and, interestingly, had not worn a rose on her jacket at the time.

Several witnesses saw Stride talking with a man before her death, and they provided descriptions of him. Constable William Smith recalled seeing the deceased with a man of approximately thirty years of age, with dark hair and mustache and dark complexion, while walking his beat around Berner Street at about 12:30 A.M. in the morning. Standing somewhere around five feet, seven inches, the man wore a dark felt deerstalker hat and a black diagonal cutaway coat with white collar and tie. Smith also remembered that the man held a good-sized parcel in his hands.

Israel Schwartz supplemented Constable Smith's story. At 12:45 A.M., while standing at 22 Helen Street, he saw a man stop and speak to a woman, whom he later identified as Elizabeth Stride, and then pull her into the street, turn her around, and throw her down. Stride screamed three times, though Schwartz did not recall that the screams sounded very loud. The attacker then called out to another man (an accomplice?) standing across the road whom he addressed as "Lipski." Schwartz walked away, but the second man followed him for a while until he began to run and lost him. Schwartz described the man who had thrown Stride down as "about thirty years old, five-feet-five inches tall, fair complexion, dark hair and mustache, full face, broad shouldered, dressed in a dark jacket and trousers and a black cap with a peak, and holding nothing in his hands." "Lipski" appeared approximately thirty-five years old, five feet, eleven inches tall, with a fresh complexion, and light brown hair and mustache. He wore a dark overcoat, a black felt, wide-brimmed hat, and carried a clay pipe in his hand.

The police identified the second woman, killed in Mitre Square, more easily, as she had some pawn tickets on her person. John Kelly, Catharine "Kate" Eddowes' lover of the last seven years, recognized her. They had lived together in a lodging house at 55 Flower and Dean Street. She was born in 1842 and orphaned at a very young age. For twenty years, she lived with Tom Conway as his common law wife, a union that produced three children, until she left him in 1880 when she tired of his physical abuse and he tired of her drinking. The next year she met Kelly, and though friends adamantly insisted that she was not a prostitute, evidence suggested that she occasional prostituted herself, especially when under the influence of alcohol. On the evening of her death, she had gone to visit her daughter in the hope of borrowing some money and had responded to Kelly's warnings about the Whitechapel killer by saying, "Don't you fear for me. I'll take care of myself, and I shan't fall into his hands." Eddowes never arrived at her daughter's home. According to the testimony of the physician who examined Eddowes' body:

> "The abdomen had been laid open from the breast bone to the pubes . . . The intestines had been detached to a large extent . . . about two feet of the colon was cut away . . . the peritoneal lining was cut through and the left kidney carefully taken out and removed. The left renal artery was cut through . . . The womb was cut through horizontally, leaving a stump of 3/4 of an inch. The rest of the womb had been taken away with some of the ligaments. The vagina and cervix of the womb was uninjured . . . The face was very much mutilated. There was a cut about 1/4 of an inch through the lower left eyelid dividing the structures. The right eyelid was cut through to about 1/2 inch. There was a deep cut over the bridge of the nose extending from the left border of the nasal bone down near to the angle of the jaw of the right side. The tip of the nose was quite detached from the nose."

In Kate's case, one witness surfaced whose account seemed to fall somewhat in line with the others. Joseph Lawende had left the Imperial Club with two friends on the morning of Eddowes' death at approximately 1:35 A.M. and saw a couple conversing at Church Passage near Mitre Square. The male appeared young, of medium height, with a small fair-colored mustache. Though he had not seen the woman's face, he later identified Kate by her clothing. Catharine Eddowes died nine minutes after Lawende and his friends saw her.

As the murders continued, hundreds of letters came pouring into police and other agencies involved in the investigation, supposedly from the real killer. However, only three of these letters have endured the test of time. The same person wrote two and coined the moniker "Jack the Ripper." The writer wrote the first letter—The Boss Letter—in red ink, received by *Central News* on September 27, 1888.

The Boss, Central News Office,

I keep on hearing the police have caught me but they wont fix me just yet. I have laughed when they look so clever and talk about being on the right track. That joke about Leather Apron gave me real fits. I am down on whores and shant quit ripping them till I do get buckled. Grand work the last job was. I gave the lady no time to squeal. How can they catch me now. I love my work and want to start again. You will soon hear of me with my funny little games. I saved some proper red stuff in a ginger beer bottle over the last job to write with but it went thick like glue and I cant use it. Red ink is fit enough I hope ha ha. The next job I do I shall clip the lady's ears off and send to the police officers just for jolly, wouldn't you. Keep this letter back till I do a bit more work then give it out straight. My knife's so nice and sharp I want to get to work right away if I get a chance. Good luck.

Yours truly,
Jack the Ripper
Don't mind me giving the trade name.

Written on the same letter, perpendicular to the rest of the writing:

> *Wasn't good enough to post this before I got all the red ink off my hands curse it. No luck yet. They say I'm a doctor now ha ha.*

The editor of the paper, believing the letter a hoax, did not forward it to police for a couple of days. When he did, the police received the letter the night after Stride and Eddowes were murdered. On the Monday following the murders, the editor received another letter in the same handwriting, postmarked October 1:

> *I wasn't codding dear old Boss when I gave you the tip. Youl hear about saucy Jackys work tomorrow double event this time. Number one squealed a but. Couldn't finish straight off. Had not time to get ears for police. Thanks for keeping last letter back till I got to work again.*

The third letter went to George Lusk, the head of the Mile End Vigilance Committee, the equivalent of a neighborhood watch. The letter had the postmark October 16 and accompanied a piece of human kidney preserved not in formalin, the general practice by hospitals, but rather in ethanol, drinking alcohol. The handwriting did not match that of the first two letters, and the signature did not say "Jack the Ripper:"

> *From hell*
> *Mr Lusk*
> *Sor*
> *I send you half the kidne I took from one women presarved it for you tother piece I fried and ate it was very nise I may send you the bloody knif that took it out if you only wate a whil longer*
> *Signed*
> *Catch me when*
> *You can*
> *Mishter Lusk*

Did the real Jack the Ripper write any of these letters? FBI profiler John Douglas claims of the first two, "It's too organized, too indicative of intelligence and rationale thought, and far too 'cutesy.' An offender of this type would never think of his actions as 'funny little games' or say that his 'knife's so nice and sharp.'" As for the third letter, the kidney it contained came from an adult human suffering from Bright's Disease, which apparently afflicted Kate Eddowes. However, there is no way today to prove the letter's genuineness.

For a month, no other murders occurred in Whitechapel, and things began to return to normal. Prostitutes began to feel safe walking the streets again, and the nighttime trade began to flourish. Then, on November 9, 1888, the discovery of Mary Kelly's horrifically mutilated body occurred in her rented room at Miller's Court. Born in Limerick, Mary had moved to London at the age of twenty-one to work in a brothel. An attractive woman, she found support from her various lovers and did not have to survive on prostitution alone. For the past year, she had lived with market porter Joe Barnett. Just before her murder, he had lost his job, and the two had argued over the fact that a homeless prostitute now stayed with them and that Mary herself had turned back to prostitution to make ends meet. Acquaintances described her as:

> "Five-feet-seven-inches in height, and of rather stout build, with blue eyes and a very fine head of hair, which reached nearly to her waist . . . tall and pretty, and as fair as a lily, a very pleasant girl who seemed to be on good terms with everybody . . . one of the most decent girls you could meet when she was sober."

However, a different picture greeted Thomas Bowyer when he looked in her window. He worked for Mary's landlord and came there to collect rent. Police surgeon Dr. Thomas Bond determined the cause of death as severance of the carotid artery, while the mutilation occurred postmortem. His examination report best describes the scene:

> "The body was lying naked in the middle of the bed, the shoulders flat, but

> the axis of the body inclined to the left side of the bed . . . The whole of the surface of the abdomen and thighs was removed and the abdominal cavity emptied of its viscera. The breasts were cut off, the arms mutilated by several jagged wounds and the face hacked beyond recognition of the features and the tissues of the neck were severed all round down to the bone. The viscera were found in various parts—the uterus and kidney with one breast under the head, the other breast by the right foot, the liver between the feet, the intestines by the right side and the spleen by the left side. The flaps removed from the abdomen and thighs were on a table . . . Her face was gashed in all directions, the nose, cheeks, eyebrows and ears being partly removed. The lips were blanched and cut by several incisions running obliquely down to the chin. There were also numerous cuts extending irregularly across all of the features. The skin and tissues of the abdomen from the costal arch to the pubes were removed in three large flaps. The right thigh was denuded in front to the bone, the flap of skin including the external organs of generation and part of the right buttock. The left thigh was stripped of skin, fascia and muscles as far as the knee."

During the reconstruction of her body, the doctor realized that her heart was missing from the scene. The TOD approximated one or two o'clock in the morning. In Bond's opinion, the murderer of Kelly, like that of all the other girls, had *no medical knowledge whatsoever.* This contradicted many other doctors who had contended the killer was a surgeon or in some kind of medical profession. The man seen with Mary on the evening of her death appeared between the ages of thirty-five and forty. He stood approximately five feet, six inches tall dressed in dark clothing.

The murder of Mary Kelly terrified the inhabitants of Whitechapel, and the nightlife in the area virtually disappeared for many nights thereafter. The Kelly murder closed the Jack the Ripper file for reasons unknown even today. The Whitechapel Killer faded into memory as a legend in the annals of true crime.

The (Un) Usual Suspects

A list of "probabilities" of a profile of The Ripper came from various eyewitness accounts and the facts of the murders:

- White Male
- Average or below average height
- Between twenty and forty years of age
- Did not dress as a laborer or indigent poor
- Had lodgings in the East End
- Possible medical expertise
- Possibly a foreigner
- Right-handed
- Had a regular job as all the murders were committed on the weekends
- Single, free to roam the streets at all hours

No one suspect emerged above all the rest, however, as the most likely killer. The files of Chief Commissioner of the Metropolitan Police, Sir Melville Macnaghten, identified three key suspects. Macnaghten retired in June of 1889.

A popular theory of the identity of Jack involved a royal conspiracy. This served as the premise for the 2001 movie *From Hell,* starring Johnny Depp and Heather Graham. According to the story, Prince Albert Victor Christian Edward, the Duke of Clarence and the grandson of Queen Victoria—known to the public as "Eddy"—lay in direct line to the throne; had he outlived his father, King Edward VII, he would have assumed it. Eddy frequently went "slumming" in the Whitechapel area, where he entered into a relationship with a shop girl named Annie Crook. He kept Annie in an apartment, she soon became pregnant, and the two married secretly. In another version of the story, the child simply is born out of wedlock. In either such case, the royal family would staunchly forbid a relationship with a Catholic woman of such low social standing. When word of the scandal reached the Queen, she demanded that the mess be "cleaned up." The job fell to the royal physician, Sir William Gull. Gull then kidnapped Annie, took her to a hospital, and drove her insane. None other than Mary Kelly cared for Alice Margaret, the child. Only Kelly, along with her friends, Polly Nichols, Annie Chapman, and Elizabeth Stride, knew the identity of Alice's true father. Afraid that the women would talk, Gull created the persona of Jack the Ripper. With the help of his coachman, he located, murdered, and mutilated each woman, one by one.

While the allure of this story is tempting, in all likelihood it is quite untrue. To begin with, while a woman by the name of Annie Crook who worked in a shop and had an illegitimate daughter named Alice Margaret existed, rumors established Eddy's sexual preference as homosexual, outlawed in England. In fact, they said he frequented a particular brothel on Cleveland Street, which catered to wealthy homosexuals. Furthermore, no ties of any kind existed between Crook and any of the Ripper victims.

In 1887, Gull suffered the first of many strokes that severely paralyzed him. This is the greatest fact discrediting this story. Though he eventually recovered, the lingering effects of the stroke at age seventy forced him to retire from the medical practice until his death in 1890, precipitated by a third stroke. As stated by John Douglas in *The Cases That Haunt Us,* The Ripper killings reflect the "work of a disorganized, paranoid offender" and not a person who "could continue functioning and interacting with people in a relatively normal way. Dr. Gull simply does not fit the profile."

Crime novelist Patricia Cornwell recently released a book entitled *Portrait of a Killer: Jack the Ripper Case Closed* in which she names British painter Walter Sickert as the man she believes "100 per cent" to be Jack. Cornwell spent a reported six million dollars of her own money to hire art and forensic experts. She purchased thirty-one of Sickert's paintings to prove her thesis. Her forensic team analyzed DNA samples from over fifty-five letters, envelopes, and stamps from Sickert himself, his wife, Montague John Druitt, as well as many of the letters that were sent to the police under the moniker of "Jack the Ripper." Investigators found *mitochondrial DNA* (linked to maternal DNA) on one of the letters sent from the killer himself (not a copycat letter). Cornwell claims this links Sickert to the killer's letter. Also, the paper of one of the supposed Ripper letters bore the same watermark as Sickert's writing paper, which he received as a gift from his father. She also claims that many of Sickert's paintings bear a striking resemblance to photographs of the Ripper murder scenes and that some of the Ripper letters contain phrases often used by the famous painter, Whistler, which Sickert had mocked when he studied under the master.

The Intrigue of the Eternal UNSUB

Why, after more than one hundred years, does the mystery of Jack the Ripper still reel us in at every turn? He is essentially no different than the many serial killers who have followed in his footsteps, and, in all likelihood, the true identity of Jack the Ripper will never become known. We attribute to Jack almost supernatural qualities to explain away our inability to detect his identity. If he were an ordinary man, we would have known his identity by now.

In essence, The Ripper represents the archetypal "boogeyman" of popular culture: He emerges from the darkness, unseen and unheard, and steals away life in the most gruesome of ways before returning undetected to the darkness from whence he came. In the annals of serial murder, Jack the Ripper looms as one of the most menacing, though guilty of only five murders, a number not comparable to many of his more modern counterparts. Perhaps it is not so much his evil deeds as our timeless fear of the unknown that has forever guaranteed him a place in both crime history and our darkest nightmares.

Comments

Due to his signature, Jack the Ripper had some knowledge of anatomy and physiology. Some letters received by newspapers contained information that only the killer could have known (adding credibility to Cornwell's thesis). During the killer's most active period, police organized frequent sweeps of the East End. Perhaps The Ripper hid among the persons incarcerated for other crimes, or they sent him to an insane asylum. Perhaps police "sweeps" took Jack the Ripper off the streets permanently, accounting for the sudden cessation of the serial killings.

Organized versus Disorganized

The Ripper crimes have elements consistent with the *disorganized type* of killer. The similarity and proximity of locations indicate a person who didn't travel far from his residence. Another element is the violence of the post-mortem mutilation. At the same time, The Ripper had enough cognizance and social skills to identify and solicit the same type of victims (prostitutes) consistently, to vary his *MO* (he killed and most brutally eviscerated the last victim indoors), and to perform the same ritualistic methods (throat slashing, followed by post-mortem mutilation).

If the killer wrote most of the taunting letters to the newspapers, it points toward a more **hedonistic and grandiose** person—a more *organized* killer who follows and collects his press clippings.

Names of Victims

Mary Ann Nichols
Annie Chapman
Elizabeth Stride
Catherine Eddowes
Mary Jane Kelly (a movie, *Mary Kelly* starring Julia Roberts depicted her life with both fictional and non-fiction elements)

Criminal Diagnosticians

Profiler John Douglas cautions law enforcement personnel on the dangers of becoming a profiler by attending a three-day seminar or by mail order. Competent profilers seek to enter the mind of criminals as *criminal diagnosticians* to discover tendencies, proclivities, fantasies, habits, and lifestyle. In the process, profilers learn how vulnerable the serial killer is to apprehension, as all criminals leave considerable information about themselves (evidence) behind at the crime scene. The behavior of serial predators appears to whisper at the crime scene, leaving important clues about his crime to those skilled enough to notice. When criminal investigators arrive at a crime scene the chase begins.

Scary Parenting

The scary, sexualized, inner world of sexual predators and the documented psychological, social, behavioral, and neurochemical influences that shape human monsters demonstrate the devastation of abusive, hateful, hurtful, and loveless parenting that perversely nurture a remorseless, cold-blooded killer. It provides the first condiment in the "recipe" for creating a child with a *psychopathic personality.*

We have answers from neuroscience, theory, and speculative logic—the investigative and theoretical tools of forensic behavioral science. Readers and students of forensic psychology and violent criminal investigation can peer deep inside the mind of violent predators. The discovery is terrifying. Usually, brain scans capture the first sign of extensive *neurological devastation.* No therapeutic treatment—nothing currently available in neuroscience or clinical neuropsychology—works to help violent psychopaths "recover." No cure exists.

The old saw from forensic psychiatrists contend "only hams are cured." Meaning, of course, human psychopathologies are not.

Psychopathy Checklist

The recognized authority on psychopathy, Robert Hare, Ph.D., produced, along with his colleagues, *The Hare Psychopathy Checklist-Revised (PCL-R),* an instrument used to determine characteristics of psychopathic personality. The most recognizable characteristic of psychopathy includes individuals who have a gift of engaging others but remain remorseless to their "prey."

Below, we have modified and condensed Hare's original psychopathy list to twenty characteristics displayed in no particular order. The *garden-variety psychopath* (non-violent type) is not a sexual psychopath; he doesn't kill and/or rape; instead, he emotionally wrecks those of whom he targets, electing to toss them away on his way to finding other prey to manipulate, control, and dominate. Some researchers refer to the non-violent variety of psychopaths as "psychopaths in suits," "white collar" psychopaths, or "industrial psychopaths." Jacobs (2006) prefers the term *hubristic psychopathy,* addressed in a later chapter.

While gender appears irrelevant in *garden-variety psychopathy,* males (mild to moderate versions) often display features of *narcissism* (inflated and self-centered self-aggrandizement) with more severe versions characterized by *grandiosity.* Female psychopaths often display histrionic characteristics (seductive and superficial features). Both genders may populate business, industry, education, and the clergy, or appear as successful community leaders.

An example of this type of "white collar" psychopathy recently gained notoriety in the form of corporate heads of *World.Com* and *Enron. Playboy* magazine, a longtime harbinger of pop culture hedonism, provided a reminder of the seductive allure of psychopathy in a pictorial display of ex-Enron female employees. Less recently, a football icon fit the same profile of narcissism mixed with psychopathy: O. J. Simpson, held civilly responsible for the deaths of his ex-wife, Nicole Brown Simpson and her friend.

According to former LA prosecutor, Vincent Bugliosi, in his book *Outrage,* O. J. Simpson has not found the murderer he pledged to find, even as he searches in vain for the killer on golf courses where he spends most of his time.

Psychopaths display the following characteristics:

- Grandiose sense of self-worth (exaggerated self-importance to the extent of ignoring other's needs); *pathological grandiosity* observed in violent types
- Glib, superficially charming
- Need for stimulation; prone to boredom
- Lack of remorse or guilt (including criminal behavior)
- Pathological liar
- Callous, lack of empathy for others
- Perhaps charismatic, personally engaging to others
- Shallow emotional responses, *blunt affect* (lack of facial expression in severe types)
- Parasitic lifestyle
- Sexual promiscuity
- Impulsive
- Lack of realistic long-term goals
- Poor behavioral controls (lack of restraint)
- Early behavioral problems (oppositional defiance)
- Irresponsible
- Failure to accept responsibility for actions
- Many stormy, short-term relationships
- Juvenile delinquency
- Conning, criminally versatile
- Cunning and manipulative

Interestingly, no *clinical diagnostic criteria* for psychopathy *per se* exist in the DSM. Since antisocial PD has replaced psychopathy or psychopathic personality, neither term appears in the DSM index (2002).

As we have noted, although serial killers get diagnosed with antisocial PD, we make the compelling conclusion that a more accurate diagnosis remains **sexual psychopaths.** These human predators rape and kill others for the *emotional "high"* of sexual control and domination, and, in some cases, even use their victims as specimens for sexual experimentation.

Robert Hare *(Without Conscience,* 1999*)* discovered that *criminal psychopaths* display thirty or more (out of a possible forty) psychopathic traits in his *psy-*

chopathy checklist. Imagine how many lesser psychopaths (with lower scores and non-violent) roam freely in society, never arrested, much less incarcerated. Nonetheless, they damage almost everyone they touch in emotional ways that linger long after the psychopath has moved on to greener pastures.

Camouflaged Psychopaths

Realistically, some highly functional psychopaths *camouflage* themselves in various societal venues—sports, business, politics, the arts, religion, and education. Psychopathic individuals litter pop culture YAAVIST Society (current North American popular culture driven by Youth, Attractiveness, Accoutrements—fashion, Visual, Isolation-prone, Successful appearing, and focus on Thin bodies). The psychology of deception runs rampant in YAAVIST Society appearing as current or retired sports heroes, business executives, politicians, and, of course, Hollywood artisans and agents exemplify.

Finally, they may hide in the ivory towers of education—the only category of society not affiliated with pop culture—yet, nonetheless, often populated with *garden-variety psychopaths* shrouded in blissful anonymity. But, they cause emotional upheaval in peers and underlings while they plot to gain promotions by convincing upper-level administrators of their worth. How convincing they can be! Studies show that they are seldom identified (or fired) until success in outing them (rarely) or taking over the very job they sought—their boss's job! Blindsided administrators may learn the hard way to pay attention to the well documented research into psychopathy.

This text does not explore the thousands, perhaps millions, of psychopathic individuals who terrorize others in the workplace by analyzing the power structure of the organization and aligning themselves with authority. They astutely observe human behavior as they plot the downfall of competitors in such crafty ways that almost no one suspects them. Consider for a moment all of the superficially glib, compulsively lying, stab you in the back narcissists you have known in the workplace. Perhaps you currently work alongside such individuals. The difference between the non-criminal types and the violent variety lies in the *profundity* of the disorder.

Criminal Profiling—Three Important Terms: Serial, *Modus Operandi,* and Signature

Serial

Although several investigators and researchers have laid claim to coining the term "serial killer," former Special Agent to the FBI, Robert Ressler (Ressler & Shachtman, 1992), deserves credit as one of the first high profile investigators to use the term. Ressler contends he chose the term "serial" after hearing a speaker at the British police academy refer to "crimes in a series"—a series of crimes attributed to the same offender. Shown before the feature attraction in movie theaters, the word conjured up in Ressler's mind the "serial films" he enjoyed as a kid. The short reel "serials" presented sequential segments as a marketing tool to entice moviegoers to come back each Saturday and watch the next installment. Popular series included cowboys and Indians, *Buck Rogers,* and the *Adventures of Zorro.*

Applying the "serial" moniker and two well-known evidentiary *tools—modus operandi* and signature—to other venues, a textbook author writing a series of books could be referred to as a *serial author.* One book follows another with a "cooling off" period between publications. The "cooling off" period refers to any breaks the author takes from writing, such as doing basic research on another book, or simply taking a mental break from writing altogether. A serial perpetrator commits a crime followed by a cooling off period that might last a few hours or perhaps ten years before he re-offends.

Modus Operandi (MO)

Legal scholars use the term *modus operandi (MO)*—the procedure or method of operation used by a perpetrator at the crime scene—to describe "how he did it." What comprises *MO* in another venue, such as authoring a college text? The book you hold in your hands provides an example. Jacobs' first effort at producing a college text occurred in 1996 after being displeased with standard adoptions. He bundled together a collection of pop culture articles relating to the course material for Intro Psych as "enhancement readings" to accompany the text. Immediately, student feedback indicated they found the articles more enjoyable and informative and preferred the handouts to the text.

Jacobs discovered that many of the standard introductory psychology texts appeared too long and too impersonal for use in group interaction and dynamics in the classroom. Also, many standard texts fall short of providing enough quality information related to pop culture, an extremely important social milieu for college students. As the years went by, he expanded the pop culture materials and wrote introductory material suitable for group interaction, application, and discussion; application suffered at the expense of documenting boring research more suitable for upper level courses. At this point, the author rewrote the articles in his own conversational style, focusing on expanding *scholarly lexica*—word knowledge—in exercises known as Word Scholar. In all, the author's first "text" contained approximately three hundred pages "shrink wrapped" in cellophane, unstapled, hand-numbered at the bottom of the pages, and sold in the college bookstore as the basic text for "Intro Psych." Soon, a publisher showed interest and custom published the work as a conventional text.

Subsequent editions *evolved* from Jacobs' original *MO*—pop culture articles mixed with standard textbook material, including shorter, more conversational-style paragraphs, punctuated by application exercises (Conflicted in Pop Culture). The ever-present focus on psycholinguistics continued—learning *scholarly vocabulary* and its documented ameliorative affect upon students' "psychological selves." Jacobs' *MO* evolved and changed, aided by more student involvement, including featuring pictures of outstanding students on the cover and throughout the text. Slowly, through experience, the author became a more experienced writer and more comfortable with student contributions and how they produced a more effective text. After co-founding a new department, Forensic Science, in 2004, a forensic science essay highlighted each chapter in the intro text, another example of the evolving nature of *MO*. Even the title of the introductory text evolved from all of the aforementioned into the current moniker: *The Real Story of Psychology* (Hayden McNeil pusblisher) 2008. The original title back in 1997 was *Psychology: The Amazing Science of Behavior.*

Just like serial crime, the *modus operandi* of textbook writing—how the text developed from inception through continual evolution—dynamic and changeable—based upon feedback and direct experiences of what "works" and what "doesn't work" (just like the *MO* of criminal behavior) continues to shape and maintain the final product.

Signature: *Cri de Coeur*

The compositional *signature* of a given violent crime represents the perpetrator's emotional "calling card." The *emotionally compelling reason* why he committed the crime in the first place: it *emotionally fulfills* him by the *nature of the crime* and *choice of victim* (victimology).

Retreating one more time to the example of textbook writing, what constitutes Jacobs' textbook signature? What displays the *one indispensable motivation* running through all titles? Discovering this one common thread discloses signature in any activity, including violent crime.

The one component in all of the Jacobs texts is his passion for *psycholinguistics* and how scholarly *word knowledge* in *Word Scholar* exercises encourages students along the path to becoming scholars. This textbook signature aids students in composing powerful *neurocognitive maps of experience,* almost single-handedly forging the *scholarly mindset.* The text does not need Jacobs' focus on scholarly words, but they represent the reason for it—the *cri de coeur*—the passionate cry from the heart—"THIS IS IMPORTANT TO ME!"

The Profundity of Psychopathy

Criminal profiling is most useful to investigators when the crime scene reflects a more *profound degree of psychopathology* and suggests aberrant behavior in the offender, as observed at the crime scene. According to Holmes & Holmes (1999), and verified by FBI statistics, criminal profiling offers the *best chance of targeting the probable offender* relative to the following crimes:

1. Sadistic torture/sadistic assaults
2. **Evisceration**
3. **Postmortem** slashing/cutting
4. **Pyromania**
5. Lust/mutilation murders
6. Rape
7. Satanic/ritualistic crimes
8. **Pedophilia**

The justification for the *efficacy* of profiling the aforementioned select group of crimes appears in forthcoming chapters. For the time being, the guiding principle of profiling predators who commit *sexually driven crimes* requires that perpetrators have (1) a depraved mind and (2) a severely flawed character defined by (3) lack of restraint and (4) emotional apathy toward victims. In a nutshell, this describes a perpetrator characterized by severe *psychopathic personality punctuated by sexual psychopathy.* The heinous nature of violent serial crimes and the societal unrest engendered by serial rapes and murders—*rapacious crimes*—requires the extraction like a bad tooth of human predators from society.

The Whitechapel Murderer—"Jack the Ripper"

The case of Jack the Ripper comprises the first subject of *Predator Profiles*—selected bios of serial killers—strategically placed in intervening chapters throughout the book. The Ripper is instructive for a number of reasons. First, although conspiracy theories abound, his identity remains unknown and probably never will become known. Patricia Cornwell's book *Portrait of a Killer: Jack the Ripper—Case Closed* presents the best guess so far; she identifies the assailant through mitochondrial DNA from a letter the police received, supposedly from The Ripper. Cornwell makes a convincing forensic argument.

In 1888, the term "serial killer" did not exist. As we have seen, over ninety years passed before the introduction of criminal profiling to the media and public scrutiny in the Atlanta child murders. The Ripper killed prostitutes, a common target of serial predators today, because of the ease of accessibility. Jack mutilated corpses as his signature after he approached them on dark sidewalks dimly lit by gas lamps—his MO. Hollywood presented two films on the Ripper, a 1970s installment with the unlikely choice of then matinee idol Tony Curtis as the Ripper. Recently, *From Hell,* starring Johnny Depp as a clairvoyant criminal investigator, followed one popular theory of Jack as a physician, specifically, a surgeon. Modern forensic science would have more than likely caught Jack and the world would have known his name.

Applied Forensic Criminology—Known Offender Characteristics

Present and future cases may reference a criminal "yardstick," the use of *known offender characteristics* in computerized databases that merge principles of psychology with forensic science. Therefore, the term for the method of capturing serial predators is **inductive,** rather than deductive. In this manner, systematized knowledge replaces speculation or "hunches" (used by the famous fictional detective Sherlock Holmes) to edge profiling ever closer to science. However, experienced profilers contend profiling is also "artwork"—it depends on speculative logic, common sense, and, sometimes, sheer luck. Nonscientific "yardsticks" of reference produce a more philosophical speculation known as **deductive reasoning.** In the words of ex-FBI profiler John Douglas, "The more you know about a predator's 'artwork,' the more you know about the 'artist.'"

Hellish nightmare accounts of dysfunctional histories, sexual obsessions, and **macabre** fascination fill FBI computers at the *National Center for the Analysis* *of Violent Crimes (NCAVC).* These comprise the *behavioral dynamics* of society's most elusive predators. From this growing database, profilers, relying on the principles of forensic science, peer inside the minds of serial rapists and killers to see what "makes them tick."

This text focuses on the *systematized knowledge* (the organized information and methodology of forensic science) used by forensic psychologists and forensic psychiatrists as well as by other criminal investigators who study serial predators. The word *forensic* means information "headed to the courtroom;" hence, a **forensic pathologist** investigates physical evidence at the crime scene, such as wound patterns, ligature (strangulation) marks, specific physical trauma of the victim, DNA traces, and related CSI evidence used in court. In contrast, a **forensic psychologist** might testify in court regarding *behavioral characteristics* of the offender suggested by the crime scene and victimology, or the feasibility of an insanity plea. Both the pathologist (M.D.-credentialed) and the neuropsychologist (Ph.D.-credentialed) may lend their expertise to courtroom proceedings.

Currently, the FBI is the largest investigative group engaged in violent crime profiling. However, retired FBI special agents offer expertise as private consultants such as John Douglas, the first full-time profiler at the FBI's *Behavioral Science Unit,* and Robert Ressler, co-founder of the modern system of profiling.

The terrorism of 9/11 led to the "retooling" of the FBI in some respects. The FBI now investigates *only the most publicized violent crimes.* The focus on terrorism in light of the current homeland *zeitgeist* represents the FBI's latest attempt to redefine itself. Despite the importance of abolishing terrorism, we shudder to think what will become of less publicized serial crimes on the already overburdened metropolitan police forces.

Multiple Murder Defined

Criminologists delineate three types of multiple murder categories. A **serial killer,** the focus of this text, murders at least three victims followed by an emotional "cooling off" period between crimes. Ted Bundy, Jeffrey Dahmer, and the fictional "Buffalo Bill," from the Thomas Harris novel, *Silence of the Lambs,* qualify as serial killers (a serial rapist rapes sequentially).

A **mass murderer** murders four or more victims in one location in one incident and in one emotional experience. From his perch high atop the Texas tower at the University of Texas, Austin, sniper Charles Whitman killed twelve students at random in the early 1960s. Along with Whitman, Richard Speck, who killed eight nursing students on a Chicago college campus, fits the definition of mass murderer. A **spree killer** perpetrates a murderous rampage in at least two or more locations in a time-compressed timeline with no emotional "cooling off."

The FBI describes serial killers as the most *cunning and sinister of all violent criminals.* Monetary gain does not motivate them, nor does "heat of passion" murder observed in involuntary manslaughter. *The driving force behind serial crime is preying upon others for perverted sexual gratification.* As we previously noted, this delineates the sexual psychopath from all other varieties of psychopathic personality. Ironically (and perversely) attracted to the police who pursue them, serial predators prove almost impossible to capture for ordinary investigative methods seldom can predict their next moves. Many are police "buffs" who impersonate police officers as a ruse to gain access to targeted victims. Some serial criminals attend news conferences related to their crimes or routinely visit their victims' cemeteries or act as pallbearers at their funerals.

Aspects of Serial Homicide

Serial homicides are almost always *intraracial;* that is, white perpetrators kill white victims, black offenders kill black victims, and so on. However, for reasons unknown, most offenders are almost always white males, twenty to mid-thirty years of age. In the initial stages of a suspected serial homicide, FBI profilers begin with age twenty-five for the *UNSUB* and then add or subtract years based on the *sophistication* of the crime scene. The first step in developing the profiling involves determining the kind of person who would feel comfortable moving around the types of victims (prostitutes or young children, for example) and choosing the disposition of their bodies.

Clearly, years of interviewing serial murderers by authorities show that no individual at any age, regardless of ethnicity or race, simply wakes up one morning and decides to kill a string of strangers. Becoming a serial predator starts with years of *systematic abuse* in severely dysfunctional families characterized by psychopathic parenting with antisocial features. Societal "safety nets"—schools and churches—fail them as much as destructive parenting. In dysfunctional milieus, *pathological personality* develops in young adolescent males headed into serial crime.

Pre-adolescents and adolescents "becoming" serial killers feel little or no remorse or guilt for killing animals, stealing, hurting other children or pets, or setting fires—some of the more recognizable "red flags" of severe psychopathy. We use the term "becoming" to denote the slowly "simmering," neurological conditions coupled with *antecedent* dysfunctional experiences and developmental glitches in the formation of severe psychopathic characteristics. Inner "demons" of rage torment the future rapist/murderer as the tentacles of dysfunction slowly squeeze away any hope of normalcy. Anger at being "different" eats away at his fragile self-esteem. As experienced profilers know, it doesn't happen any other way.

Macabre Fascination

Serial predators operate just "below the radar" of law enforcement and beyond the relentless pursuit of authorities. Some move from state to state, retreating into the background on occasion only to emerge again in different parts of the country to continue more of the seemingly motiveless killings. Yet, serial homicide *always has a motive.* Sexual psychopaths obsess over the **macabre**—with control over the victim to its last breath and death as the outcome.

Personality Disorder NOS and Dual Diagnosis

In the *DSM (Diagnostic and Statistical Manual of Mental Disorders)* (2002) terminology, a convincing argument exists to diagnose sexual psychopaths with **personality disorder NOS** (not otherwise specified), due to the presence of two or more personality disorders that drive their obsession and compulsion to kill. While satisfying the clinical classification of antisocial PD behavior, the lame diagnosis of antisocial personality completely misses features of narcissism, histrionicism, and/or borderline PD and ignores the hypersexuality and grandiosity of psychopathy. Most often, serial predators present a **dual diagnosis:** They display one or more personality disorders as well as *two or more chemical addictions,* such as pornography, sex, alcoholism, or other substance abuse dependencies. The complexity of numerous personality disorders and addictions upon cognition and emotion explains why *no single root cause* is tenable.

Serial predators addicted to the "thrilling high" of raping or killing another human being as if on a drug, compulsively seek victims in the following targeted areas:

1. Shopping malls, parks, and school playgrounds
2. City streets under cover of darkness
3. Country roads in isolated rural communities or
4. In broad daylight in congested areas
5. Anyplace, USA where victims appear vulnerable, such as children "chaperoned" by inattentive adults in a crowded mall

Like wolves who survey large herds of animals for the next meal by seeking out the weak, the elderly, the sickly, or the inattentive—any member who slows the herd—human predators similarly seek out *personal characteristics* of naïve victims to approach and fool. Most have preferences. Organized serial killer Ted Bundy favored young girls with long brown hair, split down the middle resembling Stephanie Brooks, his first love who spurned his advances. Unlike disorganized offenders, organized serial predators target a favorite physical type. It is sobering that increasingly common areas of *high statistical probability* for serial crimes across North America are apartment complexes, college campuses, nightclubs, and darkened streets where prostitutes work. Serial predators routinely cruise lonely country roads as well as busy metropolitan streets trolling for victims. Predators also find hovering around convenience stores a favorite way to snatch unattended children.

Being at the wrong place at the right time can have devastating consequences. Citizens should be able to move at will through society without fear of becoming a serial crime statistic. However, it is becoming painfully obvious that citizens—especially women, children, and parents of children—must be aware of the potential dangers of serial predators. The idiom "an ounce of prevention is worth a pound of cure" applies more than ever before. There are *no longer any safe neighborhoods.*

Families should watch children like hawks because someone else may swoop down and steal them right out of their front yards. Therefore, the ***cri de coeur***—the passionate cry from the heart underlying the presentation of this material—is *raising consciousness*. Predators exist and move freely through society. Citizens and law enforcers must embrace ways of recognition, protection, and apprehension of the monsters that hide behind human faces.

Whoever Fights Monsters

The behavior of both the hunter (profiler) and the hunted (perpetrator) are subject to the tools of behavioral psychology. With so many instances of dysfunction, profilers realize the dangers of "getting inside the head" of predators. Robert Ressler states in his book *Whoever Fights Monsters* (quoting Friedrich Nietzsche), ". . . Whoever fights monsters should see to it that in the process he does not become a monster. And when you look into the abyss (the depraved mind of the serial killer), the abyss also looks into you" (Ressler & Shachtman, 1992). John Douglas recounts his mental and emotional meltdown as a consequence of becoming "too absorbed" with serial criminals in his book, *Journey Into Darkness* (1997).

Those who study serial predators are not alarmists trying to scare the wits out of readers. Instead, we seek to educate and inform others of the dangers of serial killers whose ghastly acts taunt authorities with a "catch me if you can" mentality. *Control* is a focal point of serial predators (of the organized variety), who decide *when and where to strike*. Upon apprehension, serial predators perceive their capture as mere luck. They remain arrogant, believing themselves far smarter than the police.

The "Trigger" and Compulsion of Serial Offenses

The first murder by serial killers leaves them in an emotional "no exit" situation, perhaps feeling ambivalent. On the one hand, the "thrilling high" of finally actualizing fantasy mixes with the cold irreversible reality of being a killer. In his mind, no exit exists. Once he crosses the line from fantasy to murder, he remains a criminal forever. Whether or not he experiences a tortured psyche, he has killed another human. How does he draw back from that and live a normal life? How does he address holding back the urge to kill someone sooner or later? Often, a precipitating "trigger" that acts as the final straw leads to an escalation of murderous impulses. The two most common psychological "triggers" that launch the perverse "career" of sexual homicide are (1) loss of work (income), and (2) loss of a love interest. When the flimsy "dam" of restraint breaks, a tidal wave of sexualized fantasies and rage discharges. The pattern of serial homicide has launched, although ultimately "non-fulfillment," according to Robert Ressler, former special agent of the FBI.

Serial Killer or Wrong Man Convicted?

Could intense public outcry lead to convicting the wrong person as a serial perpetrator? Take the case of William Heirens. Seventeen year old Bill Heirens engaged in petty theft. Neighbors discovered a female murder victim near one of his burglaries.

A vicious attack upon a child whose torso washed up in a drain sewer suggested a serial killer on the prowl. The press frantically sought the killer. Was Bill in the wrong place at the right time? Did police "torture" the young petty burglar into confession ostensibly to save him from the death penalty? Was Heirens the "Lipstick Killer?" Only Heirens knows the answer.

PREDATOR PROFILE 1-2

William Heirens "The Lipstick Killer"

Ashleigh Portales
Edited by Don Jacobs

Time Span of Crimes
June 5, 1945 to January 6, 1946

Offenses Prior to Serial Killing
1. Burglary
2. Theft
3. Breaking and Entering

Quoting Heirens
"I had to plead guilty to live . . . The truth was that I was innocent, but they didn't want to hear that."

Antecedent History

William George Heirens was born on November 15, 1928 into a far from happy home. Three years later, William's younger brother, Jere, arrived. Margaret and George Heirens raised their children in near poverty, a situation worsened by the Depression. To make matters worse, George spent most of his meager paychecks, earned doing odd jobs around town, treating himself and his friends at local bowling alleys. In order to help her family, Margaret worked at a bakery and left her sons alone with babysitters most of the time. The babysitters found the children difficult to control. On one occasion, Margaret returned home to find her parlor draperies and carpet burnt from a science experiment gone awry. Another time she arrived just in time to save William from jumping off the garage roof, cardboard wings strapped to his arms. Neighborhood children remembered the boy as a loner who kept to himself, tinkering for hours with model airplanes and mechanical things and drawing. Family friends predicted "interesting things for him in the future."

Young Heirens sought solitude as his parents' fights at home escalated to physical violence. Years later, he commented about the deterioration of his home life: "Jere seemed to be able to cope with it. I couldn't." As an outlet for his mounting frustration, Heirens began stealing as he entered the seventh grade. As a successful thief, he felt a *feeling of accomplishment* that he missed in his failing family. He also enjoyed "the thrill" it produced. He typically pawned the items he stole for cash and stashed the overflow in an unused storage shed on the roof of a nearby apartment building. Soon he filled it with women's furs, men's suits, radios, cameras, jewelry, and guns, the last of which admittedly fascinated him. He marveled at and loved to study the mechanical workings of the unloaded weapons. Heirens' career as a thief continued until the age of thirteen when a police officer stopped the suspicious looking boy, frisked him, and found a newly stolen .25 caliber pistol. The officer took him to the Delinquent's Home where he stayed for the next three weeks until his hearing. There he confessed to eleven burglaries and revealed the items in his rooftop hideaway. They sentenced him to the Gibault School for wayward boys, but upon his release the following November, he resumed his sticky-fingered ways because he *desired the thrill he got from the act* even though he knew it was wrong. Soon they arrested him again, this time for prowling in the Rogers Park Hotel. The police found the front door key of another hotel down the block in his possession and allegedly beat him during the ensuing interrogation. Heirens told his mother, "It was the punishment I deserved."

The judge ordered him to St. Bede's Academy, a detention center on the banks of the Illinois River in Peru, Illinois, run by the Benedictine monks. There Heirens became a model student, earning top grades and playing on many sports teams. As a result of his intelligence, the monks pushed him to take a test, which he passed with flying colors, for admittance into a special program run through the University of Chicago. Although only sixteen years old he enrolled for the fall semester of 1945 and skipped his senior year of high school.

During their son's stay at St. Bede's, George and Margaret Heirens decided that a change of scenery might help, so they moved to a large home on a substantial lot in suburban Lincolnwood. However, life inside the sprawling house didn't change, and Heirens returned to his previous pastime: stealing, which he did mostly on his visits home. Later he said that when away at school, "I wasn't even tempted. Then I would go home, and the tensions would build, and I would find myself burglarizing to ease them." This evidence shows that family milieu profoundly affects behavior. Yet, as he boarded at the school and realized his parents couldn't afford tuition and board, Heirens found another reason to steal: It supplemented his meager income from working as an usher at the nearby Orchestra Hall and at other odd jobs. As a result, he saved enough to purchase two U. S. Savings Bonds worth five hundred dollars each. To these he added stolen War Bonds that he received via his college buddies. Once in his possession, he etched the owners' names off the bonds with a surgical scalpel, also stolen, and placed them under his bed in a suitcase. This clearly indicates how antisocial behavior can merge with psychopathy. Altogether the bonds totaled over seven thousand dollars, and Heirens periodically redeemed a few at a time, when he needed the cash. And what he usually needed the cash for was girls, another pastime he had recently discovered.

When not on one of his many dates, he spent the hours with his friend and roommate, Joe Costello, with whom he discussed philosophy and played games. As a result, Heirens' grades suffered by his second year at the university. His carefree college life didn't last much longer.

The Murders

In 1946, William Heirens attended his sophomore year at the University of Chicago. Wartime rationing had ended, and the soldiers were returning home from overseas. The peacetime economy appeared on the upswing along with the crime wave that hit when the returning soldiers could not find the jobs they had left behind when the war began. People feared the city's return to the more violent ways of another time—Prohibition—when Al Capone and his thugs ruled the streets. In response, the mayor, the police commissioner, and the state's attorney sent out one clear message to all law enforcement: ARRESTS are imminent. All this along with three recent, unsolved, and unnaturally violent murders set the stage for the undoing of young William Heirens.

The first murder, of forty-three year old Josephine Alice Ross, occurred on June 5, 1945. Divorced three times and unemployed, Josephine lived with her two daughters, Mary Jane and Jacqueline, in a small apartment on Chicago's North Side. As she faced serious money problems, she looked to the promise of her current fiancé, Oscar Nordmark, soon to become husband number four.

Josephine enjoyed visiting fortunetellers and the day before had received an especially pleasing reading, which made for a pleasant awakening on that summer morning.

Both of her daughters left for work by nine, and Josephine returned to bed. When Jacqueline returned home for lunch at 1:30 P.M., she discovered a ransacked apartment and her mother's body sprawled across the bed in her room, her throat slashed multiple times, and her head wrapped in a dress. Blood soaked and spattered the entire room, and several articles of her clothing lay in a pool of bloody water in the bathtub. Police found no fingerprints and nothing other than pocket change missing. The only lead came from two witnesses in the apartment building who saw an unfamiliar dark haired man in a white sweater and dark pants, weighing approximately one hundred ninety pounds, wandering through the building on the morning of the murder. As the victim's fiancée had an airtight alibi, the case soon went cold.

The second victim, former U. S. Navy telegrapher Frances Brown, lived near the house where Josephine Ross died. On December 10, 1945, she arrived home at around 9:30 P.M., prepared to spend the night alone, as her roommate, Viola Butler, was spending the evening elsewhere. The desk clerk in the lobby of Brown's apartment building informed her that a man had inquired about her earlier in the day, but he left when he learned that she was not home. Brown continued up to her sixth floor apartment where she laid out her clothes for the next day, phoned her mother, showered, and went to bed. The next morning the building's cleaning woman discovered her body when she who wondered why the tenant's radio was playing so loudly with the door ajar. Engels entered apartment 611 and noticed it ransacked with nothing of value taken. She followed a trail of blood beginning in the bedroom to the bathroom where Brown's naked body lay over the bathtub, her head wrapped in her pajamas, a butcher knife protruding from her neck and a gunshot wound to her head. Cryptically scrawled in lipstick on the living room wall appeared the words: *"For heavens sake catch me before I kill more. I cannot control myself."* Searching the apartment revealed one fingerprint and a bloody smudge on a doorjamb. Police determined the intruder entered the apartment via the fire escape. A neighbor told police that he had heard what sounded like gunshots around 4:00 A.M. The desk clerk confirmed that around that time a man came out of the down elevator, walked nervously to the door, fumbled with the latch, and then exited. The clerk described the man as being thirty-five to forty years old and weighing around one hundred forty pounds. Local butcher George Carraboni confessed to the crime, but his story changed so many times that police seriously doubted its validity. Though the Cleveland, Ohio police simultaneously investigated Carraboni for thirteen murders of men and women, all beheaded and dismembered, they had no evidence to warrant his arrest, so they subsequently released him. As Christmas came, the case went cold, just as that of Ross had a few months earlier.

The next murder rocked Chicago with a new wave of horror, yet although it did not seem to fit the pattern of the previous two murders in any way, the investigators eventually convicted Heirens of it, too. On the evening of January 6, 1946, six year old Suzanne Degnan went to bed on her and her sister, Betty's, last night of Christmas vacation before returning to school at the Sacred Heart Academy. The Degnan family occupied the first floor of a turn-of-the-century home, and another family, the Flynns, lived on the top floor. On that particular evening, the Degnans returned home late, quickly ate some sandwiches, and hurried off to bed. During the night, neighbors heard the Flynns' dog barking, but no one paid attention to this common occurrence.

Mrs. Flynn recalled that she heard some men talking, and that at one point she heard one say, "This is the best looking building around." Mrs. Degnan awoke briefly, thinking she heard Suzanne crying, then woke her husband. Both listened for a few minutes and then drifted back to sleep when they heard nothing else. When morning came and Jim Degnan went to wake his daughters for school, he found Suzanne's door closed, an odd and surprising occurrence since she feared sleeping in the complete dark. When he opened the door, he found his daughter's window fully open and her nowhere to be found. They searched the entire home, including the floor where the Flynns lived, with no results. The police quickly responded, and police commissioner, John Pendergast, became personally involved because of the heat they were under for the two recent unsolved murders. They soon found what appeared as a discarded tissue. Instead, the murderer had left a ransom note that the breeze from the open window in Suzanne's room probably blew off the table. The note read, *"Get $20,000 ready & waite for word. Do not notify FBI or police. Bills in $5's and $10's."* On the back, it warned: *"Burn* *this for her safty.*" Outside, they discovered a seven-foot ladder stolen earlier from a nursery several blocks away. Upright, it reached the girl's window perfectly. Investigators combed the area for any and every clue they could find until an anonymous caller urged them to look in the city's sewers.

That evening, January 7, detectives Lee O'Rourke and Harry Benoit followed that advice and noticed a sewer cover misplaced on nearby Winthrop Avenue. When they looked inside, they found the head of little Suzanne Degnan. By evening's end, police found her legs and torso in adjacent sewers. Several weeks later, they found her arms. Police then located the place where the girl had been dismembered, a basement washtub below an apartment just off Winthrop Avenue, where blood, flesh, and blonde hairs sat in the drain.

Police believed that the killer must have driven the few blocks to the site of the mutilation because people would have noticed anyone carrying a seventy pound child. A witness came forward who remembered seeing a woman carrying a large bundle in her arms in the vicinity of the Degnan home and getting into a waiting car with a bald man behind the wheel, but no one ever identified this phantom couple. None of the police suspects matched the smudged fingerprints found on the ransom note and, with each passing day, the pressure on police to find the killer grew greater and greater. The treatment of murder suspect Hector Verburgh, the janitor of the apartment building where the killer had dismembered the girl, made this most obvious. Though no solid evidence existed to corroborate his involvement in the crime, the sixty-five year old man suffered two days of intense questioning from police, interspersed with physical assaults that resulted in a separated shoulder and ten days' hospitalization. He later stated:

> "Oh, they hanged my up, they blindfolded me. I can't put my up my arms, they are sore. They had handcuffs on me for hours and hours. They threw me in the cell and blindfolded me. They handcuffed my hands behind my back and pulled me up on bars until my toes touched the floor. I no eat, I go to the hospital. Oh, I am so sick. Any more and I would have confessed to anything."

Verburg later received an award of $20,000 in damages for the abuse received at the hands of the policemen. It seemed that the investigation had reached a dead end, although police theorized that the killer must have worked as a butcher during his life due to the manner of the girl's dismemberment.

Another suspect soon entered the picture: forty-two year old Richard Russell Thomas, a drifter whom police located in the Phoenix jail, arrested for the molestation of his thirteen year old daughter. While there, he confessed to the Degnan murder. At the time of the murder, he was living in Chicago and working as a male nurse, and the Phoenix PD's handwriting expert found significant similarities between the ransom note and Thomas' handwriting. He often boasted to friends that he posed as a doctor and stole surgical supplies from the hospital where he worked. He frequently visited an automobile agency a few blocks from the Degnan home and, in fact, the sewer that contained Suzanne's arms sat directly across the street from that agency. The man's history certainly supported his claims. Among other charges, he had beat his wife, had molested two of his three children, and had served time for kidnapping and extortion. Phrases that Thomas used in the extortion document appeared very similar to those used in the ransom note. It certainly looked as if police had their man—until the arrest of local teenager William Heirens.

Confession

On the afternoon of June 26, 1946, Heirens left his dormitory and walked toward the "El" (elevated) train, intending to travel to the nearby post office. Low on cash, he decided to redeem a few of the stolen bonds to pay his college debts and to take his latest girlfriend out for a date the next evening. Along with the bonds, he tucked a revolver with all seven chambers loaded inside his coat pocket. He later claimed he brought the gun as show and that he didn't know for sure if it even worked.

At approximately 3:00 P.M., he arrived at the post office, which had closed early due to the summer months. The closed post office meant that Heirens had no money for the coming week, so he resolved to take care of the problem the only other way he knew—to steal it. He set off for his target, the Wayne Avenue apartments. That particular day he found an apartment door ajar on the third floor, but the scream of a neighbor who saw him enter scared him away. Soon other residents pursued the burglar while one called the police, who quickly blocked both ends of Heirens' escape route. He pointed his gun at the police, claiming later that he only wanted to scare them away—a grave mistake. Officers charged, and a struggle ensued that resulted in Heirens' capture.

Heirens awoke in Bridewell, the hospital attached to the Cook County Jail, with his head bandaged and hearing the words "Heirens . . . suspect . . . child . . . Degnan" whispered near his ear. One by one, he felt his fingers pressed on a cold ink pad and then onto paper. He then felt his bed moving into another room followed by the sensation of many hands poking his body, which he realized they had tied down to the bed. For hours, the painful pushing and shoving continued, along with accusations that he had killed the Degnan girl. One officer, who punched Heirens in the testicles so hard that he almost vomited, shouted to him,

> "Aren't you sorry, Bill? Tell us how you did it. You know how you did it and God knows you did it. Confess, Bill, and save yourself . . . We know you're guilty. You killed her, you sonofabitch. The game's over. You're guilty. Now tell us how you did it. Tell us, Bill."

For the next six days, Heirens remained tied down to the cot as the questions and the beatings continued. At no time did they allow him the presence of legal counsel. Two psychiatrists introduced themselves to him as Doctors Haines and Grinker and proceeded to inject him with a large dose of *sodium pentothal,* often incorrectly referred to as "truth serum."

From this point, the story's "truth" remains unclear. According to authorities, the serum elicited an alternate ego in Heirens, a shadow named "George," who slipped in and out of existence uncontrollably and who had committed the murder Heirens stood accused of. When asked George's last name, Heirens responded uncertainly but said but that it sounded like "a murmuring name." The police translated it into "Murman," which the press then decided meant "Murder Man." According to Heirens later,

> "After the use of the hypnotic drug I had the strange compulsion to take the blame for all the charges pressed against me. . . . In the beginning, it didn't have much effect, but later it overcame my own will and judgment of my innocence for these crimes."

The doctors testified that State's Attorney Tuhoy ordered the test after the interview. Tuhoy admitted at first that he had seen the transcript but that it needed to be readied for release. Later, he denied any knowledge of the examination at all. However, several witnesses claim they saw him in the room at the time of the interrogation. Conveniently, the public record never received the official transcripts of the interview, and they since have disappeared altogether. Later, Heirens asked to speak with Captain Michael Ahern about his alter-ego, George. In front of Ahern, Tuhoy, and a stenographer, Heirens admitted knowledge of "George" and that this personality may have committed the murders.

On the fifth day of his incarceration, two nurses entered Heirens' room and administered a spinal tap without anesthetic, apparently to determine the existence of any permanent brain damage. Normally after this procedure, the patient remains on his back for several hours to allow the pressure in the spine to equalize again. However, they removed Heirens from his bed after only fifteen minutes, strapped him into a chair, and placed him in a waiting police van to transport him to the jail. Heirens later stated that "the ride to the detective bureau was pure hell as the police van traveled a street with a streetcar line on it. The tracks were lined with cobblestones and each jolt to my body set off a new wave of pain." Once he arrived at the bureau, they ordered him to take a lie detector test but rescheduled it for four days later because he experienced too much physical agony to go through with it. At that time, they ruled the test inconclusive. Yet John E. Reid and Fred E. Inbau, the men who invented the lie detector apparatus, disagreed on the results in their 1953 textbook *Lie Detection and Criminal Interrogation.* Citing the questions asked and the records of Heirens' responses, they concluded that "murderer William Heirens was questioned about the killing and dismemberment of six year old Suzanne Degnan . . . On the basis of the conventional testing theory, his response card clearly establishes (him) as an innocent person."

Murderer or Scapegoat?

Convinced that he had his man for the Degnan murder, and aware that he already had told investigators to give up their inquiry into the confession of Richard Thomas, Tuhoy decided to see if Heirens' fingerprints matched the partial found at the apartment of Frances Brown. He also decided to compare it with the one on the Degnan ransom note. While Heirens lay in the hospital, police announced that his fingerprints matched the one found on the ransom note, but they did not mention that it only matched by nine points. According to the FBI handbook at the time, conclusive fingerprint identification required "twelve identical points." Detectives then turned their attention to the Brown fingerprint. On June 30, Captain Emmett Evans announced at a press conference that the print did not belong to Heirens and firmly declared, "He is not the man." Then, twelve days later, Chief of Detectives Walter Storms told the press that the "bloody, smudged" print in fact belonged to William Heirens. With that, the press dubbed Heirens "The Lipstick Killer," named for the note left on the wall in Brown's apartment. Police moved him into solitary confinement at the jail and put him under round the clock observation.

Tuhoy then decided to try and link Heirens to the killing of Josephine Ross and to determine that he indeed had written the lipstick message. Initial handwriting experts concluded that the writings on the wall and on the ransom note did not match. Undeterred, Tuhoy sought out Herbert J. Walter, the famous handwriting expert who fingered the killer of the Charles A. Lindbergh baby some twenty years earlier. After four weeks of analysis, he announced that Heirens undoubtedly had authored both, which he concluded despite some attempts made by Heirens to "disguise" his natural hand. However, in early January, after the discovery of the Degnan murder but before Heirens' arrest, a reporter had asked Walter if he thought the same person wrote the note and the message on the wall. Walter had replied doubtfully and said that in spite of a "few superficial similarities," "a great many dissimilarities" existed.

Meanwhile, Heirens finally met with his counsel, John and Malachy Coghlan, two renowned criminal lawyers, and Rowland Towle, a regarded civil lawyer. On July 1, they petitioned the release of their client into the custody of the sheriff's office and met with him at his arraignment the next day. The police charged him with several burglaries along with the murders of Josephine Ross, Frances Brown, and Suzanne Degnan. With bail set at $270,000, they transferred him into the sheriff's custody, where he collapsed from fatigue as soon as he signed in, resulting in a ten day hospitalization. While in the hospital, police raided his room at the University of Chicago and confiscated certain incriminating items, among them the surgical kit Heirens used to remove the serial numbers from stolen war bonds. Though lab tests soon ruled the instruments out in Degnan's dismemberment, the press caught wind of the story, and it spread like wildfire. The police also viewed a scrapbook on Nazi soldiers and a copy of *Psychopathia Sexualis* by social crime historian Richard von Krafft-Ebbing with interest. Krafft-Ebbing's book details various fetishes and sexual perversions and chronicles several related crimes. Heirens claimed that he had acquired both items during his various robberies and took them because he thought them interesting. The Nazi book, he said, helped him understand German culture, as he was studying the language in college. Police also searched the home of Heirens' parents and took anything that belonged to their suspect. However, lab tests revealed no evidence of victims' hair or blood on any of the items, and police knew that, so far, their evidence teetered near the circumstantial. They needed an eyewitness.

They found twenty-five year old George E. Subgrunski, a furloughed soldier who told police that he had seen a man carrying a large shopping bag heading toward the Degnan residence around 1:00 A.M. the day after the Degnan murder. He described the man as "about five feet, nine inches tall, weighing about 170 pounds, about 35 years old, and dressed in a light-colored fedora and a dark overcoat," but he couldn't describe the man's facial features accurately because of the dark street. On July 11, police showed Subgrunski a photo of Heirens, but he couldn't identify him as the man he had seen. Only five days later at a hearing, however, according to the story in the *Chicago Daily,* "Subgrunski pointed a finger at Bill and said, 'That's the man I saw!' "

As the "evidence" against him mounted, Heirens followed his attorneys' advice and considered a plea bargain offered by the state's attorney. In return for his confession to the three murders, they would spare him a trip to the electric chair in favor of one life sentence. Soon George Wright, a veteran writer for the *Chicago Tribune,* fabricated a confession supposedly written by Heirens and published it in his paper. The story spread to the other major publications in the city. The evening the story broke, Heirens heard the news on a radio report and exclaimed to the officer on duty, "I didn't confess to anybody, honestly! My God, what are they going to pin on me next?" According to Heirens, he then decided to accept the plea bargain and confess, structuring his story around the one published in the papers. He met with his attorneys to practice his official confession. He now says,

> "As it turns out, the *Tribune* article was very helpful, as it provided me with a lot of details I didn't know. My attorneys rarely changed anything outright, but I could tell by their faces if I had made a mistake. Or they would say, 'Now, Bill, is that really the way it happened?' Then I would change my story because, obviously, it went against what was known [in the *Tribune*]."

He gave his official confession before the state's attorney on July 30, but apparently he changed his mind and, before a room full of reporters, answered, "I don't know" or "I don't remember" to every question Tuhoy asked. Embarrassed, Tuhoy threatened to change the plea bargain from one life term to three if he did not get the confession. Heirens' attorneys reminded him of the electric chair and the fact that it most likely lay at his journey's end if he did not confess, and he reconsidered. On August 7, he gave a full confession to each of the three murders. They set his sentencing hearing for September 4. Surprisingly, Mary Jane Blanchard, the daughter of Josephine Ross, stated to the *Herald-American,*

> "I cannot believe that young Heirens murdered my mother. He just does not fit into the picture of my mother's death. I have looked at all the things Heirens stole and there was nothing of my mother's things among them."

She felt that someone framed Heirens. After Heirens pled guilty to all three murder charges before Chief Justice Harold G. Ward, they postponed the actual sentencing until the next day. Back in his cell that night, Heirens attempted to hang himself with a bed sheet slung over a pipe. He later commented about the suicide attempt: "Everyone believed I was guilty . . . If I weren't alive, I felt I could avoid being adjudged guilty by the law and thereby gain some victory. But I wasn't successful even at that." The next day, they sentenced Heirens to life in prison and transferred him to Stateville Prison. Before he left, Sheriff Michael Mulcahy, who had supervised Heirens during his jail stay, came to his cell and said: "You probably didn't realize this, Bill, but I'm a personal friend of Jim Degnan. He wants to know—did his daughter Suzanne suffer?" Heirens looked Mulcahy in the eyes and replied, "I can't tell you if she suffered, Sheriff Mulcahy. I didn't kill her. Tell Mr. Degnan to please look after his other daughter, because whoever killed Suzanne is still out there."

Aftermath

Today, more than fifty years later, William Heirens, over seventy years old, still sits in prison for those crimes. During the time that passed, many experts have reviewed the evidence, and they have reached some startling conclusions. Several handwriting analyses of the ransom note, the lipstick on the wall, and Heirens' own writing revealed no similarities at all. This led many to claim that, in their professional opinions, there is no way that Heirens authored either the Degnan ransom note or the lipstick message.

With the help of Dolores Kennedy, author of *Bill Heirens: His Day in* Court (1991), Heirens constantly has litigated for his freedom, but he has yet to succeed. In the meantime he has become the first inmate in Illinois to receive a college degree, completing the required 197 credit hours for a Bachelor of Arts degree. As an inmate advocate, he helps others earn their GEDs and develop other educational programs.

Though the Illinois Pardon and Parole Board dismissed his life term in 1966 for the murder of Suzanne Degnan, the two others still remain in effect. That same parole board told Heirens, "You have done beautiful things. I have never seen so much accomplishment. I think you are a good example of what rehabilitation is all about."

Recently, Chicago attorney Jed Stone and Dolores Kennedy reviewed the case and developed new, exonerating, evidence. What follows is a list of their findings, per Court TV's *Crime Library:*

- Heirens' handwriting did not appear on the Degnan ransom note. In fact, several independent experts say it belonged to Richard Thomas.
- Heirens did not write the much-publicized lipstick message on the Brown wall, and the same person who wrote the Degnan note did not write it either.
- Authorities later claimed that the Heirens fingerprint originally alleged on the "face" of the Degnan note actually appeared on the back and that no way existed to confirm it.
- The so-called "bloody fingerprint" found on a doorjamb in the Brown apartment appears as a "rolled" fingerprint like those seen on fingerprint cards at police stations and unlike those most often found at crime scenes.
- Twenty-nine inconsistencies between the confessions and the known facts of the crimes appeared in analyses of the confessions. Heirens was wrong about the basic facts about the crimes, including locations, times, and related events.

To quote William Heirens in 1995: "In 1946, I had to plead guilty to live, I was seventeen and I wanted to live. I am sixty-six years old now and I will never again admit to murders I did not commit."

Name ______________________________ Date ______________

Forensic Psych Word Scholar I

Define the following words.

1. Nihilist

2. Sexual psychopathy

3. Toxic, predatory parenting

4. "Cool-coded"

5. Forensic neuropsychologist

6. Evolutionary neuroanatomy

7. "Serial"

8. "MO"

9. "Signature"

10. Serial killer

Name ______________________________ Date ______________

Aftermath

Comment on the following in one short paragraph:

Do psychopaths have an edge over the competition in the workplace and in relationships because they are remorseless in dealing with other?

Serial Killer Characteristics

Introduction to Assessment & Profiling

The connotation of the word profiling is a mousetrap required for a special breed of violent criminals, the true human predators of society—the sexual psychopaths. Given crime scene evidence, the profile produces characteristics matching the most likely suspect, known initially as the UNSUB—unknown subject in FBI lingo.

Researching the history of *clinical psychology* for the term "profiling" finds no match, unless the term "assessment" can be substituted. To the Ph.D. clinical psychologist or clinical forensic psychologist (or MD-trained psychiatrist), *both CRIMINAL profiling and PSYCHOLOGICAL assessment are investigative tools* that seek to *narrow down characteristics* that identify and uncover diagnostic information. For Example, assessment tools seek to identify *clinical pathologies* (e.g., depression or personality disorders) through diagnosis suggesting treatment **protocol.** By contrast, profiling seeks to identify the most *likely UNSUB* through careful investigation of crime scenes (crime scene diagnostics), suggesting a sketch of the UNSUB's behavioral and psychological proclivities.

However, a compelling argument can be presented that accurate *pre-crime assessment* of psychopathic tendencies, for example, might preclude the necessity of *post-crime profiling.* Are we standing on the brink of initiating a *preventive psychology?* Must a person "act out" before society moves in to protect citizens? Have we progressed as a society to trust what brain scans tell us, or do we still need more time? It seems certain we need more time.

As we will discover, both the **assessment** of suspected mental disorders in psychiatric patients who "present" symptoms in clinical interviews versus the work of forensic psychologists and crime scene investigators seeking to construct an accurate *profile* of the likely offender function for the present to *narrow down possible contingencies.* It's how scientists and practitioners "sharpen the pencil" in diagnostics. For instance, *clinical assessment* narrows down possible areas of *psychological dysfunction* and *etiology* (causation) in patients suffering from, for example, *affect disorders* such as anxiety, depression, or other pure psychopathologies. In contrast, criminal profiling narrows down a match to one UNSUB—a sexual psychopath whose identity slowly emerges from crime scene evidence.

The guiding psychological principle that drives both assessment (in the clinical picture) and profiling (in the criminal picture) is the fact that *behavior underlies personality.* Both clinicians and profilers can document how *behavior always shows personality, habits and patterns, and cognitive mapping (ways of thinking), and always gives them away.* Today, it is the contention of the neuroscience specialty of neuropsychology that the *ultimate arbiter of behavior is in fact the brain,* a position supported by this text. What's going on in the mind of cold-blooded killers can now be observed in the brain with high resolution neuroimaging.

Domains of Information

Technically, what does it mean for clinicians or investigators to assess or profile? The clinician uses assessment as a tool to analyze *dysfunctional behavior* observed or speculated upon as *presenting symptoms* in the clinical interview. He or she then measures it against *known pathological characteristics* from the pages of the DSM—the reference "Bible" of psychopathology. Like criminal profilers, clinicians rely on *inductive* methods to discover the cause of behavior, not mere *deductive speculation.* These methods involve using observed symptoms as yardsticks for measuring presenting symptoms.

Clinicians try to connect the presenting symptoms of clients with other *domains of information,* such as the **multiaxial system** of assessment clinical psychologists use. The *axial system* guards against accidental exclusion of vital clinical information as well as preventing misdiagnosis.

In the DSM (2002), for example,

- *Axis I* documents the existence of *pure psychopathologies,* such as depression and anxiety, as clinical disorders.
- *Axis II* documents *personality disorders* and *mental retardation.*
- *Axis III* reports *general medical conditions* such as tumors or blood disorders.
- *Axis IV* summarizes *psychosocial and environmental problems* such as divorce or emotional trauma upon the presenting symptoms of the patient.
- *Axis V* reports the global assessment of functioning (GAF) scale—the general psychological condition of the client observable at any given time.

Axial information is necessary to provide clients with the most comprehensive and professional diagnoses and treatment plans.

Known Offender Characteristics (KOC)

Similar to the clinician, the criminal profiler *assesses criminal evidence* presented at the crime scene such as physical evidence left behind, as well as victimology, i.e., tell-tale characteristics of the victim—why this victim and not another? The profiler then compares the information to *known offender characteristics (KOC)* of perpetrators who committed similar crimes.

Both the profiler and the clinician seek to connect their analyses to other *domains of information.* The *criminal profiler* (with laboratory help from forensic pathologists and toxicologists) seeks to connect the crime, victimology, and crime scene evidence to a specific perpetrator from KOC just as clinicians attempt to untangle the presenting symptoms to neurochemical and societal influences from known psychopathology from DSM differential diagnostic criteria.

The absolute necessity of connecting clinical information and criminal profiling to other domains of information relies on the utilization of the concept of reductionism via determinism. Such determinism represents the connection of *cause and effect* to other domains of information as *cornerstones of science.* All possible contingencies from both crimes and psychopathology (and psychopath) enter the large end of the scientific funnel of analysis. In these processes, what comes out of the small end of the funnel defines the perpetrator in forensics and the disorder in clinical practice. For forensic science, this connection explains why the courtroom in violent criminal cases has become a *theatre of forensics.*

Presenting Symptoms

Determining a client's presenting symptoms requires the clinician to assess personality and **milieu** influences, as well as possible chemical imbalances. Such assessment leads directly to identifying disorders, dysfunction, maladaptive behavior, and neurochemical underpinnings. In fact, the ***DSM*** (2002) *compiles or profiles mental and emotional disorders* from the observation of characteristics of disorders across a wide continuum of influences. These become indexed as mild, moderate, or severe. For example, a profile exists for bipolarity, schizophrenia, or dissociative identity disorder in the DSM taxonomies. The clinician makes a clinical diagnosis if, for example, clients' presenting symptoms match the profile listed in the clusters for personality disorders.

To an experienced clinician, the presenting symptoms narrow down diagnoses. In addition to a diagnosis, *DSM* **taxonomy** generates treatment protocols such as talk therapy, prescriptive medications, and possible hospital commitment. Similarly, a **psychiatric social worker** compiles a very useful tool—the **psychosocial history** assessment—which, again, "profiles" family dynamics.

The psychosocial history documents a client's diagnosis by connecting it to other domains of information, such as:

1. Family history of dysfunction/addiction
2. Communication patterns
3. Instances of lack of nurturing
4. Personality disintegration
5. Societal and psychosocial influences

Criminal profilers assess many of these *psychological influences* as they meticulously piece together predator profiles from crime scene evidence.

These techniques are similar to observing "presenting symptoms" of psychiatric patients.

Serial Killer Characteristics

Profilers have uncovered the following general characteristics of serial killers from careful evaluation of paper-and-pencil inventories, face-to-face interviews, and crime scene evidence. Serial killers tend to:

- Be white males, twenty-five to thirty-four years of age

- Have an average IQ
- Often possess charisma, or a charming personality (aka *possess psychopathic charm*)
- Be victims of abuse as children and/or be illegitimate
- Select *specific types* of victims
- Select *vulnerable victims* they can control and dominate
- Prefer to kill "hands on" such as strangulation and/or stabbing
- Display *sexually sadistic fantasies*
- Admire police work and may impersonate officers as a ruse
- Operate in all parts of North America with a higher incidence in Southern states
- Have frequent mental illness in family histories
- Have parents which often have criminal backgrounds
- Have a familial history of alcohol and/or drug abuse
- Have experienced *serious emotional abuse* during childhood
- Be sexually *dysfunctional as adults*—no satisfying, mutually satisfying, give-and-take experience in mature sexual relationships

Seven Underlying Psychological Perspectives in the Study of Violent Offenders

The following **psychological perspectives** provide the paradigmatic underpinning in the analysis and identification of **psychopathy.** Tangentially, they constitute the underpinnings of the mousetrap—the criminal profile—so effective in catching them. The following principles that provide important perspectives in understanding psychopathy *across* ***the continuum*** appear in subsequent chapters.

1. Behavioral Psychology

Behavioral psychology **(behaviorism)** focuses on learned behavioral *patterns and habits* in various social milieus (social contexts of learning) as *formative influences* in the developmental programming of all behavior, including psychopathy. FBI profilers trained in psychology contend that "behavior lies behind personality; a knowledge of *behavioral psychology* is essential in connecting the behavioral dots—habits and patterns—of human behavior, specifically psychopathic personality.

Psychopathic parenting, with features of *antisocial behavior,* appears in severe physical and/or sexual abuse that emotionally and physically disfigure childhood and adolescent behavior. "Red flags" of dysfunction often contribute to the McDonald Triad such as **enuresis** (beyond an inappropriate age of eight or nine years), cruelties to pets and peers, and fire-starting **(pyromania).**

Most notably, both physical abuse and emotional abuse can rewire the brain producing *irreversible neurological brain damage* in the brains of violent psychopaths. Startling neurological brain imaging presented in Chapter Five documents **"cool-coded"** prefrontal lobe damage, as well as damage to other brain regions such as the midbrain limbic system's amygdala, and areas in the temporal lobes and the cerebellum. Modern sophisticated brain scanning technology makes such damage discernible to a jury of peers.

Parenthetically, the general term *behaviorism* refers to the distinctly North American perspective in psychology rooted in *direct observation, cause and effect,* and **empiricism**—laboratory proof, analysis, and reporting. Behavioral psychology seeks *empirical verification* exemplified by the rigorous laboratory forensic evidence extracted from crime scenes by **forensic pathologists.** Today, *forensic evidence drives ninety-nine percent of criminal investigation and criminal prosecution.*

The Myth of the Phenomenological Self-Report

The behavioral approach to understand violent, criminal behavior has looked less favorably upon the more speculative **intrapsychic** hunches (speculative "self-reports" also known as phenomenological reports) popularized by 1970s researchers. Most notably, these appear in various *psychodynamic* perspectives focusing on unconscious motivation, courtesy of Sigmund Freud (1890).

Interestingly, Robert Ressler and John Douglas obtained similar phenomenological reports from incarcerated violent perpetrators encouraged to talk about aspects of their crimes. Enjoying the **cathartic** experience, the perpetrators gloated one more time about their crimes. In the process agents benefited by learning more about "what made them tick." The *self-report* based on direct face-to-face interviews with incarcerated offenders accounts for almost one hundred percent of *known offender characteristics.* Contrary to strict behavioral psychology, self-reports from incarcerated criminals prove uncannily accurate, invaluable, and indispensable in developing criminal profiling. This explodes somewhat the myth of the fallibility of phenomenological reports relative to violent predatory crime.

2. Forensic Psychology

Forensic psychology is characterized by testimony from **forensic psychologists, psychiatrists, and neuropsychologists**—psychologists who specialize in criminal behavior and the startling new evidence from high-resolution brains scans that have created much debate in the courtroom. These highly paid professionals provide *expert insight* extracted from crime scene evidence, victimology, and brain scans. Forensic psychologists suggest the state of mind of the perpetrator, as well as strategies in (or countering) *insanity pleas.* With neuroscience firmly planted in the courtroom, first-degree murder can now become mitigated to second-degree murder (hence a sentence of life in prison instead of a death sentence) due to compelling forensic evidence such as "cool-coded" brain scans.

Forensic psychology has deep roots in both criminology and behavioral psychology. **Forensic neuropsychology** has deep roots in biology, neurology, and medicine.

3. Cognitive-Behavioral Perspective

The cognitive-behavioral perspective focuses on the relationship between *aberrant thinking, focus, and motivation* produced by powerful *neurocognitive maps* of behavior based upon biology (nature) and experiences (nurture).

This perspective, more than any other, explains why *addiction to hardcore violent pornography* upon a neurologically damaged brain (a rewired brain) is central to the expression of sexualized violence. Connecting the two perspectives of cognitive (thinking) and behavioral (acting) creates a perfect metaphor for the interconnectedness of mind-body, the current model of **holism** in behavioral neuroscience. Our term for this inter-connective holism of **br**ain, m**ind**, and bo**dy,** is BRINDY.

4. Psychopathology (Abnormal Psychology) Perspective

Psychopathology targets dysfunctional family relationships, and pure psychopathologies, such as **DID** (dissociative identity disorder). Examples of pure psychopathologies also include anxiety, depression, bi-polarity, and paranoid schizophrenia. In diagnosing the violent perpetrators of serial crime, abnormal psychology most often documents such severe **personality disorders** as due to **antisocial personality.** As we have noted, this disorder appears more correctly characterized by **psychopathic personality** with features of sexual psychopathy. Our view rests upon New School research into psychopathy. Often, a history of precursor "red flags" in childhood and adolescence (such as **Conduct Disorder** or **Oppositional Defiant Disorder**) appears in both severe psychopathy and antisociality. Discussed at length in upcoming chapters, features of antisociality (and narcissism, histrionicism, and borderline PD) appear as features that **exacerbate** sexual psychopathy.

5. Developmental Psychology

Developmental psychology targets unsatisfied emotional "crises" brought about by *incompetent parenting* (with strong elements of emotional detachment and/or ambivalence). This appears especially true in the tradition of Erik Erikson's classic *psychosocial stages* of "lifespan" development. More severe dysfunction in self-image and in relationships occurs due to *psychopathic parenting with antisocial features.*

Primate studies showing insufficient **tactile stimulation** (i.e., Harry & Margaret Harlow and John Bowlby) and concomitant effects of emotional scarcity, lack of attachment, and lack of bonding is known to *retard* brain development as neuropsych envisions the skin—our largest sense organ—to be the outer bark of the brain. Stunted emotional behavior and severe neurological deficits in the development of the cerebellum—the brain region most affected by lack of tactile stimulation and motor stimulation before age two—rewire neurological systems. This makes conventional *talk therapy* completely ineffective as a treatment protocol for the mitigation of violence. How can a person be rehabilitated if he or she were not "habilitated" in the first place?

6. Neuropsychology and Addictionology

Compelled versus Choice

The compatible disciplines of **neuropsychology** and **addictionology** identify powerful *neurotransmitters* and *neurohormones* underlying normal thinking (cognitive), emotions (affective), and behavior. They also indicate how the identical chemistry becomes *perverted in psychopathy* through cortical rewiring due to compulsive behavior characteristic of addiction. **Neuropsychology** is the study of behavior at the tissue (cortical) level of the brain relative to **neurotransmitters**—chemical messengers of the brain—and **neurohormones**—chemical messengers that target remote cells of the body—and their impact upon behavior. In addictionology, developmental insults of physical, verbal, and/or sexual abuse add insult to injury in targeted regions of the brain.

Understanding addiction and its effect upon neurological systems provides insight into *compelled* behavior versus so-called *choice* behavior due to the devastation of neurological systems. The seat of addiction in the brain is, of course, the midbrain limbic system, specifically a region in the hypothalamus—the **medial forebrain bundle** (MFB)—with brain-wide influence as well as the nucleus accumbens (NAcc) and ventral tegmental areas (VTA)

Severe neurological abnormalities in neurocognitive *mapping* are exacerbated with addiction to sexually explicit images, including hardcore violent pornography, which authorities document as the single most devastating influence in sexual psychopathy.

7. Evolutionary Neuroanatomy

The **evolutionary neuroanatomy** of brain development no longer can be ignored in the development of psychopathy, according to neurologist Paul McLean's **Triune Brain Paradigm.** The *reptilian brain* (the brainstem) and its connections to the midbrain limbic system clearly exert influence.

With few exceptions, serial killers are not psychotic, meaning at the time of the commission of the crime, they were not insane. As noted previously, they know exactly what they are doing. One notable exception is serial killer Richard Trenton Chase, the "Vampire of Sacramento."

PREDATOR PROFILE 2-1

Richard Trenton Chase "The Vampire of Sacramento"

Jeremy Crabb, Amanda Roderick, Brandon Strickland, Kimberly Bunt, and Heather Mitchell, PA.*

Span of Time Crimes Committed
From December 29, 1977 to late January, 1978–a total of 4 days.

Childhood

Richard Trenton Chase was born on May 23, 1950 and committed suicide on December 27, 1980, living to be thirty years old. According to family reports, he was a mischievous child who grew up in a household of anger, mental illness, and alcoholism. Richard's mother was an alcoholic and a drug addict who displayed features of schizophrenia; his father was a compulsive disciplinarian who used physical abuse as a way to solve problems. As a child, Chase seemed to live in a *fantasy world.* Later, he became a firestarter (pyromaniac) who showed no remorse.

As he grew older, his crimes escalated to more serious offenses. He was a loner, showing signs of antisocial behavior. No one ever perceived him as a "ladies' man." Rather, they ridiculed him for sexual impotence. Often he found dead animals, brought them home, and likely cannibalized them. Once, he injected rabbit blood into his veins, which made him very ill. Eventually, psychiatrists diagnosed Chase with *paranoid schizophrenia* with features of *somatic (bodily) delusions.*

Physical Description of the Offender

Chase was a white male twenty-seven years of age, 5 ft. 11 in. tall, weighing approximately 145 lbs with an undernourished look. He appeared homeless, with an unshaved beard, long brown hair, and sunken eyes.

Prior Offences to Serial Killings

Most of his prior offenses were petty thefts, public intoxication, drinking while driving, possession of narcotics and resisting arrest. He was nothing more than a petty criminal. The killing began soon after his arrest for carrying a gun without a license and for torturing animals.

Modus Operandi (MO)

Chase qualified as a *disorganized* killer. He walked from house to house checking for unlocked doors and believed that a locked door meant "unwelcome." He made no plans. After finding an unlocked door, Chase entered with the intent to kill. When he found victims, he usually shot them in the head and proceeded to the next level of his sexualized crimes.

Known Signature

Postmortem, Chase raped and mutilated the body of the women he killed. This mutilation extended far beyond overkill. Chase consumed human organs and drank victims' blood. On

*PA denotes Psychology (department) Assistant

some occasions he extracted the organs and entrails to consume later. Often, Chase brought buckets with him to the crime scene to fill with victims' blood.

Current Status

Richard Chase committed suicide with a drug overdose in his cell on December 27, 1980. His body held enough medication to kill three men.

Comments

One of the few serial killers to suffer from mental illness, Chase believed he had a disease that turned his blood into powder, which gave rise to his *blood fetish*. He referred to his fetish as "The Soap Dish Disease." In a fit of mania, Chase once ran through a hospital complaining that someone stole his organs. Chase remained unremorseful for his serial crimes.

Futuristic Application

A notable difference can be delineated between *clinicians* and *profilers* in their methods of *pre-crime versus post-crime analysis.* In clinical assessment, the patient presents symptoms in real time, during therapy. In criminal profiling, the crime scene speaks to investigators and profilers who seek to reconstruct the events of the crime.

If authorities could generate a pre-crime *assessment* and a post-crime *profile* of the typology of a serial killer such as Ted Bundy, forensic psychologists (psychologists who testify in court about the validity and meaning of crime scene evidence) might be dumbstruck with the psychological similarities between the two documents.

Using speculation alone, an effective *assessment* would indicate tendencies toward *antisocial* behavior that could be validated by scores on a psychological test. Such tests could include an **inventory test** like the *Minnesota Multiphasic Personality Inventory (MMPI)* and a **projective test** that explores the deeper dynamics of personality, such as the *Thematic Apperception Test (TAT)* or *Rorschach Inkblots.* Pre-crime criminal assessment would predict propensities for future criminal behavior accurately, at least so goes the theory.

Will a time ever come when forensic psychologists, acting as "behavioral police," intervene on the basis of an assessment analysis to prevent future crime?

Department of Pre-Crime?

Academy Award winning director, Stephen Spielberg presented the notion of the Department of Pre-Crime recently in the film *Minority Report.* If authorities are ninty-nine percent sure that an individual has the *requisite psychopathic personality,* antisocial personality, or neurological damage to commit murder, rape, or serial crime, why would they hesitate to intercede before the crime?

This is precisely what happens in profiling once profilers link a particular crime to a targeted offender. Authorities seek to capture and arrest the suspect, therefore intervening before the body count rises.

Will a time come when the United States requires preventive assessment of all citizens, say on their eighteenth birthdays, as a screen for *criminal tendencies?* Would the **futurist** require this document prior to the issuance of a driver's license? Could this assessment prevent crime and reduce the use of a criminal profile?

Today, the FBI devotes more time to terrorism and less to serial crimes. Would a Department of Pre-Crime within the Department of Justice impinge upon our constitutional rights? If citizens had to answer an inventory of criminal tendencies, might it resemble that devised by our criminal profiling students?

From the Desk of Student Profilers—

Targeting Psychopathy in a Pre-Crime Assessment Questionnaire

1. Have you ever "acted out" by hitting or harming another person? No. Yes. If yes, under what circumstances? Explain. Have you had thoughts of harming another recently? If yes, explain.
2. Are you currently taking prescribed medication? For what medical/ psychological condition? Discuss any and all mental illness in your immediate family.
3. Do you agree or disagree: I do not always feel compelled to live by the rules of society. Explain.
4. Have you ever had an incident of road rage where you were the attacker? Explain.
5. Have you ever received a diagnosis of a mental or emotional disorder? Have you ever experienced confinement in a psychiatric hospital? No. Yes. Explain.
6. On a scale of 1–10 (with 10 being the highest), to what degree do you value your life and the lives of others? Explain.
7. On a scale of 1–10 (10 being the highest), how active is your fantasy life? Give two examples of your sexual fantasies. Do you believe any of your fantasies are deviant?
8. Do you harbor animosities against individuals with different ethnicities or race or gender other than yours? Explain.
9. Have you ever experienced arrest? No. Yes. Explain.
10. Are you quick to anger? No. Yes. Explain.
11. Are you currently registered as a sexual offender? No. Yes. For what offense?
12. Were you abused (sexually, verbally, or physically) as a child? No. Yes. Explain.
13. Has a court ever required mandatory attendance for you at an anger management class? No. Yes. Explain.
14. What is your opinion of men who hit women? What is your opinion of serial rapists and serial killers?

What's Eating Jack?

Let us return to our analysis of the words "profiling" versus "assessment." We can see what the FBI had in mind by developing the criminal profile when we add the word "criminal" or the words "violent crime" to the word "profiling." The difference is considerable, and this addition moves criminal profiling closer to the auspices of criminal justice and criminology. Popular culture provides a helpful analogy through Stephen King's fictional character, Jack Torrance, from the novel *The Shining.*

"*Jack Torrance picks up the ax and buries it in the chest of Dick Hollorann.*"

Frustrated novelist Jack Torrance takes his psychological pathologies a step further by killing an Overlook Hotel employee and then stalking his own wife, Wendy, and son, Danny, with an ax. Had Jack escaped the snowbound lodge, he would have made a fitting subject for criminal profiling. Had he not committed criminal offenses—one homicide and two attempted murders—his behavior would have remained non-criminal, although admittedly pathologically unstable and in need of assessment, diagnosis, and treatment.

It's a scary thought that many individuals in society are "Jack Torrances"—slowly "simmering" in sexual psychopathy, yet remaining undiagnosed as a functional dysfunctional. Their state is similar to functional alcoholics who remain psychosocially productive but struggle with alcohol addiction.

Corpus Delicti—Mens Rea, Actus Reus, and the Corpse

Criminal perpetrators "act out" by going beyond crime's cognitive component—the legal concept of *criminal mental intent* known as **mens rea**—or literally "guilty mind." Thus, they cross the invisible line between thinking about and acting out a criminal act—the physical part of the crime known as **actus reus**—or literally "guilty action."

Both *mens rea* (mental) and *actus reus* (physical) comprise what criminal attorneys recognize as elements of ***corpus delicti***—the *body of evidence* prosecuting attorneys need to win their case. (Interestingly, *corpus delicti* is often misrepresented in **pop culture,** especially in movies, as "the body of the dead victim," which is not true.) *Corpus delicti* refers to *rules of evidence* regarding the mental state of intent as it merges with the physical act to produce the dead victim. Chapter Five will investigate both *mens rea* and *actus reus* as elements of *modus vivendi*—the sexualized elements of serial crimes.

Severity of Rapacious Behavior

In the case of the serial criminal, no treatment exists to reverse his rapacious behavior. This is not because the person is just "bad" or "evil." In most cases, the reason lies with *horrific abuses from antisocial parenting* and *irreversible neurological brain damage.* What else explains their blatant disregard for human life? Due to damaged brains, they perceive the world far differently than those with normal brains. Society must incarcerate every serial killer for life with no possibility of parole or else provide capital punishment. Once serial killers get a taste of rapacious crime, they commit more and more. Serial killers never commit suicide unless cornered by authorities with no hope of escape. Obviously *self-aware* of their heinous acts, they choose to end their own cowardly lives themselves.

Narcissism

Many serial killers possess features of *narcissistic personality disorder.* Suicide is entirely out of the question when perpetrators display such **narcissistic** traits. Vincent Bugliosi, Charles Manson's former prosecuting attorney, observed this in his book *Outrage* (1995), as he recounted the O. J. Simpson trial. Convinced of Simpson's guilt and **narcissistic personality disorder,** he correctly observed (along with other forensic psychologists) that Simpson would not harm himself and never would have shot himself in the now famous LA freeway chase. Almost to the man, serial killers (of the organized typology) are narcissistic and arrogant and believe they are intellectually superior to pursuers.

Health professionals *assess* psychiatric patients with psychological dysfunction in an effort to help them attain mental and emotional health. Their focus is on *liberation and regaining self-esteem and respect,* criminals, on the other hand, experience incarceration or "warehousing" in prisons.

Profiling: Connection Between Autopsies and Psychology

As we have seen, the Atlanta child murders provided the launch pad for media and law enforcement scrutiny of the criminal profile. Although contemporary historians cannot agree on the date of the first criminal profile, numerous works of literature in recorded history brilliantly display killers' tortured minds.

- One example is Cain and Able in the Old Testament.
- Or perhaps Othello, driven to murder his wife in Shakespeare's tragedies, best exemplifies a tortured mind.
- Then there are the mystery fiction novels of Sir Arthur Conan Doyle that show detective Sherlock Holmes and Dr. Watson piecing countless puzzle pieces together in their efforts to identify criminals. Holmes even had a functional forensic lab in his Baker Street residence consisting of his magnifying glass and collection of chemicals.

Outside of fiction, perhaps autopsies provided the initial setting for criminal profiling. Some investigators suggest this possibility by pointing to the postmortem exam performed by Dr. Thomas Bond, M.D., a police surgeon, on Mary Kelly, Jack the Ripper's last victim. Dr. Bond used wound patterns from Kelly and other victims to surmise the killer's *modus operandi* (MO). Today, we view the Ripper crimes as personifying the use of MO and signature. MO is the killer's "operational methods," and signature represents that which *fulfills his sexual and psychological needs.* Investigators using these two methods can *link murders to the same perpetrator.*

However, not all profiles represent the results of physical evidence gathered at crime scenes or postmortem exams. World War II provided the profile of a war criminal and mass murder. The CIA asked a psychiatrist, Dr. Walter Langer, to provide a profile of Adolf Hitler. Since it was unrealistic to assume Hitler would die of natural causes or experience capture, the profile's essence focused on "what made him tick." In the unlikely event of his live capture, the profile suggested ways to interrogate him, a common practice today. Langer's profile indicated that Hitler saw himself as the "savior of his country," and that if his capture seemed certain, he would commit suicide, which he did. Today, it is common practice in law enforcement to profile leaders of *military campaigns* and those who claim responsibility for *terrorist attacks.*

Profilers Seek Answers Beyond Persona

The following is a list of pertinent questions a profiler seeks to answer:

1. What are his behavioral and psychological characteristics?
2. Based on the nature of the crime (related to MO and signature), what makes him tick?
3. What might be some of his most recognizable physical characteristics? Based upon what the crime scene suggests, did he have to carry the victim from place to place? Is he physically strong or weak?
4. What does the nature of the crime reveal about his fantasies? Sadistic? Sexually perverse?
5. What did he leave behind that suggests the psychosocial influences seemingly at work on his tortured psyche?
6. Where does he live?
7. Where kind of work might he perform? Does the time of the murder suggest he is a shift worker? Did he leave any evidence from his shoes or clothing that suggests his line of work?

PREDATOR PROFILE 2-2

Edward Theodore "Ed" Gein
"The Plainfield Ghoul"
"The Mad Butcher of Plainfield"

Ashleigh Portales

Time Span of Crimes
December 8, 1954 to November 16, 1957

Date of Death
July 26, 1984, age seventy-eight, of respiratory failure, at Mendota Mental Health Institute

Quoting Gein
"I had a compulsion to do it."

Preferred Prey

Though initially sexually satisfied with the body parts he pilfered from corpses, Gein's deviant desires eventually craved live victims. This led to murder. All of Gein's women, either alive or dead, were in their early fifties, with stocky build and full of figure. They all had strong personalities in life that overpowered those around them. Gein's victims were obvious stand-ins for his mother, Augusta Gein, who bred her youngest son in complete ignorance of and isolation from the opposite sex.

Mothered into Madness: The Birth of an Oedipus

On August 27, 1906, George and Augusta Gein saw the birth of their second and final child, a son they named Edward Theodore, or Ed for short. Throughout the pregnancy, Augusta prayed that the child would be a girl. As a result of her strict German Lutheran upbringing and marriage to George, a perpetual drunk and flamboyant womanizer, she felt deep resentment and aversion to all things male. Ed's father George and his older brother Henry *spent more time in the local bars than with his family or at work,* a fact that led his wife to despise him. Orphaned at the age of five, George's fervent religious grandfather raised him on a farm near LaCrosse, Wisconsin. As soon as he was old enough, he left for the city where he discovered alcohol and bounced from job to job before marrying Augusta in 1899. Their marriage never was a union of love. The exact opposite of George, Augusta was *fanatically religious* and a *harsh disciplinarian* who showed very little affection. Due to her husband's frequent absences, she ran the family's grocery store practically single-handedly. When her second son arrived, she vowed she would not allow him to become one of the loathsome men that she felt surrounded her in life.

In 1913, when Ed was seven, the Gein family closed the grocery store in La Crosse and moved to a dairy farm approximately forty miles east where they spent one year before finally settling on an isolated acreage just outside the small town of Plainfield. Here, Ed's *domineering mother* kept her boys busy from sunup to sundown with farm work. For Ed's first sixteen years, school was his only contact with the outside world. But even there the ever-present *domination of his mother kept him from naturally integrating into society.*

She raised him never to look at a woman, the main element of sin in her eyes. In addition, whenever he brought home a male friend, Augusta quickly extinguished the relationship, believing the boy a moral threat to her son. She constantly quoted scripture and reminded her sons that all boys were sinners made in their fathers' image. As a result, Ed withdrew completely from other children and became a *social recluse.* Schoolmates recalled him as "shy" and "feeble." Cut off from human interaction, Gein retreated into the world of horror comics and books about violence. His favorites, which he secretly devoured, were volumes on Nazi concentration camp experiments and human anatomy textbooks, where he concentrated most on the images of the female body. The results of this pastime appeared in idle conversation when Gein unexpectedly would make some macabre comment that left the person at an utter loss for words. This continued to happen well into adulthood.

Around him, the home atmosphere continued to deteriorate. As Augusta's hatred of her husband grew ever deeper, she began to berate him to the point where he responded with physical violence, while Ed and Henry looked on helplessly. After each attack, Ed's mother would get down on her knees and pray for her husband's death.

In 1940, her prayers were answered. The loss caused Ed and Henry to take extra jobs to supplement the family's meager agricultural income. With his father gone, Gein drew even closer to his mother and deeply admired his older brother. That adoration ended one day when Henry suggested to Ed that the degree of his attachment to his mother might not be healthy. Soon after, Henry died under mysterious circumstances. In spring 1944, a fire broke out near the Gein farm, and Ed and Henry went to extinguish the blaze. During the fight, they allegedly became separated. Acting as a concerned sibling, Ed later led a search party to the exact spot where his brother's body lay. Though there was obvious bruising on Henry's head, they attributed his death to asphyxiation from smoke inhalation.

Shortly thereafter, the one great love and hate of Ed Gein's life also passed away. Augusta Gein suffered her first stroke soon after the death of her eldest son. Ed took the next twelve months nursing her back to some semblance of health, but another stroke soon followed, resulting in his mother's death in December 1945. Ed was thirty-nine and *all alone in a world he barely comprehended.* In blatant rejection of his mother's death, Ed sealed off her room in the farmhouse, creating a makeshift shrine to her memory. Yet the behaviors deeply ingrained in her son from birth continued dictating his life, making it *impossible for him ever to engage in a healthy relationship with another living woman.* The dead, however, were another matter.

Looking for Love in All the Wrong Places

Ed never could gain the love of the one person he most desired to please: his mother. In the wake of her death, and without her strict rule and heavy judgmental hand to control his life, he felt helpless. So, as quickly as he could, he sought a replacement. Afraid of rejection by a live partner, Gein found solace in the quiet of the cemetery and decided to choose a lover from among its inhabitants. Since his mother also resided there, he reasoned, the women in this hallowed place surely were worthy of his attention. Ed began to rob the graves of the dead, but not in the traditional sense of the term.

When a woman that Ed knew in life died, he waited until the night after the funeral to return to the graveyard, dig up her body, and remove the parts he wanted. He then returned the rest of the body to the coffin and replaced the dirt over the grave. Other times he took the entire body back to the farmhouse in his truck, leaving behind only the empty coffin. In those cases, he often *performed necrophilic acts on the bodies before dismembering them.* He used the body parts to construct various objects that he kept around the small part of the farmhouse he still lived in and stored different organs in his refrigerator for later consumption. At one point, he even exhumed his mother's corpse and preserved several parts of her body.

Behind all this activity lay Gein's curiosity about the female sex. In the deepest part of his psyche, he harbored a *secret desire to be a woman.*

In his private quest for gender transformation, he used the skins of various bodies to piece together what he called a "woman suit," which he donned on the nights of a full moon, complete with a salt-preserved vulva placed in his underwear. Then he danced in his pasture by the moonlight. Yet, soon the souvenirs from his dead victims did not satisfy his carnal desires, and he turned to the land of the living.

He chose for his first victim thrice-divorced Mary Hogan, born to German parents in 1900, and owner of the local bar, Hogan's Tavern. Rumors circulated in the God-fearing town that Mary had ties to the Mob, formerly ran a prosperous brothel in Chicago, and used the money to buy the bar. On December 8, 1954, patrons entering saw a blood trail leading from the door to the back door across the parking lot where it ended in tire tracks and a spent cartridge from a .32 caliber rifle. A search revealed no trace of Hogan, who seemed to have disappeared into thin air. About a month after her disappearance, Plainfield resident Elmo Ueeck recalled that he often had seen Gein sitting alone at the back of Hogan's Tavern staring longingly at its proprietor. He suggested to Gein that, if he had played his cards right, Hogan might be back at his farmhouse making him dinner instead of missing and presumed dead. Always the oddity, Gein replied, "She isn't missing. She's back at the farm right now." Thinking it another of Gein's feeble attempts at humor, Ueeck shrugged off the comment, as did several other Plainfield residents whom Gein repeated the story to.

On the opening day of deer season three years later, practically every man in Plainfield was out in the surrounding woods celebrating the annual event, including Frank Worden, the son of the fifty-something owner of the hardware store, Bernice Worden. Bernice, a devout Methodist of spotless reputation within the town, was minding the store alone that November 17, 1957, when Ed Gein came in a little after 8:30 A.M. to have an empty glass jug filled with antifreeze. A few weeks earlier, he had asked Worden awkwardly to go ice skating, an invitation she had lightly ignored, figuring it a joke. Since that day, she often had noticed Gein staring at her from his pickup truck or from across the street. After obtaining the antifreeze, Gein left the store, but he returned a few minutes later under the guise of wanting to trade in his old shotgun for a newer model. Picking up a hunting rifle off the store rack, he slipped a cartridge from his pocket into the gun behind Worden's back and shot her. Gein then loaded her body into his truck and returned to his farmhouse outside of town.

Later that afternoon, his neighbors, teenaged siblings Bob and Darlene Hill, knocked on Gein's door and asked if he would give them a lift into town to buy a new car battery. Gein quickly stepped out of the house to meet them, his hands covered in blood. He explained that he was dressing a deer, which puzzled the Hills who knew of his distaste for blood and hunting in particular. Gein agreed to help his neighbors and, after returning to the house to wash up, he drove them to town and then accepted the invitation of their mother, Irene, to stay for dinner. Around the same time, Frank Worden returned from hunting and noticed the lights still on in his mother's store though the door was locked. Puzzled, Frank went home and got his spare key then returned to the store. What awaited him inside shocked him. Someone had torn the cash register from its place on the counter, and a large pool of blood lay on the floor near the back of the shop. Frank phoned Sheriff Art Schley who soon arrived with one of his deputies. "He's done something to her," he told them as soon as they walked in.

"Who?" they asked.

"Ed Gein," he replied. His mother's mention of Gein staring at her lately along with his memory of Gein asking if he intended to go hunting the next day had run through Frank's mind ever since he discovered his mother missing.

Meanwhile, back at the Hills' home, Gein just was finishing his dinner when a neighbor entered and told them about Worden's disappearance. Gein commented that "it must have been someone pretty cold-blooded." Irene Hill jokingly asked Gein, "How come every time someone gets banged on the head and hauled away, you're always around?" Gein simply grinned and shrugged his shoulders, then happily complied with Bob Hill's request for Gein to drive with him into town so they could see what was going on. As they walked out to the car, Sheriff's deputies arrived in search of Gein and ordered him into their squad car to return to the station for questioning. Officer Dan Chase began by asking Gein to recount the events of his day and then had him repeat the story a second time, which resulted in many inconsistencies. When Chase voiced his suspicions, Gein blurted out,

"Somebody framed me!"

"Framed you for what?" Chase asked.

"Well, about Mrs. Worden."

"What about Mrs. Worden?" asked the officer.

"Well, she's dead, ain't she?" replied Gein.

"Dead!" exclaimed a surprised Chase. No one knew if Worden was dead or alive as she still was missing. The police only said that she was missing.

"How d'you know she's dead?"

"I heard it." Gein lied. "They told me in there."

Hearing of Gein's apprehension, Sheriff Schley headed over to the man's farmhouse with Captain Lloyd Schoephoerster to see if they could uncover anything. Nothing could have prepared them for what they found.

Inside the House of Horrors

Schley found the door off the kitchen open and entered by flashlight, as the farm had no electricity. Stepping inside, Schley felt something brush against his shoulder and spun with his flashlight to see what it was. Before him, cast in the beam of his small light, the headless corpse of a woman hung from the ceiling, completely trussed, dressed, and skinned like a deer. The officers radioed for help and returned to the house where they discovered Gein had hung the body by piercing the tendons of one ankle with a sharpened tree branch and slitting the other foot below the heel and tying it to the pole with a wire. The head and throat were missing, and Gein had cut out the anus and genitals. Filth covered the surrounding room; rotting food, dirty dishes, and soiled rags of clothing lay everywhere. Empty cans and rusty farm tools littered the floors. Sand filled the sink, and a long row of dentures lined the mantle.

Soon police cruisers filled the land surrounding the farmhouse, and police brought in a generator to illuminate the premises for search. Bathing the house in artificial light revealed even more grisly horrors. The ensuing search revealed:

- two shin bones
- four human noses
- an empty quart can converted into a tom-tom by human skin stretched over both top and bottom
- bowls made from the inverted halves of human skulls
- nine death masks made from the well preserved faces of women, four of which hung on the walls around Gein's bed
- ten female heads with the tops sawn off above the eyebrows
- bracelets of human skin
- a purse with a handle of human skin
- a knife sheath made of human skin
- a pair of leggings made from human skin
- four chairs with human skin stretched over the seats
- a shoe box containing nine salted vulvas, of which Ed's mother's was painted silver
- a hanging human head
- a lampshade of human skin
- a collection of shrunken heads, including that of Mary Hogan
- a shirt made of human skin
- two skulls topping the posts of Ed's bed
- human lips strung together
- a refrigerator full of human organs
- a full "woman suit," complete with mask and breasts, composed entirely of human skin
- Bernice Worden's heart in a pan on the stove
- Bernice Worden's head in a paper bag, hooks driven through the ears with a string between them in preparation for hanging on the wall
- Bernice Worden's bowels wrapped in an old suit

Eventually police concluded that Gein's menagerie of horror included the bodies of at least fifteen different women.

The Master of Horrors

With the search of his farmhouse underway, Ed Gein sat in jail under interrogation for the next twelve hours without saying a word. The next day, he broke his silence. He confessed to shooting Bernice Worden and to mutilating her body, stating that he could not recall certain specific details of the crime as he felt "in a daze at the time." Under further questioning, he said that he thought it all was an accident. He then described how he dressed the body and drained the blood into a bucket, which he then buried in a hole in the ground. When asked if he thought he had trussed up a deer, he replied, "That is the only explanation I can think was in my mind." He then told police that, to the best of his knowledge, he killed no one besides Worden, and that the rest of the body parts in his home had come from the bodies he dug from their graves. He confessed that he had known most of the women when they were alive and that, after he took what he wanted, he returned the grave to what he cheerfully described as "apple pie order." When the police addressed the subject of necrophilia, Gein adamantly denied the charges, shaking his head and crying, "No! No! They smelt too bad!"

Later he admitted to killing Mary Hogan as well, though he claimed not to recall any of the details. He described his many instances of grave robbing, detailing how he lifted the casket lids with a crow bar, sometimes only removing the head by sawing across the neck and snapping the spinal cord. Other times he removed additional parts or even the entire body, then replaced the lid and refilled the grave. Throughout his many interrogations, Gein remained calm and cooperative, literally void of all emotion or remorse.

On Thursday, November 21, police formally charged Gein with the murders of Hogan and Worden. He pleaded insanity, and the judge committed him to the Central State Hospital for the Criminally Insane in Waupun, pending psychological testing. Eventually they ruled him schizophrenic, a condition that often has roots in childhood emotional traumas. Deemed unfit to stand trial, he received indefinite commitment to the hospital on January 1, 1956.

Aftermath

Outrage consumed the town of Plainfield that the man who caused such horrors in their lives would not stand trial for what he had done. Though the judge assured them that the ruling did not mean he never would stand trial in the future, they became even angrier when they learned that authorities had decided to auction off the Gein farm and various items. That auction never took place because on the evening of March 20, 1956, the farmhouse went up in flames while Plainfield's residents stood by and watched. Though the cause of the fire remained undetermined, when Gein himself learned of the incident, he commented that it was "just as well."

In May 1960, dogs digging on the property uncovered another pile of bones, bringing Gein's catalog of bodies to its final total of fifteen, including Mary Hogan and Bernice Worden. In January 1968, hospital authorities informed Judge Robert Gollmar that they believed Gein was now mentally fit to stand trial. Though the Judge felt it would be a waste of time and money, he allowed the trial to proceed. The trial lasted just a week the following November and brought forth a verdict of "guilty" of murder but "not guilty" because of Gein's obvious insanity. Thus, the court remanded him back to the care of the Central State Hospital.

In February 1974, a petition filed on behalf of Gein requested his immediate release, claiming that he had recovered his mental health during his sixteen years of his incarceration. Judge Gollmar ordered a fresh battery of psychiatric tests and the release of the results in a hearing. Waiting for the hearing to begin, Gein chatted with reporters outside the courthouse, revealing his desire to take a trip around the world when freed. However, four different doctors presented Judge Gollmar with overwhelming evidence that, in their professional opinion, he never should release the blatant psychopath.

They contended that though he could appear well organized on the surface, his psychosis waited shallowly beneath "ready to be reactivated under the right conditions." Thankfully, the judge rejected Gein's petition and returned him to the hospital.

In 1978, authorities transferred Gein to the Mendota Mental Health Institute, where he died of respiratory failure on July 26, 1984. They buried him in an unmarked grave in the Plainfield cemetery in the plot of land directly beside his mother.

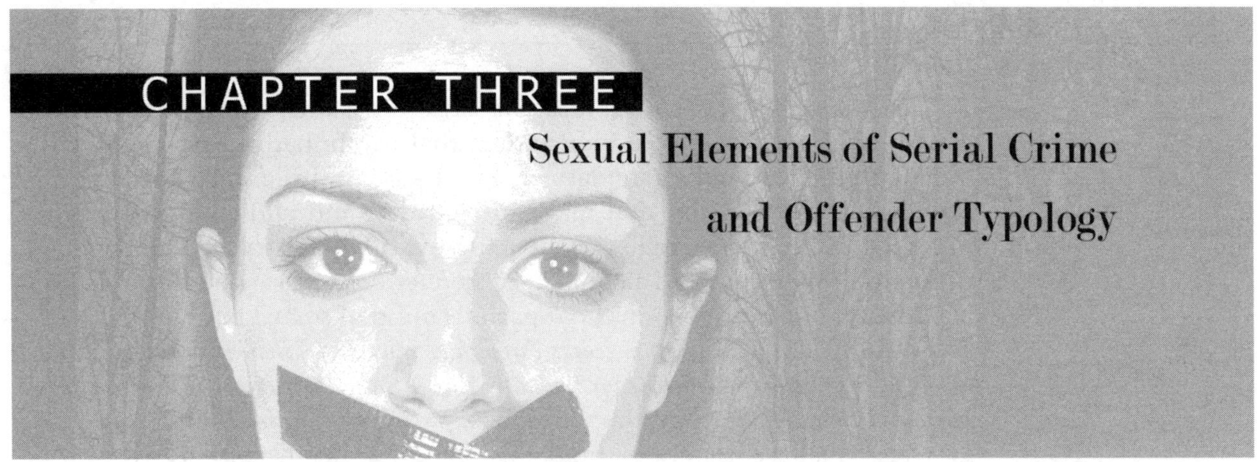

Sexual Elements of Serial Crime and Offender Typology

Manie sans delire . . .
"By the nineteenth century, *abnormality in the human mind* was being linked to criminal behavior patterns. Philippe Pinel, one of the founders of French psychiatry, claimed that some people *behave abnormally even without being mentally ill.* He coined the phrase *manie sans delire*—"MADNESS WITHOUT CONFUSION OF MIND"—to denote what eventually was referred to as psychopathic personality."

—*Criminology,* L. Siegel (2003)

Mens Rea—Criminal Motive, Fantasy, and Eroticism

Chapter Three presents the *five sexualized components of serial crime* from start to finish beginning with an all-consuming obsession with erotic fantasy forged from an addiction to hardcore pornography, or perhaps sexual repression from a dominating mother (for those with Freudian leanings), to the aftermath that follows the victim's death. Next, the two types of serial killers—the organized and the disorganized will be analyzed.

In behavioral psychology, "reading" the mind of a killer is an *inferential* task accomplished by observing behavior, the nature of the crime, and choice of victim—all against the yardstick of *known offender characteristics.* Taken together, profilers must project personality and psychological proclivities into a composite reflecting the most likely UNSUB.

Interestingly, Freud was the first maverick personality theorist to use the words "mind" and "personality" interchangeably. Moreover, he preferred using the term "character disorder" over personality disorder to describe longstanding personality deviance. In actuality, Freud's preference better describes the personality of sexual psychopaths. It seems the word c*haracter reflects more profoundly who a person really is* as character is closely aligned to *"moral character"*. Believing a person's *character* to be emotionally deviant, then "once a pedophile, always a pedophile" or "once a sexual psychopath, always a sexual psychopath," reflects the ingrained nature of character. Modern neuropsychology may scoff at rehabilitation, but more correctly, it remains true to the profundity of brain wiring and the superhuman effort required to rewire it, yet again, in rehab.

The *sexualized components of rapacious crime—"the thrilling high"* experienced by perpetrators is the best evidence of sexual psychopathy. This chapter is a prelude to the role addiction plays in sexual serial murder and rape. The cortical regions of the brain that make intoxication and addiction possible are presented in Chapter 4, followed the pathways of chemistry in Chapter 5. Addiction to illegal drugs and alcohol in subsequent chapters show how the torch of addiction burns brightly in the rapacious mind of sexual predators

As we will show, *Addiction works through emotions to "compel" behavior, while a person who is neither addicted nor mentally ill continues to rationally "choose" what course of action to take.* In psychopathy, *feelings* trump *cognitive restraint.*

The sexualized aspect of predatory behavior begins with *mens rea*—the criminal motive—a confusing mix of anger, frustration, fantasy, and perverted erotic desire.

In criminal law, *mens rea* is the legal term that describes the mental "fuse" that ignites the observable behavior of crime.

The word **motive** derives from the Latin *motus* (pp. of *movere*) meaning "to move" and clearly is a verb of action. Motive implies emotion or desire operating on will (cognition), which causes action. In this definition, *emotion "wags the tail" of thinking to produce behavior.* Clearly, *emotional thinking* motivates the killer. This sounds like an oxymoron, yet it is similar to the concept "heart over head."

The predator is deep into sexualized fantasies, anger, frustration, and perverted eroticism when he selects a victim (an organized offender) or crosses the path of an unfortunate victim (disorganized offender). His perplexing neurochemical cocktail drives—*compels*—sequential rape and murder. The serial killer's murderous thoughts gravitate around *emotion*—the feeling that comes with complete domination, degradation, and sexual experimentation with the victim. Sadly, it's the only thing that "jazzes" him in his otherwise depressing life.

Following our discussion of the sexualized motivation inherent in criminal intent (*mens rea*), we will look at these remaining elements: the physical act of murder, *actus reus,* and the corresponding components, namely *modus operandi* and signature. We also will look at the aftermath, the time after the victim's death.

Manipulate, Dominate, and Control

The study of motivation always has intrigued psychologists. From the classic studies of Abraham Maslow's *hierarchy of needs* and McClelland's *achievement motivation* to the celebrated **Hawthorne Effect** of Elton Mayo in his "Western Electric Studies," researchers have sought insight into human motivation. Blue chip companies pay **I/O psychologists** six and seven figure amounts to motivate employees to work toward higher profits against fierce competition. In forensic psychology and criminology, knowledge of motive (and *signature*) often separates the exceptional profiler from the average one.

Sexualized motivation may not be obvious at the crime scene, but it's the first thing on the mind of an experienced profiler.

Sexualized *Mens Rea*

Sexual fantasies and the behavior leading to the kill (similar to sexual foreplay), jazzes the serial predator's perverted emotional life like nothing else. Yet, ultimately, it is only temporary. As Bundy noted, the "brutal urge" always comes back stronger.

Sexualized fantasies, spawned from an addiction to **pornography, voyeurism,** and/or *fetishism,* and enhanced by alcohol or drug addiction (what we have termed poly-addiction) provide the emotional "fuse" of *mens rea.* Upon capture, serial killers often act arrogant and unremorseful. They are not sorry for committing crimes; rather, they are sorry they got caught. It is a sad irony that serial murderers try in vain to gain a measure of control by dominating, humiliating, and ultimately killing others when their personal lives are full of chaos and havoc.

The more serial killers succeed in victimizing others, the more addicted they become to their scary, inner world of sexual exploitation. This world consists of an addiction to hardcore, violent pornography, or in the alternative, features an all-consuming sexual repression leading (as in the case of Ed Gein) to pathological curiosity. According to the FBI's study of serial homicide, an addiction to hardcore pornography is the most common thread uniting all serial killers. As sexual tension builds from his full-blown porn compulsion, the serial predator eagerly plots the next murder in order to use what he learned from his last one. Compulsively watching pornography keeps sexual tension high. The organized serial killer becomes **obsessive-compulsive** in his macabre fascination with sexual exploitation through rape and murder.

Through journal entries and/or "trophies," the killer keeps souvenirs of the crime. These may include articles of the victim's clothing, a bracelet, or a shoe. Sometimes in bizarre instances of cannibalism, he keeps a body part (Jeffrey Dahmer favored the victim's head). By crossing the line and committing the first homicide, most serial killers feel compelled to kill again and again, in the same way the addict seeks his drug of choice.

A few offenders do realize the severity of their crimes and become overwhelmed and apprehensive. They appear begging to get caught. However, this is extremely rare.

Isolation Drives *Mens Rea*

As Chapter Two mentioned, the developmental psychologist Erik Erikson targets the crisis of *intimacy* versus *isolation* as a major influence in young adults (ages twenty to mid-thirties), which is the age range of most serial predators. If we merge Erikson's theory with brain neuroscience, we may theorize that the neurologically deficient brain is on a collision course with outrage due to *lonely, isolated, and emotionally disconnected feelings.*

Perhaps Erikson would agree that the serial offender is unable to bridle predatory thinking because his sexual fantasies are too **dissonant** for his damaged brain and lack of experience to handle. He may be overwhelmed with feeling that no one around him understands, that he is in *isolation.* Perhaps his only comfort comes from the quick "fix" of compulsive masturbation and fantasy. There's always the fantasy.

In Predator Profile 3-1, we meet Richard Ramirez, "The Night Stalker." His one adult mentor was his "pot"-smoking older cousin, a Vietnam veteran who delighted in recalling his sexual conquests and brutality during his tour of duty. Ramirez began his life of crime as a voyeur and a rapist, finally becoming a serial killer.

PREDATOR PROFILE 3-1

Richard Ramirez "The Night Stalker"

Ashleigh Portales

Time Span of Crimes
June 28, 1984 to August 31, 1985
Offenses Prior to Serial Killing
Possession of narcotics
Theft
Breaking and entering
Assault
Attempted rape
Quoting Ramirez
"Even psychopaths have feelings.
Then again maybe not."

Demon Child

Ricardo Levya Ramirez, aka Richard, was born in El Paso, Texas in 1960, the youngest of five children. Richard's neurological state was under fire from the very beginning. His *hot-tempered father* frequently used Richard as a punching bag as the outlet for his rage, and a male teacher inflicted additional abuse. Richard often tried to avoid his father by hiding in the local cemetery. At times he even spent the night, because he *felt at peace in this macabre refuge.* According to forensic psychologist Dr. N. G. Berrill of John Jay College of Criminal Justice, one method to overcome one's fears is "to identify with what's frightening you. One way to do that is to become a frightening person yourself." Richard soon did just that.

In school, Richard was somewhat of a *loner.* Because he suffered from *epilepsy never properly treated,* he had to give up football, a staple of popularity in West Texas culture. Despite this disappointment and the ridicule he received for his thin feminine figure, Richard *determined to become famous. He longed to be noticed, to stand out in a crowd, and to make a difference.* He began to learn how to make a difference, albeit a negative one, from his older cousin, Mike, who served as Ramirez's substitute father. After serving as a Green Beret in Vietnam, Mike returned from the exotic land covered in tattoos and toting a treasure chest of war "trophies." He introduced thirteen year old Richard to *marijuana* and constantly bragged about raping and murdering a number of Vietnamese women and children during his tour of duty. And he had the pictures to prove it. He showed his cousin numerous Polaroids of himself in sexually dominating positions with his victims. Later photos showed him killing those same people. He told Richard that "killing made [him] feel like a god . . . There was nothing more powerful."

Ramirez soon became *desensitized to the images, which became eroticized in the developing pubescent boy's deviant sexual fantasies. He began perceiving people as objects whose sole use was suffering and degradation for his pleasure.* He idolized Mike for his ability to kill without qualm. In addition to his "war memorabilia," Mike also passed his knowledge of *predatory hunting* on to Richard. The two spent many a night in the desert observing and sneaking up on animals. Once Richard mastered this skill, Mike taught him how to kill his prey with a knife or gun and, ultimately, how to mutilate the corpse. Richard's final instruction from his cousin came one night when Mike's wife began to nag him about his marijuana use and lack of a job. Annoyed, Mike drew his revolver from his place on the couch and shot her in the face, killing her instantly.

Mike told Richard, spattered with the woman's blood, to leave, but later Richard returned with his father to survey the scene and never confessed he had witnessed the crime.

Richard later confessed that he had *begun to feel a connection with the dead, which bordered on the mystical. He had a growing desire not only to frighten people but to mutilate them, degrade them, and strike waves of fear in the hearts of those around them.*

In order to fulfill this desire, and to live up to the legacy of his role model Mike, Richard dropped out of high school his freshman year and began cleaning hotel rooms when he wasn't smoking pot. This provided ideal access to women, and one day he surprised a guest, tied her up, and attempted to rape her. Fortunately her husband interrupted the assault and turned Ramirez over to the police. In a blatant miscarriage of justice, the judge committed Richard's sentence to probation because he was only fifteen. The budding killer was on the loose.

In 1978, when Richard was eighteen, he moved to southern California where he continued his *parasitic lifestyle* of stealing to supply marijuana and LSD and to sustain his steady diet of convenience store junk food. His teeth rotted from such large amounts of sugar, creating a horrible halitosis. This fit perfectly into Ramirez's idea of becoming a "demon," a notion he acquired from the Church of Satan, which he discovered upon his move to California. The church's "sacred" *themes of dominance, power, and control beckoned to Richard.* Its teachings fit Richard's temperament perfectly, and he turned his weakness into strength there. Satanism pervaded California culture, and "satanic panics" swept the country. Police had arrested teachers at the McMartin preschool not far from where Richard lived and accused them of being a "ring of Satanists corrupting children."

Throughout the country, serial murders took place as satanic sacrifices. According to Katherine Ramsland, Ph.D.:

> "During that decade [the 1980s], Robert Berdella killed six men in Missouri for satanic purposes, Antone Costa killed four women in Cape Cod in rituals, Thomas Creech admitted to 47 satanic sacrifices, and Larry Eyler buried four of his 23 victims under a barn marked with an inverted pentagram. Nurse Donald Harvey, suspected in the deaths of 47 patients, admitted to a fascination with black magic, and Leonard Lake, who had teamed up with Charles Ng for a series of torture murders, was affiliated with a coven of witches. One killer targeted homeless men, ringing his victims with a circle of salt . . . A former associate of John Wayne Gacy named Robin Gecht inspired a group of three men known as the Ripper Crew in killing an estimated eighteen women. They would murder a victim, sever her left breast with a thin wire, clean it out to use for sexual gratification, and then cut it into pieces to consume. Ostensibly, they were worshipping Satan, and eating the flesh was a form of demonic communion."

A teen even murdered his parents in their bed to demonstrate his allegiance to Satan. Because people feared Satan and his followers, Richard aligned himself with the Prince of Darkness so they would fear him, too. He began wearing dark clothing, displaying pentagrams, and prowling the streets at night. He adopted the Australian rock band AC/DC's song "Night Prowler" from their Highway to Hell album as his personal anthem. The cultural zeitgeist was ideal for a sexual predator to emerge, and Richard Ramirez was ready to do just that.

Demon Unleashed—Hell on Earth

Ramirez relied on his predisposition to breaking and entering to gain access to his victims. On June, 28, 1984, his black-clad figure blending with the night, he removed the screen of an open window and entered the ground floor apartment of seventy-nine year old widow Jennie Vincow. Vincow's son lived directly above and found her body the next morning. The amount of *overkill* was astonishing. She had suffered numerous stab wounds, as well as sexual assault, and he had gouged her throat with so much force that she appeared nearly decapitated. Ramirez often employed this technique to *heighten his sexual experience.* He ransacked her apartment, stole her valuables, and left her body sprawled across the bed.

Ramirez had gotten away with his first murder, and the empowerment spurred him to even greater degrees of violence. After savagely bludgeoning two elderly sisters, he carved a pentagram into one of the women's upper thigh. Such satanic symbols became the calling card of the Night Stalker. He left pentagrams in lipstick on the wall, mutilated victims with them and with the AC/DC insignia, and removed eyes in ritualistic fashion. If he encountered a boyfriend or a husband, he removed the threat with a single shot to the head from a .25 caliber pistol. Clearly, the *women were the targets.* Amazingly, many of Ramirez's victims survived, though he believed them dead when he left. Thirteen, however, lost their lives before police apprehended him.

Demon Returned to Hell

On August 31, 1985, while Ramirez attempted to hijack a car, the female occupant began to scream. Onlookers intervened and overpowered him until the police arrived. He then began to *boast about his crimes* in true psychopathic fashion. "I am a minion of Satan," he claimed, "sent to commit the Dark One's dirty work." In his confession, Ramirez told police, "I love to kill people. I love watching them die. I would shoot them in the head and they would wiggle and squirm all over the place, and then just stop. Or I would cut them with a knife and watch their face turn real white. I love all that blood . . . I told one lady to give me all her money. She said 'no,' so I cut her and pulled her eyes out." Ramirez carried his satanic flamboyance over to the courtroom. At a preliminary hearing, he displayed a pentagram tattooed into his palm for the press. The story caught like wildfire, making Richard an instant celebrity. *He loved it.*

Basking in his notoriety, he played to the crowd every chance he got. Upon entering the court room at trial he shouted, "Hail Satan!" He donned dark sunglasses, and tried to intimidate jurors with long, menacing glares. The demonic persona convinced the jury, and they convicted Ramirez on thirteen counts of murder. When his lawyers informed him that he could receive the death penalty he replied, "I'll be in hell, then, with Satan."

When, in fact, he did receive the death sentence, he went to San Quentin's Death Row. There he told guards he hoped people would write books about him, like they had about Ted Bundy and Jack the Ripper. He further confided his thrill that a movie would depict his crimes. Ramirez wasn't alone in loving his fame. Throngs of women corresponded with him, many proposing marriage. Richard accepted one proposal, and a marriage ceremony took place on Death Row, where he still resides.

Did Ramirez honestly believe in Satan and commit his heinous acts to serve the Dark Lord? Psychology tells us the answer is probably no, yet Ramirez's milieu provided the ideal cover to kill and acquire the timeless notoriety he desired.

Actus Reus—The Criminal Act

Secret pornographic archives provide the means to the sexualized fantasy (*mens rea*) that ignites serial killers, but the physical act, the *actus reus*—the act of murder—brings law enforcement into their orbit. Parents teach children from a young age that they are free to think any thoughts they want without consequences, but that acting upon them is another story.

In criminal law, the term *actus reus* denotes the physical component or the criminal act that constitutes a serious crime against a person (**malum in se** crimes from British common law). In sexualized crimes, we see *actus reus* characterized by violations of the victim's body by the perpetrator's penis, fingers, hands, knives, ligatures, or other objects that inflict torture, pain, and/or death.

In murder by firearm, actus reus is the simple act of *pulling* the trigger; in a stabbing death, it's the slashing, plunging, or thrusting *motions* of the knife. In ligature (garrote) strangulation, it's the *motions* required in tying the knot or pressure on the tourniquet that result in death. In sexualized crime, perpetrators *seek to prolong actus reus* as long as possible. Ironically and eerily, the sexual behavior of serial killers during *actus reus* is comparable to the *foreplay phase* of the human sexual response. Sustaining this stage as long as possible prolongs the killer's erotic fantasies. It's a vital part of the *addiction factor* of serial rape and murder.

Fantasy and Erotic Cognitive Maps

According to sexual researchers (Hendrick & Hendrick, 1992), consenting partners must *feel safe* in sexual encounters to experience full and comfortable intercourse or sexual pleasuring. The serial killer intent upon sexual exploitation replaces the intimacy of consensual sex with control, domination, degradation, and manipulation.

Because *serial killers are sexual perverts,* they seek victims to sexually exploit. Their powerful, sexually perverted *cognitive maps* inexorably lead them to murder (*actus reus*). By contrast, normal, healthy, neurological development produces balanced neurochemistry and hormones. These produce far different erotic "maps" than those of the neurologically damaged brain of the sexual psychopath. Violent, hardcore *pornographic eroticism* is the predator's single most damaging component.

In strikingly similar ways, the *four-stage model of sexual arousal* by classic sex researchers Masters and Johnson offers insight into the serial killer's sexualized crimes.

The Human Sexual Response and Sexualized Crime—Striking Similarities

According to Masters and Johnson, the first stage of the human sexual response is the **excitement phase** prompted by erotic thoughts, sights, or physical contact. The most marked aspect of the excitement phase in males is vasocongestion of blood in three spongy cylinders—two *corpora cavernosa* on the top of the penis and one *corpus spongiosum* on the bottom—resulting in penile erection.

An extended, multi-layered period of the excitement phase is not unusual among serial rapists and killers. First, erotic fantasy fueled by pornography is partially responsible for targeting a victim. Then, the search begins. Sexually jazzed, they drive around in neighborhoods, perhaps unfamiliar to them, all the while searching, fantasizing, and anticipating. When the serial predator targets a victim, the excitement phase may peak and become almost unbearable, as he has spent vast amounts of time in preparation. If he fears being seen and captured, he may masturbate in his car. If possible, he may brush up against her in a crowd for a brief moment of **frottage.**

The **plateau stage** of Masters and Johnson is equivalent to feeling that the sexual act is imminent. Physiologically, males experience intense muscular tension and full erection. As the fantasy of the serial rapist or killer transitions into confronting the victim, the plateau stage compels him to move forward and capture her with actus reus. In the plateau phase, rapists and killers often *take as much time as possible preparing the victim* for rape or murder, once they have incapacitated victims with handcuffs, ligature or **rape kit.** Just like lovers in a sexual frenzy, the killer begins his sexual siege.

According to Masters and Johnson, even though it lasts no more than fifteen to twenty seconds, the **orgasmic phase** is the summit of the human sexual response cycle. Orgasm, ejaculation, and the release of muscular tension characterize it. The serial killer's orgasmic phase—the expenditure and release of sexual tension—may occur during any or all of the following:

1. rape
2. fondling
3. masturbation
4. stabbing
5. ligature strangulation
6. beating
7. vaginal/anal intercourse
8. sodomy
9. sadism
10. fetishism
11. overkill (mutilation)

Many serial killers have admitted to criminal investigators the dread they experience in killing victims, because death ends the *highly anticipated* plateau and orgasmic phases (unless they prize necrophilia). For example, Jeffrey Dahmer experimented with using "living zombies" as sex slaves by pouring acid into holes he drilled into victims' skulls in the hope of keeping them alive for continued sexual experimentation.

According to Masters and Johnson, the final stage, or aftermath, following orgasm is the **resolution phase.** Physiologically, the body returns to its prearousal state. Muscle tension subsides, and all physiological responses return to normal, such as heart rate, breathing, and blood pressure. In Ted Bundy's words, the "brutal urge" to touch, to control, to penetrate the victim in *sexualized signature,* finally subsides.

Halpern and Sherman (1979) suggest that, given an array of choices, normal sexual partners prefer to touch in the resolution phase. In contrast, the serial killer experiences increasing sexual tension as he kills his victim; following rape or murder, he may continue to brutalize the corpse (overkill). In instances of **necrophilia,** he continues to have sex with the corpse long after the victim is dead. He may return days later and continue.

Modus Operandi (MO)

Step-By-Step Procedures within *Actus Reus*

According to Jacobs (2003), *Modus operandi* (MO in police and FBI lingo) is the third component of *modus vivendi*—the sexualized components of serial crime. Part of actus reus, MO, reflects typical physical methods and procedures the killer uses as he proceeds, step by step, with the crime. In contrast to MO, signature *personalizes* the state of *emotional attachment* to the crime for a specific perpetrator.

MO is a legal term that denotes the perpetrator's *physical procedures step by step* during the commission of the crime. MO is "how" he did it.

In rapacious crimes, MO is a function of learning, experience, and modifications employed from feedback at the crime scene. Hence, it is not surprising to find MO *routinely modified* as perpetrators gain more experience and confidence. Organized serial predators are very astute observers; they pay particular attention to every detail of the crime; they are meticulous to a fault with crime scene feedback. According to Ressler and Douglas, this is not the case with the *disorganized* offender whose MO, if any, is haphazard.

Dynamics of MO

MO exists as a dynamic, hence changeable, quality that develops over time. For example, a serial offender may learn from direct experience during his first crime how difficult it is to prevent his victim from screaming, so he uses what's available at the crime scene, such as a sock or a piece of cloth that he forces into her mouth.

Subsequent crimes may show the offender "graduated" to a **"rape kit"**—consisting of duct tape, rope, or other supplies needed to render the victim helpless. At best, MO shows the evolution of a given predator's step by step *actus reus.* Hence, it is important to note that, due to its evolving nature, *MO is not by itself an accurate gauge* in apprehending serial predators.

Shaping Criminal Behavior

Behavioral psychology uses the following terms to explain how experience provides feedback to "shape and maintain" learned behavior:

1. **Successive approximations**
2. **Shaping**
3. **Chaining**

Small steps and chaining evolve into shaped behavior, which is the final product of behavior, and become evidence of MO within *actus reus*.

Behavioral psychologists view the *small steps* taken by the serial criminal during the commission of his crime as comprising *successive approximations* towards the final end result of death. For example, the amount of time a predator watches a targeted victim (*voyeurism*) and carefully documents her "comings and goings" helps him decide the best time and best place to commit the crime. As serial killer Ted Bundy admitted in an interview, "the victim may not know the perpetrator, but the perpetrator certainly knows the victim."

The offender's careful analysis may disclose the best way to enter a targeted residence is through a window in the back of the property away from heavily traveled streets. Wearing a ski mask to prevent detection, or deciding on the best way to surprise the sleeping victim are methods of accomplishing the crime that comprise his *successive approximations*—the summation of all the small steps used. When the *small steps* of MO are combined into a sequence, a seamless "chaining" of behavior produces *shaping—the behavioral sequence* that produced the crime scene.

Unless investigators observe the entire process, they never know every specific step the killer takes in committing the total crime (including its aftermath) with certainty. However, due to the accuracy of statistical analysis and crime re-construction with computer models, investigators usually gather enough information to construct an accurate profile. Upon being apprehended, the offender often addresses the evolution of procedures that characterized his MO.

Experience: The Best Teacher

MOs may remain relatively unchanged in one aspect of the crime but "evolve" in other aspects. For example, wearing a ski mask to prevent identification and entering a window away from a busy parking lot late at night may remain the baseline MO for a serial rapist who enters a victim's bedroom and gags her with a piece of clothing. However, committing his crimes in single family dwellings instead of apartment complexes may not change. If he's comfortable attacking victims in a home with less exposure to other people instead of an apartment with "paper-thin walls," he likely will remain in his "comfort zone."

Authorities realize the extent of most offenders' development of MO and the immense amount of time they devote to each crime, from fantasy to aftermath. Imagine how differently serial predators' lives could have turned out had they used all the energy expended in rapacious behavior to build productive careers or healthy relationships. The glaring problem here, of course, is that they don't know the first thing about normalcy.

Signature

"Calling Card" within *Actus Reus*

In contrast to MO, **signature** (or **personation**) is the predator's emotional *cri de coeur*—"*cry* from the heart." It denotes his emotional connection to his victim and the nature of his predatory behavior at the crime scene. Signature is *ritualistic behavior* at the crime scene and suggests the *underlying emotional theme* or motive. Although unnecessary to the commission of the crime, signature is the offender's personal, sexualized, "imprint" exemplified in verbalization or behavior that is *typical and distinguishable* from all other offenders.

For example, a serial rapist's signature may be a typical, distinguishable explicit vocalization at every crime scene, such as intimidating and humiliating his victims with very graphic, degrading, and vulgar talk. He may force them to say something vulgar during each incidence of rape, such as "I know you've fantasized about this before." Signature appears when he forces his victim to perform the same sexual act in a certain sequence. For instance, he might force her to perform oral sex and at orgasm ejaculate on her stomach or breasts. This specific signature allows crimes to be linked together as forensic evidence against a targeted offender. As we have seen, behavioral psychologists refer to behavior "linked" together in a certain sequence as *chaining* (or chained behavior).

An entirely different signature occurs, for instance, when a serial rapist enters for a burglary, and as soon as he disables his victim and gains control by handcuffing her, he morphs into the "rapist mode." He may force her to masturbate herself at gunpoint while he masturbates himself and then ejaculates into her vaginally.

Sexualized Signature: The Fantasy Element

Perhaps the best justification for characterizing serial crimes at the crime scene as sexualized is the presence of signature. Unchanging (but perhaps "escalating") signature remains a constant and enduring part of the offender's *behavioral dynamics,* linking the graphic sexual behavior or sexual overtones to the psychological or "fantasy elements." Sometimes, however, unexpected interruptions or unanticipated victim behavior may prevent signature from showing up at all crime scenes.

In serial homicides, a killer's signature may be bloody attacks seen as **overkill.** He may slash his victims' bodies with multiple stab wounds or other mutilations, or carve unnecessary "graffiti"—initials, words, or symbols—into his victims' bodies. Hence profilers use the term "overkill."

Signature is whatever aspect of *actus reus* is not necessary to kill the victim. For example, after death killers place some victims in special positions with certain body parts exposed (usually a breast) in curious **tableaux mortido** (death pictures). They situate these in meticulous detail, evidence of their need to dominate and control, specific traits found in serial offenders.

Analysis of crime scene evidence shows that offenders spend considerable time at the crime scene, as though they are prolonging their murderous fantasies for as long as possible. This is similar to the way normal lovers seek to prolong sexual encounters by touching, cuddling, or sleeping together.

Forensically, signature provides the "stake in the heart" that convicts the serial killers, as signature analysis links crimes together to reveal the "smoking gun." Experienced field agents in local law enforcement and the FBI can attest that the odds of two different offenders fitting the same MO and signature while operating in the same area at the same time are remote.

Experienced profilers, such as Ressler and Douglas, suggest that connecting the "how" of actus reus (the MO) and the "why" of actus reus (the signature) targets the "who" (the serial killer).

It is instructive for the forensic psychologist to analyze, understand, and document *crime scene dynamics* in as much detail as possible. These are the elements of *modus vivendi* covered thus far—*mens rea, actus reus, modus operandi, and signature.* The aftermath of the crime—the time subsequent to death—is the fifth factor of *modus vivendi.*

In behavioral psychology, the **law of frequency** states that the more frequently a person (e.g., a serial killer) responds to a given stimulus (stalking and killing a victim), the more likely he will make the same response to the same stimuli in the future (continue serial killing). In other words, this behavioral law accounts for one of the most recognizable idioms in pop culture: "Practice makes perfect."

The *law of frequency* explains why serial killers practice what they fantasize; they seek perfection in the macabre practice of death. Another prescient behavioral law, the **law of recency,** states that the sooner a person responds to a given stimulus, the more likely it is that he will respond to it again—very soon.

Aftermath

The aftermath is the period following any ruinous event. In the aftermath of serial rape or murder, the perpetrator may perform **necrophilia**—sexual intercourse with the corpse—or other sexualized behavior that demonstrates the macabre "relationship" the killer envisions with his victim. For example, serial killer Jeffrey Dahmer anticipated lying next to his victim and later cannibalizing some part of the victim's body as a way to feel "connected"—a way to make it a part of his body forever.

Numerous killers return to perform sexual acts with the corpses or parts of the corpses. When the perpetrator relives the rape and/or murder, he savors the sexualized feeling and extends the "thrilling high." He may revisit the crime scene, a shallow burial spot, or a cemetery to experience the sexualized elements associated with fantasy and the crime itself one more time.

Contrasting *Modus Operandi* and Signature: A Student Profiler's Perspective

Profilers define MO as any action or series of actions performed by a perpetrator as a necessary means to access and/or isolate the victim. Here the actualization of the mental, fantasy part (*mens rea*) takes place. Serial killers can refine, improve, or even discard any aspects of MO as required to achieve their goal of acquiring, disabling, and raping and/or killing their victims. The following are examples of *modus operandi:*

1. Use of **ruses** (romancing, flirtation, asking for help)
2. Stalking, or otherwise locating a *particular type* of victim
   Targeting a prostitute, hitchhiker, student, or ground floor resident
   Specificity of body build, hair color, gender, age, and race
3. Choosing a certain location;
   May be significant to the perpetrator
   Easy access to the victim
   Permits escape from the crime scene
   Satisfies requirements of signature
4. Use of restraints, force, or a weapon to incapacitate, restrain, silence, and/or kill the victim

Signature

The definition of signature is any action or series of actions performed by a perpetrator that is *not necessary* to raping or killing a victim during and/or after the physical commission of a crime. *Emotionality* enters *actus reus* and becomes *personalized with signature.*

Signature is unique to each killer, such as:

1. Verbalizations made to the victim
2. Apologizing to the victim
3. Ante-mortem or post-mortem sexual activity with or without the victim (rape, masturbation, necrophilia)
4. Taking of trophies (body parts, locks of hair, jewelry)
5. Ante-mortem or post-mortem markings, mutilation, or dismemberment (overkill)
6. Washing of the corpse
7. "Staging" of the corpse
8. Returning to the crime scene or grave of the victim
9. Insertion into the investigatory process

Paradigm of the Homicidal Triad

Experienced profilers commonly observe the so-called **homicidal triad** in sexual psychopathy. This behavioral triad indicates the presence of "red flags" or the potential ability for violent crimes. It *shows lack of restraint* and a tendency to act out. A *cavalier disdain for structure and responsibility* is inherent in the triad, which includes:

1. **Enuresis** (en-yur-e-sis) or bed wetting at an inappropriate age
2. Cruelty to small children, animals, or pets
3. **Pyromania** (fire starting)

Besides low self-esteem, anger, frustration, and blaming others for severe childhood abuses and/or neglect, many serial predators (sexual psychopaths) have frightening histories of *compulsivity and ritualism such as lying, cheating, and compulsive masturbation.* As they enter late adolescence, they begin to display criminal "careers" as pyromaniacs, cat burglars, stalkers, Peeping Toms, rapists, and fetishists before eventually (in their early to mid-twenties) committing murder.

Organized Predator Poster Boy: Ted Bundy

The "poster-boy" predator for the *organized offender* is the serial killer Theodore "Ted" Bundy. Applying his personal vita to the following delineation of organized versus disorganized offenders is instructive.

In terms of IQ (Intelligence Quotient), the organized offender is almost always average to above average in intelligence. Bundy had been a law student at the time of his offenses and possessed a "gift of gab" along with above average academic skills. In contrast, the disorganized offender is usually below average in intelligence and may not have a high school diploma or a GED.

The organized offender is *socially competent* and sophisticated, meaning he has a grasp on the way society works and possesses the requisite interpersonal skills to maintain at least a persona of "normal" social relationships. Bundy made friends easily and possessed strong manipulative skills. The judge who presided over his murder trial berated Bundy for his offenses (paraphrased): "I would have liked to hear you argue in court someday . . . you would have been a good lawyer . . . but you chose the wrong path." The organized offender is proficient at acquiring and keeping skilled jobs, as was the case with the offender's father. Bundy was adopted into a home with a stepfather who had a stable work record.

The disorganized offender is socially incompetent, displays socially inadequate behavior such as disturbed or nonexistent social relationships, and is a "job-hopper," an unskilled laborer following in the footsteps of his father's spotty work history.

The organized offender often has success in sexual relationships with girlfriends, wives, or ex-wives. Bundy had girlfriends and even married one of his admirers who witnessed the trial. While in prison, conjugal visits by his wife produced a daughter.

The disorganized offender is sexually incompetent. When interviewed on death row, disorganized offenders report never having experienced a mutually satisfying sexual relationship with the opposite sex. The disorganized offender bases sex on control, domination, degradation, and/or abuse.

A history of inconsistent discipline characterizes the organized offender's childhood. Researchers documented this fact in Bundy's parent-child relationship. On the other hand, harsh discipline characterizes the disorganized offender's history of discipline.

During the crime's commission, the organized offender's mood is somewhat stable, enhanced with the abuse of alcohol. Bundy was in a state of intoxication during the commission of his crimes. The disorganized offender's mood is anxious with minimal use of alcohol during the commission of the crime.

A precipitating stressor or "trigger" for the crime exists for organized offenders while minimal situational stressors exist for disorganized offenders. In Bundy's case, having to leave law school due to finances was the "trigger."

Organized offenders live with a partner (a wife or girlfriend), while disorganized offenders live alone. Due to a total lack of transportation or a fear of mechanical breakdown, the disorganized offender lives close to the crime scene, while the organized offender displays a wider range of mobility with dependable transportation. Bundy drove a VW beetle in good repair.

The disorganized offender displays minimal interest in the media coverage. The organized offender follows news coverage avidly (true in Bundy's case). After the crime, the organized offender may leave town or change jobs. Bundy left town and eventually moved from state to state. In contrast, while the disorganized offender never strays from home, his behavior changes radically, as observed in increased drug and/or alcohol abuse.

Individuals with full-blown psychopathic personalities marked by lack of empathy have many sexual relationships with people they perceive as "things to be used." The difference between the garden variety non-violent psychopath and the violent sexual psychopath is telling. The sexual psychopath, a sexual pervert for life, kills and becomes known to the world as a serial killer. Perhaps one of the most perverted was Ted Bundy.

PREDATOR PROFILE 3-2

Theodore "Ted" Bundy

Ashleigh Portales

Original Name
Theodore Robert Cowell
Time Span of Crimes
January 4, 1974 to February 9, 1978
Execution Date
January 21, 1989, Florida

Offenses Prior to Serial Killing
1. Minor burglaries
2. Shoplifting (usually jewelry)

Quoting Bundy
"I'm the most cold-blooded son of a bitch you'll ever meet. I just liked to kill. I wanted to kill."

Victimology and Modus Operandi

Bundy preferred *young girls with long brown hair parted down the middle.* These young women served as physical and emotional stand-ins for Stephanie Brooks, Bundy's first real girlfriend who dumped him because she considered their relationship a mere fling. Bundy was in love with her, and her rejection pushed him over the edge he had been teetering on for a long time.

Bundy preferred college campuses where he could stalk his victims throughout the United States. He often feigned injury by using crutches and dropping his books to elicit help from his targeted prey. When they bent over to retrieve the books, he would knock them unconscious with his crutch, handcuff them, and load them into his VW Bug. Bundy showed forethought by removing the passenger seat so his victims could lay conveniently beside him in transit without threat of discovery.

Making of a Monster

Ted Bundy was born Theodore Robert Cowell on November 24, 1946, the *illegitimate child of* Louise Cowell. After Ted's birth at a home for unwed mothers, Cowell returned home to live with her parents. She decided to raise Ted as her parents' son, not their grandson (making Ted her brother). If the truth came out, the family's shame would be too extreme. Ted's "mother" suffered from chronic depression and eventually received shock therapy. His "father" (his real grandfather) was a maniacal man who verbally and physically abused everyone in his household. Both "parents" fueled their respective problems with constant alcohol abuse. When Ted was four, his "sister" removed him from the abusive household to marry John Bundy, who adopted Ted. Thus, "Ted Bundy" was born, and his "sister" quickly became his "mother." Even though she and her husband had four other children, Ted *never bonded* to her and *never felt like a part of the family.* He never connected with his siblings or with his stepfather, who led a Boy Scout troop.

Bundy was highly intelligent and did well in school even though he felt *isolated* from classmates. He never dated in high school. Curious about the cause of his social obscurity, Ted sought out his birth records and learned the truth of his parentage. *Ted realized his entire life had been built upon lies.* As his world turned upside-down, his psychopathic nature materialized.

Ted enrolled in college where he excelled at studies in law and psychology. With a promising political career, he served as an assistant director of the *Seattle Crime Prevention Advisory Committee.* He wrote a pamphlet instructing women on rape prevention. He engaged in charity work and campaigned for the U.S. Republican Party. He proudly wore the armor of the psychopath: he was engaging, manipulative, and possessed the "gift of gab."

Yet when Stephanie, the love of his life, abruptly ended their relationship, Bundy's rage for "being different" exploded. The straw that broke the camel's back occurred when he had to leave law school for financial reasons. Years of pent up rage, barely camouflaged beneath the slick persona of psychopathy, exploded in a string of murders unrivaled in American crime. In Ted's eyes, women (his mother, grandmother, and girlfriend) had ruined his life, and payback was long overdue. Suddenly he realized he "hated women," so he felt completely "entitled" to his actions. He felt no remorse when he killed, which allowed him to pass lie detector tests, for he believed his victims "had it more than coming."

The Monster Unleashed

For many years a predatory monster lay camouflaged behind the façade of the "boy next door." Exacerbated by alcohol, marijuana, and other mitigating circumstances, the savagery (stalking and rapacity) of sexual psychopathy began to prowl. On January 4, 1974, Bundy broke into the home of Sharon Clarke and severely beat her face and sodomized her. Miraculously, Clarke survived, but subsequent victims were not so lucky. Spurred by his self-proclaimed "brutal urge," the sexual sadist side of his predatory lifestyle roamed college campuses in constant search of his next girl.

A wild animal like a wolf loves the thrill of the kill, even if only to secure his next meal and to survive another day. Ted Bundy was no exception except he killed exclusively for the thrill and the aftermath. After killing his victims, Bundy hid the bodies in secluded spots in the woods time and again, returning to them at various stages of decomposition in order to immerse himself in his erotic fantasies and relive the crimes. He could only find in death the emotional and sexual connections he longed to have with women in life. In a macabre attempt to make his trophy women more attractive in the murders' aftermaths, he would bring along a makeup kit for aesthetic value. After committing *necrophilia* (sexual intercourse with the dead body), he would lie beside his "lover" like a "bed partner," sometimes all night in the woods. It may have been in the feral wilderness that Ted, the monster, felt most at home. True to the model of an *organized serial killer,* Bundy closely followed media coverage of the crimes.

The Monster Caged

An officer who pulled Ted over for driving a stolen car apprehended him in Florida in 1978. Yet he escaped twice before he ever came to trial. After each escape, his killings became more frenzied and violent, with greater degrees of *overkill.* When he finally faced justice, he taunted the press and police with a *cavalier attitude* and constantly displayed *inappropriate affect,* such as smiling when discussing some brutal aspect of his sadistic crimes.

Inappropriate affect characterizes the neurological damage often found in sexual predators. Relying on his knowledge of the law, Ted fired his lawyers and chose to represent himself. This allowed the monster yet another chance to savor his crimes as he grilled police witnesses for specific grisly details and in-depth accounts of their observations at crime scenes. Often during this time, observers saw Bundy closing his eyes and smiling.

Despite his obvious guilt, Bundy acquired a bizarre notoriety and a following composed of swooning young women who made an audience in the courtroom. Many of them never realized they were just the kind of girl Ted Bundy preyed upon—naïve and immature and ignorant of the importance of looking beneath a pleasing exterior. Even the grisly testimony of Carla LaRonch, a young woman who barely escaped Bundy's attempted kidnapping, did nothing to deter his naïve throng of admirers. One of the "pitiful princesses" married Bundy while he was in prison, where a conjugal visit produced a daughter.

Although found guilty and sentenced to death, Bundy delayed the execution process by offering information about additional women he had murdered in exchange for temporary stays of execution. He continued his manipulative, controlling ways until the end, when he granted a final interview the day before his execution to Dr. James Dobson of "Focus on the Family" radio fame. At the time, Dobson was lobbying before the Supreme Court for harsher pornography regulations. Bundy claimed he was just a normal "American boy" ensnared by the evils of hardcore pornography, and he attempted to convince Dobson that he was more valuable to law enforcement alive than dead so he could continue to be an advocate for the restriction of porn.

Later, when asked about his crimes, Bundy replied, "Murder is about domination. When they take their last breath, you're God."

Bundy's last desperate effort to prolong his miserable life with the Dobson interview proved unsuccessful though, as the state of Florida executed him on January 24, 1989 in the electric chair. He was forty-one. By that time, his cache of *known victims included at least thirty-five women.* Only Ted Bundy knew the true total, and he took that with him to his grave.

There is strong evidence that Bundy may have committed his first murder at age fifteen, when a young girl in his neighborhood disappeared. At the time, no one suspected young Ted, and though he never became connected to the young girl's murder, in hindsight it seems highly likely that he was the perpetrator.

Crime Scene Analysis

In 1978, FBI special agents Robert Ressler and John Douglas modified the Teten-Mullany *Applied Criminology Model* of serial killers into the *Organized/ Disorganized Model.* This model remains influential today. From 1979 to 1983, FBI agents hatched the *Criminal Personality Research Project,* the dream and literary "child" of FBI special agent Robert Ressler. The undertaking was truly a landmark for the FBI, who at the time, ironically, showed limited interest in what motivated murderers, rapists, and child molesters (incidentally, all of these offenses were not against federal law.) The study took dead aim at the *psychological and behavioral characteristics* of violent criminals related to their backgrounds, specific crimes, crime scenes, and victimology.

FBI special agents obtained this information firsthand by entering correctional facilities initially before or after "road schools" where agents taught FBI techniques to local law enforcement agencies. The agents spoke with offenders individually, augmenting personal interviews with homework and poring through stacks of forensic evidence such as court transcripts, psychiatric assessments, and police reports. The exhaustive study resulted in the creation of a list of crime scene protocols known as *Crime Scene Analysis (CSA).*

The current dichotomized *organized versus disorganized model* follows the presentation of the six steps of the CSA paradigm. These are:

1. ***Profiling Inputs.*** CSA's initial stage is evidence gathering. This includes all *crime scene materials* gathered at the crime scene, such as photographs of the crime scene and/or of the victim. Evidence includes comprehensive background information on the victim, autopsy reports, and forensic information relative to the "psychological autopsy" of the crime scene, such as postulating what occurred before, during, and after the crime. Profiling Inputs are the CSA's foundation. Any errors and/or miscalculations in this *evidence gathering* stage can lead investigators in wrong directions.
2. ***Decision Process Models.*** *Logistics* is the best word to describe CSAs second stage. A *logical* and *coherent pattern* must emerge from this stage, an emerging picture of the perpetrator suggested by the crime scene. Was a serial perpetrator responsible? Or, does the evidence point to a single instance of a crime (the offender's one and only crime)?
3. ***Crime Assessment.*** *Crime reconstruction* best describes the third stage. What are the sequence of events and the behavioral characteristics of victim and offender? What "role" did the victim play? What "role" did the offender play? Analyzing this stage allows investigators to piece together the emerging criminal profile gradually.
4. ***Criminal Profile.*** The actual profile starts to occur with the fourth protocol, which includes background data, behavioral characteristics, and the perpetrator's physical description. This stage provides suggestions based on personality type for the most effective ways to interview the offender if apprehended. The goal at this stage is *identification and apprehension.*
5. ***Investigation.*** This is the *application* phase where law enforcement agencies receive the actual profile to aid in the perpetrator's apprehension. New information continually modifies the original profile, sometimes on a day to day basis.

6. ***Apprehension.*** The last stage of CSA is crosschecking the profile with the apprehended offender. This stage has built-in difficulties in the event the offender never is caught, police arrest him on another charge, or he ceases criminal activity.

The Phenomenon of Staging

Staging refers to actions by the perpetrator or other persons that *purposely alter the crime scene* prior to the police's arrival. Investigators know staging takes place for at least two reasons:

1. The perpetrator purposively manipulates the crime scene in an attempt to direct the *investigation away from the most logical suspect—himself.* Perpetrators can attempt to steer the investigation to other more "logical" suspects by appearing overly cooperative or ostentatiously distraught.
2. A family member may try to protect the victim and/or the victim's family. This occurs predominantly when a family member or friend discovers the body first in *rape-murder crimes* or autoerotic fatalities (such as **autoerotic asphyxiation**). Perpetrators of rape-murder crimes often leave the bodies in curious *tableaux* that are shocking and/or degrading. A husband may cover his wife's nude body, or a wife may cut her husband down from a noose following his accidental death from an autoerotic episode. Hence, staging in such instances would attempt to leave some dignity to the victim. The wife may try to make her husband's death in an autoerotic event look like a suicide and may even go as far as to write a suicide note. Middle aged and older males may have a stroke or heart attack and die while masturbating to pornographic movies. Wives sometimes remove the pornographic videos from the residence as well as any evidence of their husbands' autoerotic behavior. These are instances of staging non-criminal related events.

"Red Flags" of Staging

"Red flags" of staging alert investigators to potential discrepancies and aid in preventing a "misdiagnosis" of the crime scene. These include *inconsistencies* between the crime scene analysis and the evidence, such as point of entry, point of exit, time of day, and the position of the body in reference to context in the overall picture.

Suppose a homeowner returns home from work and interrupts an attempted burglary. The startled burglar kills the man in his own home. An inventory of the crime scene discloses that the burglar did not steal anything of value, such as easily accessible jewelry or expensive electronic equipment. Use *skepticism* and *speculative logic* to analyze the following *alternative theories* about why the man was killed:

1. Unfortunately, the man startled the burglar, which led to his death; the burglar simply fled the residence without stealing anything.
2. The "burglar" was not a burglar at all.
3. The murder involved the wife.

4. The wife and the "burglar" staged the whole "burglary attempt" Because of a secret affair. She and the burglar staged the crime to kill the husband and make it appear to be a bungled burglary.

Skepticism

Psychologically, the experienced profiler approaches a crime scene with philosophical **skepticism.** This "mindset" withholds opinions until investigators have gathered and analyzed all the facts. An example of the failure to use skepticism (educated doubt) occurred in the recent events surrounding the sniper attacks in Virginia (2002). Nightly news "talking heads" with very limited information attempted to profile the individual. When police eventually captured the snipers, the profiles proved misleading and wrong. On a nationally syndicated program, experienced FBI profiler Greg McCrary later spoke out against hasty attempts to profile offenders on limited information. In the snipers' case, McCrary noted there simply was not enough evidence to produce an accurate profile.

Psychology of Sexual Sadism—Psychopathology of Compliant Victims

The pioneering work of former FBI agent Roy Hazelwood provides accurate insight into the minds of sexual sadists and their compliant victims. A compliant victim is a female co-offender who helps "recruit" other victims for her sadistic "lover." The definition of **sexual sadism** is that it is a sexual perversion where gratification is obtained by the infliction of pain of a physical, sexual nature with accompanying mental anguish.

In many cases of sexual sadism, the sadist *keeps victims alive as long as possible* to prolong his perverse sexual appetites. The *fait accompli* of serial sexual sadism is rape or murder. The word "sadist" comes from the exploits of the *Marquis de Sade* in the nineteenth century, who delighted in inflicting sexual cruelties on his "lovers." According to Hazelwood, the sadist follows five steps in "creating" his perverse "female Igor"—the **compliant co-offender.** The steps are:

1. Through astute observation of **body language,** the sadist identifies a *vulnerable* co-offender—a naïve, dependent, immature, and therefore controllable female. Often, such a compliant person has a diagnosable **dependent personality disorder** and already displays behavior consistent with co-dependency, i.e., being a "doormat" for others. She may have had abusive parents or an abusive relationship with a boyfriend or ex-husband. In any event, the sadist appears as the "saving" persona, her "rescuer."
2. The sadist charms her with his "smooth talk" and seemingly gentle nature. He may lavish her with gifts, offer physical protection, financial support, or whatever he perceives as the "legitimate" answer to her problems. The victim perceives him as a loving and caring "nurturer," worthy of her love. Women often "fall" for the demeanor and **persona** (i.e., social "masks") that psychopathic individuals use to deceive.

3. Soon, she is totally dependent upon him and under his emotional "spell." He then encourages her to *engage in perverse sexual practices* that she most likely considers deviant or at the very least "kinky." The *small steps* he uses to lure her into his perverse world eventually lead to the *shaping* of full-blown perversities that evolve into habitual sexual practices. This shaping of sexual perversity accomplishes two control mandates: First, it demolishes her fragile will and "esteem," along with any sense of normalcy regarding values. And, second, she becomes *isolated from others*—the *fait accompli* of sexual sadism. After a relatively short time, the co-offender becomes a sexual "slave." With a compliant co-offender, the sadist perversely "gets his cake and eats it, too."
4. The sadist uses domination, manipulation, control, and physical punishment for lapses to become the center of the "recruit's" universe. The compliant co-offender feels hopeless and depersonalized, which eventually will play a central role in victimization. Sadly, she is worse off in the sadist's hands than in any prior dysfunctional relationship.

Through his use of *mind control* and *physical punishment,* she complies with his every demand, partly to avoid his wrath. He has succeeded in "making over" her cognitive map. As mentioned earlier, powerful cognitive maps provide the blueprint for self-concept, feeling, and behavior. The sadist has changed her fragile (or non-existent self-esteem) into the persona of a "bad," "stupid," "inferior," or "inadequate" "depersonalized" slave. She is now a pure example of **co-dependency.**

Signature Sexual Offenders—Sadist Signature Killers

Criminologist, homicide detective, and true crime author Robert Keppel (1997) presents his own paradigm to explain sadistic homicide offenders. His term is *signature killer* because of the unmistakable presence in all crime scenes of the killer's sadistic "calling card."

As mentioned earlier, according to Ressler and others, signature relates to motive, to emotionality, and, ultimately, to sexualized "non-fulfillment." The *addictionologist* knows why serial predators feel *compelled* to continue *criminal behavior;* they operate out of a full-blown addiction where every aspect of the crime is sexualized.

According to Keppel (1997), the basis for understanding even the most minor sex offenses (in his words "Sex Crimes 101") is the realization that *anger expressed through control drives serial killers.* To analyze serial crimes, Keppel uses the following categories to describe the perpetrator's *psychological dynamics* manifested at the crime scene:

1. ***The Anger-Retaliation Signature.*** This signature often displays *overkill* against the victim as an anger-retaliation symbol. The killer chooses to *retaliate against the real source of his anger* by using a symbolic victim. Examples of serial killers who follow this typology include Arthur Shawcross and John Wayne Gacy. According to Shawcross, he murdered women because his mother rejected him, while Gacy murdered "lost boys" who sought consolation from

him as retaliation against his alcoholic father who never expressed genuine emotion and love. (Serial killers are so dangerous because they are not what they appear to be. They may pose as roofers or service technicians while canvassing victims door to door. They may return several months later in what appears to be a random, chance occurrence, or they may dress as a clown (as Gacy did) to entertain children.) The anger-retaliation killer seldom kills his own mother; he chooses someone like her. He chooses victims who represent domineering women in his life, whom he believes are responsible for his troubles—unless the killer is homosexual and seeks to destroy young males. According to Keppel, examples include mothers who were overcontrolling, promiscuous, physically or sexually abusive, or who inspired fear and terror in their children, or fathers who rejected their sons.

2. ***The Picquerism Signature.*** The serial killer who is a **picquerist** is a sexual deviant who becomes sexually aroused by biting the victim or by penetration of the skin through cutting, slicing, or stabbing with a long-bladed knife. In rare cases, picquerism may involve sniper activity. Victims are not victims of chance. The killer may stalk his victim for weeks or months, choosing those who fit his preferred type. Picquerist crimes are particularly gruesome due to deep and violent stab wounds. *Knife penetration and the control of every aspect of bringing his victims death* drive this type of signature homicide. After six picquerist murders near San Diego, California, a twenty-five year old black male, Cleophus Price, became identified as the serial killer.
3. ***Sexual Sadism Signature.*** According to Dr. Richard Walter, a forensic psychologist at Michigan State Penitentiary, the three Ds of sexual sadism are *dread, dependency, and degradation.* Prolonging the sexual "high" in each stage by inflicting as much pain and misery as possible provides the killer with *modus vivendi*—sexualized feelings related to sadism, such as breaking his victim's *will to resist.* Delaying the victim's death prolongs the sadist's desire for psychological terrorism. If death comes too fast, the serial sexual sadist feels cheated.

The Sexual Addiction Factor

More importantly, according to Jacobs (2003), *signature* provides evidence of the *addiction factor* that highlights serial rapes and murders. Due to many factors relative to learning and neurochemistry, offenders become addicted to the feeling—the "thrilling high" generated in the brain's *pleasure pathways.* The same pathways explain addiction to any drug, such as alcohol, marijuana, cocaine, or MDMA (ecstasy). Some serial killers dread killing the victim because *restraining the victim* and controlling when the victim dies produces more of the "thrilling high" than murder.

The relentless *addiction factor* that compels serial killers to go from one kill to another is what Bundy called the "brutal urge." It only recedes when the killer feels "spent" in aftermath. The same frenzy that describes his addiction may describe two lovers tearing off each other's clothes and having consensual sex in a wild display of shared passion. The parallels are striking to the *organized* serial offender. This factor is startling news to so-called "experts" who inject decision into the motivation of serial psychopaths. They are ignorant of the psychological ramifications of addiction and neurochemistry that lie behind emotion. Feeling, not cognitive decision-making, drives serial crime.

The most important ingredients for forensic neuroscientists who study serial rape and homicide to address are:

1. The driving force of erotic *feeling* relative to the killer's sexual fantasies as he *contemplates* the crime
2. How he *feels* (as *mens rea* accelerates) *preparing* to commit the crime
3. How *feeling* carries him through the physical perpetration of full-blown *actus reus* exemplified in MO and signature
4. How he *feels* "spent" from endorphin release in aftermath

To the neuroscientist, every step from imagery to debauchery to aftermath is due to neurochemistry and neurohormones driven by fantasies and deviant cognitive "mapping," which reach climax in MO and signature. This view explains why some serial killers experience revulsion at the memory of the crime the next morning when alcohol (or other drugs) wears off. But, as Bundy explained in his last interview prior to lethal injection, the brutal urge always comes back.

The following three chapters provide compelling details of the neurochemical underpinnings of rapacious behavior. These offer evidence of the strong addiction component of the serial killers' "decision" to rape or murder. They show that such a "decision" to act out arises largely from *compulsion* generated by *chemical addiction.* Compulsion by addiction makes more sense to the neuroscientist than a rational, cognitive decision to kill—the so-called "choice" some presume that serial offenders make because of rage, anger, or retribution for past ignoble influences.

Paraphilias

The essential features of **paraphilias** are recurrent, intense sexually arousing fantasies, sexual urges, or behaviors generally involving the following:

1. Non-human objects
2. Sadism: Causing the emotional suffering of another
3. **Masochism:** Causing emotional humiliation
4. **Pedophilia:** Sexualizing children

Voyeurism

A serial killer may show signs of a sexual deviance known as voyeurism early in his developmental stages. The act of observing unsuspecting individuals, usually strangers, who are naked, who are in the process of becoming naked, or who are engaging in sexual behavior is known as *voyeurism.* The act of looking or "peeping" (i.e., a "Peeping Tom") is to *achieve sexual excitement,* and generally the killer seeks no sexual activity with the observed person. Convicted killer Richard Ramirez began his serial killer "career" by observing hotel guests in various stages of nudity.

In its severest form, peeping constitutes an exclusive form of observing sexual activity, and its practitioners are known as *Peeping Toms.* Voyeurism causes clinically significant distress or social or occupational impairment in the voyeur, unless the voyeur is a sexual psychopath. Voyeurism is designated 302.82 in the DSM, IV, TR (2002).

Frotteurism

Frotteurism involves touching or rubbing against a non-consenting person. The behavior usually occurs in crowded places where the individual can escape more easily. He rubs his genitals against the victim's thighs or buttocks or attempts to fondle her genitalia or breasts with his hands. While doing so, he usually fantasizes an exclusive, caring relationship with the victim. Most acts of frottage occur when the person is fifteen to twenty-five years of age, after which the frequency generally declines.

As a teenager, Jeffrey Dahmer often fantasized lying next to a nude male and listening to his heart beat, a fantasy that fueled his murderous rampage against homo-sexual males. After strangling his victims, Dahmer often laid next to their corpses to fondle them. Later, he often cannibalized them.

Such fantasies, sexual urges, or behaviors cause clinically significant distress or impairment in social, occupational, or other important areas of functioning. DSM, IV, TR (2002) designates it 302.89.

Sexual Sadism/Sexual Masochism

The earlier section on *Signature Killers* discussed the sadist involved in **sexual sadism.** Infliction of psychological or physical suffering (including humiliation) upon the victim gives the sadist sexual excitement. Such fantasies, sexual urges, or behaviors cause clinically significant distress or social or occupational impairment in all perpetrators except for serial offenders. On the other hand, the **masochist** involved in sexual **masochism** experiences real acts of humiliation, such as beating, binding, or other suffering. These fantasies, sexual urges, or behaviors cause clinically significant distress or social or occupational impairment. Tethered to control and domination, sexual masochism surfaces in a serial killer when his acts humiliate others. The DSM, IV, TR (2002) designates sexual sadism as 302.84 and sexual masochism as 302.83. Sadomasochism combines characteristics of the two paraphilias.

Sexual Disorder NOS

Sexual Disorder NOS (Not Otherwise Specified) is a category for coding a sexual disturbance that does not meet the criteria for any specific sexual disorder and is neither a sexual dysfunction nor a paraphilia. Examples include: marked feelings of inadequacy about sexual performance, or other traits related to self-imposed standards of masculinity or femininity; *a succession of lovers whom he feels are objects to be used;* and persistent and marked distress about sexual orientation.

Name ______________________________ Date ______________

Forensic Psych Word Scholar III

Define the following words.

1. *Manie sans delire*

2. *Malum in se*

3. Rape kit

4. Successive approximation

5. Shaping

6. Personation

7. Paraphilias

8. Overkill

9. Staging

10. Aftermath

Name ______________________________ Date ______________

Aftermath

Comment on the following in one short paragraph:
What one factor do you feel best explains why serial killers are sexual deviants for life? What happened? Nature or nurture or both?

UNIT II

Readings in Psychology of Deception

CHAPTER FOUR

POD Readings *Au Courant*
Showcasing Eclecticism in Forensic Psychology & Neuropsychology

Written, Edited, & Compiled by Don Jacobs
& Top Student Scholars in Forensic Psychology
Weatherford College

1 Science Can Be Dead Wrong

A Case in Point: David Reimer aka Bruce, Brenda, & Joan

Don Jacobs

Science can be as powerful as electricity—lighting up an entire city, or as a towering thunderstorm of hubris, destroy everything it touches. In considering a course of action—a good place for students to start is *to ALWAYS do your own research* in whatever issue or problem you face, including launching a new relationship. Yet, is this always enough? Are second or third opinions sufficient? Can doing our own research touch all of the perceived bases of our problem? No, it cannot, especially when dead wrong information is used as the basis of an irreversible decision.

Today, doing our own research is even more important AND THAT WAS BEFORE the Human Genome Project 'cracked' the codes of most of our genes. What does the future hold for cloning and genetic tinkering? Are we truly living in a Brave New World, a world our grandparents would scarcely recognize?

With medical technology more powerful than ever and Big Pharma—the big pharmaceutical firms—dictating medications that promise cures for a plethora of conditions—conditions ranging from erectile dysfunction to restless leg syndrome, and from eating disorders and chemical dependency, to panic attacks and schizophrenia.

That SCIENCE CAN BE DEAD WRONG may be the best place to start. When science is wrong, it can be especially devastating, since we tend to believe such a powerful discipline. Let's begin with an intensely private person—David Reimer—a person who became known around the world as the real life person behind the celebrated *John/Joan case*—a case that prompted an award winning article in *RollingStone* Magazine, then a book and, now talk of a screenplay and movie. The year of the article appearing in *RollingStone* Magazine was 1998; the year of the events that would forever change David Reimer was 1966, when he was just an infant.

Science Can Ruin a Life

What follows is the true story of a boy who was born Bruce—then became Brenda due to a botched circumcision—then David as his elected name when finally told the truth; but became known as John, and then Joan in academic articles. David Reimer was the real flesh and blood person.

On May 4, 2004, David Reimer committed suicide in Winnipeg, Manitoba in Western Canada. He shot himself in the head while sitting in his car in a parking lot. He was just thirty-eight years of age. He left behind an estranged wife, adopted children, and a world of pain. As an infant, a botched circumcision forever changed his life.

Born Bruce Reimer in 1965, David (the name he adopted when finally he learned of his early infant trauma) suffered a botched circumcision when he was just eight months old. Most of his penis was burned off by an apparatus not intended for circumcision; reconstructive surgery at the time was too primitive to restore his burned penis. David's well-meaning parents sought answers from local physicians and then a famous one-of-a-kind sexologist.

After seeing Dr. John Money on TV, Bruce's distraught parents sought the advice of the famous sexologist of prestigious Johns-Hopkins University. At the time, Dr. Money was the authority on *sex reassignment surgery* in infants and children. The doctor persuaded Bruce Reimer's parents to have their son completely castrated and raised as a girl. Money believed *it mattered little what gender an infant was born, it only mattered how he or she was raised.* Therefore, Bruce's remaining genitalia was surgically removed. With hormonal therapy, Brenda (as he became known) was raised a girl.

For Bruce, this decision proved to be a complete emotional disaster. As a teenager, Brenda NEVER FELT feminine, nor did her behavior resemble the mannerisms society expected from female behavior—he consistently exhibited boyish 'rough and tumble' play observed in all MALE mammals. The contradictory feelings David experienced with his MALE BRAIN, now cocooned in a surgically reconstructed female body, would prove to be both humiliating and emotionally devastating throughout his teenage and young adult life.

As a teenager, Brenda experienced panic attacks, depression, anxiety, and suicidal ideation—all conditions that prompted a psychiatrist to insist Brenda's parents inform their daughter of the circumstances of the failed circumcision. Brenda, at age fourteen, refused to see Dr. Money again, and threatened suicide if she were forced to go. Finally, Brenda's father felt he had no choice but to tell his son the truth. He recalled that his son did not become emotional or angry. He just sat there somewhat dumbfounded. Yet, David recalled the 'world lifted off my shoulders' with the news. He knew there must be a reason for the way he felt. Soon after, David began living as a male.

Through annual visits to Dr. Money's Psychohormonal Research Unit, Brenda's 'progress' and 'adaptation' as a 'fully-functioning' female were *consistently misreported.* This blatant misrepresentation of the 'success' of sex reassignment, produced a flurry of similar infant sex reassignment surgeries to thousands of other children destined to be failures. Subsequent follow up *longitudinal studies* on sex reassignment surgery confirmed the overwhelming failure rate of this now discontinued practice, a practice that reeked of child abuse.

If David's experiences were not enough, the Reimer family seemed cursed in other ways. Both his mother and his twin brother Brian suffered from cyclical depression. At age sixteen, Brian attempted suicide by drinking drain cleaner. He became an alcoholic in his young adult years, eventually losing his job, his wife and children, and his own life in 2002 by a toxic cocktail of antidepressant meds and alcohol.

Award-winning writer John Colapinto discovered David's story and wrote an article for *RollingStone* Magazine. Later, he would write a New York Times best-selling book, *As Nature Made Him* (hardcover 2000, paperback 2006). David cooperated fully with Colapinto, in hopes the experience would prove cathartic. It may have been beneficial at the time, but it also recaptured horrendous memories. Colapinto split the profits on the sales of the book fifty-fifty with David. In a short while, the profits and an impending movie deal finally brought a measure of financial relief to David's *unrelenting fear of his own future.* Yet, nightmarish memories continued to plague him as he brooded at home with no work, or social outlets to distract him.

After making a series of ill-advised investments, David was soon financially strapped again. His stay-at-home wife, Jane, soon felt compelled to find work to help the family survive. David felt threatened by her sudden independence and his depression deepened. No matter how much his wife pleaded and assured him they could make it through the setback together, David, feeling alone and forever fearful of his future, one day stormed out of the house in tears. Apparently, he had finally reached the final rung on his emotional ladder.

After leaving his beloved mate, David briefly visited his parents but drove home while Jane was at work and the kids were in school. There, he found a shotgun and sawed off the barrels. With suicidal resolve, he drove to a nearby grocery store parking lot and ended his life.

Since about 1997, controversy had descended upon the Twins Study often used by Money to describe his most celebrated case (in his own mind) when Milton Diamond, a professor of anatomy and reproductive biology at the University of Hawaii, reported that the reassignment surgery had been in reality a complete failure. "John had never identified as a female or behaved in typical feminine ways. In fact, John had been miserable the entire time." The paper was a bold move for a fellow scientist to challenge the seemingly bulletproof reputation of an esteemed scientist. Indeed, Money's theories had influenced the writing of all textbooks at the time and continuing for years after the news hit academic circles. Could his theory that nurture trumped nature in scripting sex and gender be true? Or, did he fraudulently twist his findings?

In the annals of medicine or psychology, it is not possible to find a more persuasive example of *medical hubris* from a self-absorbed 'expert' who built his theories of gender and sex upon castles of sand. John Money was in influential in gender studies from the 1950s-1980s. He died in 2006 never once admitting he could have been wrong.

LESSONS LEARNED. What did the medical/psychological community learn from the experiences of David Reimer? Praise for *As Nature Made Him* offers clues:

> "As object lesson in medical hubris and close-the-ranks collusion, and in the tragic results when ideology trumps common sense in thinking

about sex and gender . . . " Deborah Tannen, author of *You Just Don't Understand*

"Makes a convincing case that gender has less to do with signals we send and receive from the world than with ineradicable messages encoded in every cell of our brains and bodies." *Elle Magazine*

"Colapinto's book is a stinging and overdue indictment of the 'sexual reassignment' of infants like baby Bruce and those born with male and female sex organs." *Cleveland Plain Dealer*

"This is a mesmerizing tale that manages to balance an engrossing look at what happened to Brenda with a persuasive argument that biology, not environment, determines sexuality." *San Antonio Express*

Ancillary Assignment: Google: John Money Wikipedia, David Reimer Wikipedia, and Milton Diamond Wikipedia

2 Scott Peterson: A Diagnostic Evaluation

Ashleigh Portales

Scott Peterson's Problems

On Christmas Eve, 2002, Scott Lee Peterson murdered his wife, Laci, and unborn son, Conner, and dumped their bodies in the San Francisco Bay. The decayed remains of mother and son washed ashore a day apart from each other, Conner on April 13, 2004 and Laci the morning after (Ablow, 2005).

It soon became apparent that pictures released to the press of the smiling couple told little of the true story of life inside the Peterson home. At the time of the murders, Scott was one month into an affair with Amber Frey, a masseuse who believed he was a widower. She was only the latest in a string of girlfriends Peterson had deceived since his 1997 marriage to Laci. Referring to his casual use of multiple women for his own pleasure, Dr. Keith Ablow, the forensic psychiatrist hired by Peterson's defense team, states that Peterson "walked among us as an emotional vampire feasting day-to-day on the life force of others, particularly women" (p. 17).

Peterson, who had already shed his wedding ring and dyed his hair, was arrested five days after the body of his son was discovered, April 18, 2004. The contents of his vehicle, which he had purchased under his mother's name, confirmed that he was preparing to weave a new web of lies in another place. Along with nearly $15,000 U.S. dollars and a significant amount of Mexican currency, Peterson had packed four credit cards issued in the names of various family members, his brother's driver's license, his entire wardrobe (including fourteen pairs of shoes!), four cell phones, camping equipment, and twelve tablets of Viagra (Ablow, 2005).

Scott Peterson was put on trial for the murders of Laci and Conner Peterson in June of 2004. In November of that same year, he was convicted and sentenced to death. Today, he awaits the culmination of his sentence, residing on San Quentin's Death Row (Ablow, 2005).

Biopsychosocial Model of Scott Peterson

Biological Factors. Evidence exists that the genetic predisposition for failure to develop strong emotional ties to one's own offspring ran strong in the Latham (Scott's maternal) line. A family history of abandoning children dates back to at least his grandmother. In December of 1945, Scott's grandfather was murdered. His wife promptly gave custody of the couple's four children to the local Catholic orphanage. Among those children was Scott's mother, two year old Jacqueline (Jackie) Latham, who would grow up to give two of her own children up for adoption and consider surrendering a third, stopped only by the chastisement of her pediatrician, before giving birth to Scott. While Scott and his older brother were growing up, his parents shipped the older brother off to live with relatives in another state when he displeased them, reinforcing the idea that children were expendable (Ablow, 2005).

Scott most likely received the genetic vulnerability for antisociality from his father. Before he met Jackie, Lee Peterson had a history of buying things he could not afford as a means of masking his impoverished past. He was known to use fake names and identities when collection agencies got too close. A relative of his (quoted in Ablow, 2005) referred to Lee as "not someone comfortable with real emotion—his own, or anyone else's" (p. 35). It was also rumored that Lee did not particularly care for children and may have ended his first marriage to escape those he already had (Ablow, 2005).

Social Factors. Scott Peterson's socialization toward both Antisocial and Narcissistic Personality Disorders began shortly after birth when he came down with pneumonia severe enough to require the infant's complete isolation in a climate-controlled plastic chamber. Such early separation of an infant from his mother has been shown to cause emotional detachment later in life (Ablow, 2005). After his recovery, Scott came home to a family where perfection was not the goal, but the norm, according to Dr. Keith Ablow (2005):

> *Scott Peterson must have learned early on that being anything but the perfect child would not be tolerated by his parents . . . He was in a perpetual state of unconscious panic that his mother (who had given away two other children and who had considered giving away her third) would abandon him and that his father would do nothing to save him.* (p.45)

Despite the checkered pasts of all involved, Jackie referred to her family as "The Brady Bunch" and objectified Scott as her "Golden Boy." Though Lee and Jackie did their best to paint a picture of perfection, some of those close to the family were able to see through the guise. One such source spoke regarding the murders of Laci and Conner, as quoted in Ablow (2005):

> *"If this could happen to any family, it doesn't surprise me one bit that it happened to this one—because of the disconnectedness from reality. Jackie says*

> *everything with a smile, but it's a predatory smile . . . And if you ever . . . made any other wrong move, you knew you were going to fall out of her good graces. Then you would be nothing to her, dead in her world."* (p.48)

Scott had been socialized to perfection, to truly being "golden." But his own life was not going exactly in the direction he had planned, and the pressures were mounting. He had not been able to perform at a large university and was forced to settle for a smaller school closer to home, purchasing the diplomas from his ideal school from a counterfeit source online. His dreams of a high paying job had been forfeited for selling fertilizer. He had been forced to buy a home in a less prominent city and to borrow the money for the down payment from his parents. His wife was pregnant with a child he did not particularly want or care for and she was pressuring him to buy a bigger home far beyond their means. He had a girlfriend he wanted but could not have; divorce would shatter his perfect image. The more Scott Peterson was left to live life on his own, out from under his parents' watchful eye, his "golden" glow began to fade and he struggled to maintain the lustrous shine his mother insisted he had (Ablow, 2005).

Psychological Factors. From the beginning of his life, Scott Peterson was told he was "golden" and he believed it. He could not understand why the golf team at Arizona State University did not beg him to play (Ablow, 2005). He had never experienced this type of rejection. He thought every woman should be attracted to him and could not comprehend why women would not continue seeing him upon discovering that the person he had pretended to be did not exist (Ablow, 2005). Scott saw himself as far above average and did not understand why the world did not cater to his "golden" identity the way his mother and father always had.

Scott Peterson felt trapped. Trapped in a marriage to a woman he did not really love. Trapped in the impending role of fatherhood brought by Conner's impending birth. Trapped in an identity he could no longer maintain.

Scott Peterson felt angry. Angry at his mother and father (though he would never in a million years voice such defiance to them) for stealing away his individuality and fashioning him into a puppet whose strings they controlled. Angry at the world for not recognizing who he was and what he was entitled to and for not giving him all the splendors of life that he felt he deserved. Angry at his wife for robbing him of his freedom, of his chance at a life with Amber Frey, and for forcing him to be accountable for his actions. Angry at his son, who was not yet even born, for tying him down to a family and all the responsibilities that go along with it. Angry at himself for not being able to perform at the level expected of him.

Most importantly, Scott Peterson felt no emotional connection to anyone. People were simply a means to his pleasure and when they ceased to be such they were expendable. Others could be manipulated, deceived, used, and discarded as quickly as they could be replaced. Scott was ready to assume a new identity, just as he had done for countless women many times before. The only thing left to do was to eliminate the obstacles that stood between him and the new life he felt he deserved.

Multiaxial Assessment:

Axis I: Clinical Disorders

V71.09 No diagnosis

Axis II: Personality Disorders and/or Mental Retardation

301.7 Antisocial Personality Disorder

301.81 Narcissistic Personality Disorder

Axis III: General Medical Conditions

None

Axis IV: Psychosocial and Environmental Problems

Primary Support Group: Familial love and support are superficial at best and highly conditional. It was never made secret that children who did not please their parents were expendable and would be eliminated from the family. Meanwhile, Scott was labeled the "Golden Boy" and forced to grow up dominated by parents who molded him to their whims. Conflict was also brewing within Scott over the impending birth of his first child, an event he did not welcome. Additionally, he was having an affair at the time of the murder, and had had several previously.

Occupational Problems: Scott was extremely dissatisfied with his work as a fertilizer salesman, feeling it was beneath him and his status as "golden."

Housing Problems: Due to financial constraints, Scott and Laci were forced to buy a home in Modesto, California, an area which did not match the status Scott felt he deserved.

Economic Problems: Scott and Laci were struggling to pay their bills. Laci was working as a substitute teacher to make ends meet. Scott's parents had floated the down payment on his house, his country club membership, and the cost of installing a pool in the backyard. Added pressures came from Laci who, with the coming addition of a child, wanted a bigger house far beyond the couple's means.

Interaction with the Legal System/Crime: Scott was arrested for and convicted of the murders of his wife and unborn son and sentenced to death.

Axis V: Global Assessment of Functioning

55* (Scott was in a constant state of emotional disconnect and had marked difficulty maintaining a healthy relationship with his wife, evidenced by multiple affairs)

Antisocial Personality Disorder

Antisocial Personality Disorder is characterized by "a pervasive pattern of disregard for and violation of the rights of others" (American Psychiatric Association (APA), 2000, p. 706). Individuals displaying this disorder are typically very apathetic towards the emotions of others, have no regard for the law, repeatedly lie for personal gain, are impulsive and

* While a man who murders his wife and child clearly poses a danger to others, suggesting a lower GAF score, even in the commission of the murders Scott was clever and cunning enough to plan how to hide the bodies, clean up after himself, and construct plausible alibis for both his family and his mistress, indicative of a higher level of functioning.

often violent, and behave irresponsibly (APA, 2000). Several of these criteria apply to Scott Peterson.

Failure to conform to social norms with respect to lawful behaviors (APA, 2000, p. 706): There is no greater disregard for the law and the rights of others than the taking of another's life. This is exactly what Peterson did when he murdered his wife and unborn son on Christmas Eve, 2002 (Ablow, 2005).

Deceitfulness (APA, 2000, p. 706): Peterson lived his life in a web of lies he began weaving at least in high school, when he led his fellow golf teammates to believe he had won a golf scholarship to Arizona State University when he had in fact never been contacted by the school and was later told to try out for a walk-on position. He bolstered this lie by purchasing fake ASU diplomas online. He also used deceit as a foundation for his romantic relationships. He created false identities to facilitate several affairs, both before and after his marriage to Laci (Ablow, 2005).

Consistent irresponsibility (APA, 2000, p. 706): It was highly irresponsible of Scott to risk losing his marriage over several affairs. It was also irresponsible to squander the money he had to borrow from his parents on frivolous things such as a country club membership and a backyard swimming pool, especially in light of the fact that he had a baby on the way. Peterson also spent substantial amounts on his girlfriends, showering them with lavish gifts, dates, and even trips (Ablow, 2005).

Lack of remorse (APA, 2000, p. 706): In the event that one of Scott's girlfriends caught him in a lie, he never once apologized or even acknowledged that he had hurt them in any way (Ablow, 2005). Moreover, from the moment Peterson made the claim that his pregnant wife had disappeared, he never shed anything but crocodile tears for the news cameras. His lack of emotion was noted by many people, including the officers charged with his care. In fact, during the police transport following Peterson's arrest for the murders of his wife and son, DNA confirmed that the badly decayed bodies found in the San Francisco Bay were those of Laci and Conner. When Scott was told this information, his eyes did not even water and, only one hour later he requested the detectives get him a double cheeseburger, fries, and a vanilla shake—a meal which he consumed in its entirety (Ablow, 2005).

The same apathetic Peterson sat in court, sometimes smiling, but never showing negative emotion, even as his death sentence was being read (Ablow 2005; Associated Press 2005). The complete absence of guilt or remorse in any form continued to be on display when Peterson arrived at San Quentin State Prison in California, where he will be held until his death sentence is carried out. During the long intake process, an officer suggested to Peterson that he take a nap, to which he replied, "No, I'm just too jazzed" (Associated Press, 2005).

Narcissistic Personality Disorder

The key to Narcissistic Personality Disorder is "a pervasive pattern of grandiosity, need for admiration, and lack of empathy that begins by early adulthood and is present in a variety of contexts" (APA, 2000, p. 714). The individual with this disorder has an extremely inflated view of himself and believes others should view him in the same light. Several criteria for this disorder can be seen in Scott Peterson.

Grandiose sense of self-importance (APA, 2000, p. 717): Peterson wholeheartedly bought into the "Golden Boy" label he was given by his mother. His entire family catered to it and he was genuinely shocked when the rest of the world did not follow suit. Scott greatly exaggerated his achievements by faking multiple college diplomas from a university he had attended for only one semester, and by leading others to believe he was recruited by the golf team there (Ablow, 2005).

Preoccupation with fantasies of unlimited success (APA, 2000, p. 717): Scott ideally envisioned himself as a professional golfer or an international businessman whose work took him around the globe. This stood in stark contrast to the reality that was his life.

A former girlfriend of Peterson's related to Dr. Keith Ablow (2005) that she felt:

> "Scott must have resented the move to Modesto. It wouldn't go with what he was aiming for—living in a bigger city and having adventures and money and all that" (p. 112).

The international jetsetter he desired to be is exactly who he portrayed himself to be to his last mistress, Amber Frey (Ablow, 2005).

Requires excessive admiration (APA, 2000, p. 717): Never content with just one woman Peterson made passes to the bartenders at his wedding and to his sister's babysitter just days after Laci's disappearance (Ablow, 2005). This trend continues in prison. According to the warden at San Quentin, he is "flirtatious with the women and lights up when he sees female correctional officers" (KNBC News, 2005). He also boasts about a nickname given to him by one of the various women who write to him: "Scottie True Hottie" (KNBC News, 2005).

Interpersonally exploitative (APA, 2000, p. 717): Scott used his many relationships with women to satisfy his own personal desires for sex and admiration. When planning his flight from murder charges, he also used his relationships with his mother and siblings to acquire false licenses and credit cards, without their knowledge, to further the success of his escape (Ablow, 2005).

Lacks empathy (APA, 2000, p. 717): At no time during his romantic relationships did Peterson ever acknowledge that he had hurt either his wife or his mistresses. He also showed no compassion for the pain of others regarding the disappearance of his wife and child. In a secretly recorded phone call between himself and Laci's mother before his arrest, she rants at him for killing her daughter and pleads to know where her daughter is. Throughout the heart-wrenching dialogue, Peterson's voice remains flat and monotone, completely void of emotion (Ablow, 2005).

Arrogant, haughty behaviors or attitudes (APA, 2000, p. 717): From his cell on death row, Peterson posts messages to supporters via the Canadian Coalition Against the Death Penalty. The condemned man oozes arrogance as he writes about a scholarship fund established in his wife's name. "It hurt when my donation from jail to this fund was almost rejected . . . any donation to it is a wonderful way to carry forth the memory of my wife" (Peterson, 2005a). He later extends accolades to his mother-in-law concerning "rumors" that profits from a book she is writing will go toward the same charity.

Peterson (2005a) writes:

> "Some people have done things to profit off of my wife and son having been taken from me and murdered. The profits from this possibly going to charity would counter this disgusting trend. If the rumor is true, what a wonderful act."

Possible Interventions

Personality disorders are extremely difficult to treat because the disordered individual does not recognize a problem with themselves. The general consensus among clinicians favors incarcerating the severe Antisocial to prevent further harm to society. The only hope seems to lie in identifying at-risk children and training their parents in the areas of problem recognition and alternative parenting strategies (Durand & Barlow, 2006). Yet in the case of Scott Peterson, whose own parents were obviously disordered themselves, it is doubtful this type of intervention would have worked. Parents who thought of their family as perfect were not likely to have solicited or accepted help from the psychiatric community.

Narcissistic Personality Disorder is just as resistant to treatment as Antisocial Personality Disorder for largely the same reason; the individual fails to realize a problem exists. While some attempts at cognitive therapy have been made, results are inconclusive (Durand & Barlow, 2006). Scott Peterson is not likely to consent to any such type of treatment given the fact that he has yet to admit to killing his wife and unborn son.

Unfortunately, sometimes dysfunction and disorder perpetuate from generation to generation. Such is the case of Scott Peterson. For him, incarceration is undoubtedly the best course of action. Perhaps it is the only way to put an end to the "psychological perfect storm" that has decimated the Peterson family (Ablow, 2005, p.8).

3 Minding the Twenty-First Century Brain

Don Jacobs

> "The biochemistry of the brain is no longer a complete mystery. Some psychiatrists go so far as to say that 'for every twisted thought there is a twisted molecule.' This may not turn out to be true, but it seems reasonable that what poisons the brain poisons the mind."
>
> Mary Kilbourne Matossian *Poisons of the Past (1989)*

Neuroimaging & Neuropsych: Getting Over the Threshold

"To enhance public awareness of the benefits to be derived from brain research, the Congress, by House Joint Resolution 174, has designated the decade beginning January 1, 1990, as the "Decade of the

Brain" and has authorized and requested the President to issue a proclamation in observance of this occasion. Now, Therefore, I, George Bush, President of the United States of America, do hereby proclaim the decade beginning January 1, 1990, as the Decade of the Brain. I call upon all public officials and the people of the United States to observe that decade with appropriate programs, ceremonies, and activities."

Filed with the Office of the Federal Register, 12:11 P.M., July 18, 1990

The years from January 1st 1990 to the end of 2000–the much celebrated Decade of the Brain—came and went. But, who knew? What programs, ceremonies, and activities were held? Ceremonies aside, *what did transpire* was a surge in activities and programs involving sophisticated neuroimaging (brain scan) tools—PETs, fMRIs, and SPECTs—among others, as neuroscientists began to observe how *cerebral tissue functions in real time.* This alone was huge. Research scientists began to map quality of blood flow, indirectly showing metabolic activity, and regional specificity of brain function. Also, imaging showed *cerebral connectivity*—the networking empire of MIND—the perfect metaphor for the holistic brain. Neurochemistry contained in the axon vesicles—dopamine, norepinephrine, and others including serotonin—comprise a life force greater than the Freudian concept of libido. It's behind our personality and how we feel, and ultimately the fuse to observable behavior.

Thousands of volunteers and patients were willing to lay motionless in massive steel cocoons to get this revolutionary data. In the process, the brain became illuminated in regal reds and mellow yellows—both showing normalcy in form and function; a brain with the 'blues'—indicative of slow or disrupted metabolic functioning—showed up 'cool-coded' blue.

Would a picture book of damaged and dysfunctional brains soon emerge within pages of the Diagnostic & Statistical Manual of Mental Disorders (The DSM, 2002)—the bible for psychiatrists, psychologists, and numerous mental health practitioners in careers across the continua from social workers to psychiatric nurses?

Wouldn't such unprecedented insights from this decade and continuing, give practitioners a leg up on our magnificent three pound organ of behavior? Would the list of those that would benefit be headed by practitioners in such diverse careers as education, criminal justice, marketing, law enforcement, business, criminology, psychiatry, sports, psychology, and the military? What about every parent and teacher?

Controversy was a sure bet to enter the picture with such dramatic promise. It did. In an article in *Neuropsychiatry,* published in February 2001, *Why Don't Psychiatrists Look at the Brain?* Dr. Daniel Amen, medical director of the Amen Clinics, stated

"psychiatrists are the only medical specialists who rarely look at the organ they treat. The odds are that if a patient is having serious problems with feelings (e.g. depression), thoughts (e.g. schizophrenia), or behavior (e.g. violence), psychiatrists will never order a brain scan."

No one had to be a mental health specialist to guess where Amen was headed with his attack: One, psychiatrists most likely reach for the prescription pad to prescribe psychoactive medication; two, psychologists engage in talk therapy (psychotherapy) indicated by presenting symptoms from the (as yet picture-less) DSM; or three, they use any of a number of eclectic procedure such as ECT—electroconvulsive therapy or some other *au courant* treatment *du jour.*

Would brain images eventually accompany the DSM? Would this pass muster with the American Psychiatric Association? For all its early hype, brain scanning was treated by mainstream psychiatrists as unwelcomed interlopers. Were the results just not specific enough? As one might expect, neuroimaging has its share of doubters. Even though about five hundred neuroimaging studies are currently being published a year and with numerous commercial imaging clinics opening in all parts of the country including Texas, some experts are of the steadfast opinion that 'technology has been oversold as a psychiatric tool.' Yet, other researchers remain optimistic. It's getting to the next level of a more 'standardized application' apparently that has to happen. The scans have to show a consensus of opinion—what we call 'getting over the threshold' in this article.

Yet, we have to face facts. We know what a schizophrenic brain looks like. With scans, we can diagnosis a depressed brain, an anxious brain, and an autistic brain. Incredibly, we can peer inside the mind of killers and see prefrontal and temporal lobe damage. Might this measure of 'seeing is believing' be exactly what attorneys need to segue into forensic legal proceedings? This kind of high stakes drama in popular culture might kick-start the next step: inclusion of neuroimaging into the media. How else do things get done in the USA? Anyone care to look at Paris Hilton's brain?

Incredible as it sounds, the brain's metabolic activity—a cardinal sign of *function versus dysfunction*—remains largely unobserved by anyone other than neuroscientists in research institutions. From the time of Philippe Pinel, psychiatry has remained like an auto mechanic who fails to run the diagnostics on a sluggish car and installs a new transmission instead. 'Presenting symptoms'—the car doesn't move forward or backward—dictates a new transmission.

The point is: every option mental health professionals choose is intent upon *changing brain functioning.* Would an orthopedic surgeon start an incision to repair a torn ACL just because his patient was limping? Wouldn't he or she first want to see if the ACL is, in fact, torn and in need of repair? Amen continues, "but (psychiatrists without referencing scans) don't know which areas of the patient's brain work well, which areas work too hard, and which do not work hard enough." Precisely the point.

Today, we live in the age of neuroscience, well past the journal articles (nine years past and counting, in fact) that characterized the Decade of the Brain. Yet, a seven to ten year lag is expected from journal articles to textbook inclusion. Doing the math puts some of the findings in textbooks slightly past due with a deluge expected by 2010. Want proof? The newly published 2008 version of the seventh edition of David G. Myers' psychology textbook *Exploring Psychology in Modules* has ONE reference to neuroimaging on page 50–51 of the 620 page book. The image shown is the difference between a schizophrenic brain and a normal brain. No Geidd Study, Volker or Blair Studies. Why?

Psychology is a neuroscience even though it is often treated otherwise by individuals who should know better. Imaging technology is evolving so fast that it is on a par with high definition television.

> "With increased resolution, we'll be able to do more sensitive and more precise work, and I would not be surprised if *anatomy alone based on volume* will be a diagnostic feature," according to Dr. Jeffrey Lieberman, chairman of the psychiatry department at Columbia University Medical Center and director of the New York State Psychiatric Institute. "We have gained enormous amounts of knowledge from thousands of imaging studies, we are on the threshold of applying that knowledge, and now it's a matter of getting over the threshold."

The fact is neuroscientists do know which areas work well and which areas do not, even with the problem of idiosyncratic brains—the fact that not every brain is exactly the same. So, what do we need to get over the threshold? What we need for a picture book with the DSM is what it already has: a *continuum of severity* indexed mild, moderate, and severe. Variations in scanning would fit 'severity continua' like a glove.

In cutting-edge psych classes, even undergrad neuropsych students take the liberty of suspecting something's not right when they are asked to interpret 'cool-coded' scans. Truth is: e*xperience in interpretation* over the years has yielded the signature of scans with 'habits and pattern' formations required of the behavioral sciences to fashion theory into fact. This is the difference in analysis between ushering in the new science of imaging from the 1970s to (almost forty years later), a considerable *predictive science* today.

But, why all of the reluctance to put neuroimaging right in the center of psychiatric and psychological diagnoses and treatment where it clearly belongs? After all, NIDA—the National Institute of Drug Abuse—and many other federal agencies display impressive websites with the latest findings directly from neuropsych.

We have plenty of neuroimaging equipment in all major hospitals and in university research centers. Enough, in fact, that we could transport patients (and inmates) across town for a neuroimaging appointment without much hassle. In modern times, might parole boards be interested in viewing a brain scan from a convicted felon about to plead his case for being freed back into open society? What about serial killers or pedophiles? Still remaining is a certain hesitancy even from those who have published many imaging studies.

> "I have been waiting for my work in the lab to affect my job on the weekend, when I practice as a child psychiatrist," said Dr. Jay Giedd, chief of brain imaging in the child psychiatry branch at the National Institute of Mental Health. He has done MRI scans in children Monday through Friday for 14 years. It hasn't happened."

Again, the threshold has not been crossed. What must happen for neuroimaging to get over the threshold of acceptance and move forward into mainstream practice and the DSM picture book? If it's just time, we can wait. *But, there's so much available right now we can use.* The remaining parts of "Minding the twenty-first century brain" over the next few months will focus on what neuroimaging and neuropsych has given us that has already PAST THE THRESHOLD. Stay tuned.

This introductory installment is one of three planned in my series of "Minding the Twenty-First Century Brain." Part II will address puberty and adolescent brain development (sure to be popular with parents and educators); Part III will cover addiction, violence, and the certainty of chemical imbalances. The articles simply 'scratch the surface' of what neuroscience knows from neuroimaging studies, but hopefully it's enough for practitioners and the public to demand more. *More is better,* as application of this knowledge in everyday careers can make a vast difference in outcomes.

Written in non-technical language, the thrust of the articles is directed at the ones who need insight the most—individuals, parents, teachers, counselors, and psychotherapists—the practitioners that represent *issues of home, relationships, and school*—the very backbone of society.

4 Pandora's Hox

The Hox Gene Paradigm

Don Jacobs

One outcome of the unprecedented human gene 'tagging' project, *The Human Genome Project (HGP),* and the decades of research preceding it, confirmed what scientists hypothesized at least since the 1980s: *gene expression is modified by individual experience.* Apparently, biology (nature) is never immutable, nor is learning (nurture) *per se.* Learning may be nothing more than an exercise in turning some genes "on" or "off." Sounds simple enough. In the age of neuroscience, *nature appears to be designed for nurture* so what happens to us in the womb, experiences in the environment, and especially in *personal milieus* (family, peers, careers, social interactions) across the lifespan is just as important as the existence of genes themselves. Enter *hox genes* and supporting cast—the promoters, the enhancers, and the transcription factors—soon to be addressed.

The 'Versus' Controversy

Traditionally, the "versus" controversy in philosophy first, science second ("nature versus nurture") created a *false dichotomy* for the biological sciences and hence, the behavioral sciences, and gave the impression that one was more important than the other. Mathematician Francis Galton, Charles Darwin's cousin, ignited the "nature versus nurture" controversy in 1874, five years before Wundt established psychology as a science of human behavior.

A kind of intellectual masturbation ensued for over a century involving Descartes' "Cartesian Principles" of dualism that fanned the flames of the "versus" controversy. The father of North American psychology, William James, took a swing at the controversy by arguing humans "possessed more instincts than animals, not fewer." Philosophical luminaries such as British Empiricists John Locke and David Hume jumped on the "experience" bandwagon, while Rousseau and Kant argued for the immutable "human nature" biological side. Hence, an indelible line was drawn in the sand over which one of the "versus" was more formative in human behavior. From the standpoint of behavior, it penetrated everything from philosophy, science, personality theory, moods, and to the causes of criminality.

Behaviorist John B. Watson, a proponent of learning theory, stood on the shoulders of Ivan Pavlov and used the lab and "Little Albert" to show how a conditioned stimulus (CS) worked by pairing a neutral stimulus (a rat) to a startle response (a loud noise) and how it conditioned fears and phobias in humans through learning (or through nurture). Following Watson's famous "stimulus guarantee" challenge, suddenly it mattered greatly the type of stimuli parents put around their children.

According to Watson:

> "Give me a dozen healthy infants, well-formed, and my own specific world to bring the up in and I'll guarantee to take any one at random and train him to become any type of specialist I might select—doctor, lawyer, artist, merchant-chief, and yes, even beggar-man and thief, regardless of his talents, penchants, tendencies, abilities, vocations, and race of ancestors" (Watson, 1966).

Interestingly, the movie *The Village* by M. Night Shyamalan, Watsonian behaviorism is in evidence in an attempt by the elders to create a utopian society set amid eighteenth century nature worship. The red clothed creatures—"Those We Don't Speak About"—are intended to keep the naïve villagers from entering the "real world" outside the confines of protection within the animal conservatory.

Even with all the carefully crafted non-violent stimuli within the village with the conspicuous absence of money, guns, drugs, and TV—and the menacing stimuli beyond the perimeters—"the creature"—still cannot prevent a murder. Apparently, the power of nature (the feeling of jealousy over a lost love) cannot mitigate wholesome non-violent nurture.

Radical behaviorist B. F. Skinner (1969) demonstrated the power of learning (or conditioning, hence learning by nurture) through positive and negative reinforcement, punishment, or ignoring behavior, *through consequences*—"what follows what we do." Again the Shyamalan movie shows villagers the consequences of disobedience (crossing into the forbidden zone of the woods with hairless corpses of livestock lying dead around the village) presumably, a warning by the "creatures" not to cross the boundary.

On the flip side, Chomsky showed how humans must come equipped with an innate "processor" for language since human babies cannot learn language merely by trial and error—a clear argument of biology and genes (nature). Otherwise, so could chimps, who by nature, don't have the "processor" and do not babble in humanoid sounds at the operant level—the level required for behavioral reinforcement of double syllables (again, a nurture experience).

The controversy between the behaviorists and the geneticists raged on. But, that was then and this is now. At least all of the aforementioned theorists made famous for their stand on nature versus nurture were, at least, half right. In science, being half right is far better than being completely wrong as observed in myth and other systems of false belief and other embarrassing instances of *parataxic logic,* courtesy of H. S. Sullivan. Felt the bumps on your head lately?

The *Human Genome Project* (HGP) taught us that genes are *active participants* in our lives from "womb to tomb." Somehow, experience intertwines with nature; it doesn't gather at the top like water "versus" oil. With this news, everything comes out truly *individualistic* as a function of perception expressed as *nature via nurture* (not "versus"). Early theorists in psychology predicted as much in *a priori* perceptual studies but couldn't prove it. Behavioral scientists may work around individuals wearing straightjackets but it doesn't mean their minds are harnessed like one.

Both nature and nurture have been actively in the spotlight of general psychology since Wundt laid down principles of structuralism versus William James who countered with functionalism, a continuation of the "versus" controversy. The Gestaltists somehow knew they were destined to oversee the exchange of the "versus" controversy in psychology for more plausible "via" but couldn't prove it either. Behavioral scientists are compelled to measure things even if it means putting things on a continuum and dealing with percentages of the truth.

Science 'versus' Beliefs

With new understanding of gene expression from the HGP, we must think "outside the gene." Translation: we must open our minds. Open up our minds? That calls for a revamping of our belief systems and personal "addictions." Scientists have long known *behavior and belief systems outweigh scientific fact.* Belief versus fact? Yet another "versus" controversy we humans love to nurture. Pardon the momentary diversion: If we listen to scientific facts alone, why do millions of people still smoke those disgusting cigarettes, cigars, and pipes? What do the facts say? You're killing yourself and others (with secondary smoke) and headed to an early grave! The response: "So . . . that happens to everyone else!" Why do millions of teenage girls give birth to children every year? What do the facts say? Use protection or you'll have nine months to get your act together. The Bush Administration and the Far Right swear they have the answer and they have put the money where their mouths are at the tune of pumping millions of federal dollars into abstinence, baby! That has worked about as well as the Reagan Administration's 1980s failed drug policy: Just say "No" to drugs! So, by implication our teens Just say "No" to sex! With billions of brain cells crying "Yes! Yes! Yes!" In the heat of the moment (or at point of purchase), we just say NO. Right? Wrong. Alicia, help! We need your *Clueless* expertise!

Jacobs (2008) observed in an essay summarizing the so-called Triune Brain—our three-brains-in one: the vast majority of teenagers are "parked" in the midbrain (emotions, sex, and stimulation) while their parents are "parked" in the dorsolateral prefrontal cortex (cognition, strategizing, and weighing consequences). Here lies the generation gap and the uphill climb to make teenagers understand things cognitively. In the celebrated case of John/Joan, the boy who lost his entire penis to a botched circumcision, we discovered another quirk of nature. Since John's chromosomal sex was XY his "Brindy"—his brain, mind, and body—was male. No amount of hormone injections or talk therapy could have ever changed "John" into "Joan," since his brain was male (hard wired male brain). The celebrated sex expert at the time, Dr. John Money of prestigious Johns-Hopkins, was anything but "on the money." If fact, he was dead wrong trying to prove, with a combination of Watsonian and Skinnerian behaviorism, that: "it doesn't matter what you're born, it ultimately matters only how you're raised." The failed experiment showed us the real score: trying to make John into Joan was a complete disaster. Nature can be expressed through nurture and vice versa, if and only if, both are "turned on" to the idea so to speak.

Freud assumed *a priori* that "eventually, everything (in a person's psychological make-up) will be shown to have biological implications." Presumably, this insight occurred after he gave up cocaine, not before.

Pandora's Hox

Back to the Hox Gene Paradigm. In the 1980s when animal geneticists opened up the common house fly genetic "code" they found a small group of genes they called "hox genes." They quickly discovered that hox genes set out the plan of the body. Researchers found hox genes in the rat, monkey, and yes, in humans. Like all genes, hox genes are capable of being switched "on" or "off" (meaning to become active or inactive) in different parts of the body at different times. Geneticists discovered that special DNA called *promoters* were the "switchers" that control the on/off "button" on the hox.

Subtle *variations in the genome of species* suggested the genome is less of a blueprint and more of a *receipt for taking "orders" from milieu.* Hox genes take a certain amount of time in a certain sequence for the "order" to be expressed.

Realistically, the project devoted to "time and sequencing" may take another five years to ascertain, say around 2015. Another DNA element, called an *enhancer,* plays a role but only if "transcription factors" engage both the promoters and enhancers producing enzymes that copy DNA to RNA. Confused yet? Regardless, slight changes in promoters can have huge consequences in gene expression, apparently determining a myriad of physical characteristics from where the head, thorax, and limbs go to apparently emotional ones, such as how resilient, how addictive, whether we're introverted or extraverted, or how co-dependent or independent we eventually become.

Parenting

Opening and closing hox gene "windows" at critical times is nature's way of nurturing a constellation of human and animal characteristics such as language and task acquisition, personality and yes . . . love, gender, and criminality. For example, childhood mistreatment by incompetent or "toxic" parenting has long been postulated as a blueprint for antisocial behavior. Reported by a news magazine recently and according to research at London's King College with over four hundred New Zealand males as subjects, who have been followed since birth, antisocial parenting does indeed produce antisocial behavior but *only in special instances. The difference lies in the gene promoters.*

Apparently, the subjects with so-called "high-active genes" were immune to deleterious parenting—no matter how bad the parenting was they were resilient and grew up to be normal, not antisocial. In contrast, males in the sample with low-active genes (as well as being reared in dysfunctional homes with bad treatment) were responsible for "four times the number of crimes such as rapes, assaults, and robberies."

While hox genes seem to be "at the mercy of our behavior, not the other way around," they are "responders" to experiences and well as senders in the interchange between nature and nurture. They are "both the consequence and the cause of behavior," not the "versus" controversy that produced the false dichotomy scientists once held so dear.

Now, we know why the same parental conditions can produce two completely different kids with entirely different temperaments. Apparently, one child's hox was active while the other was inactive. We still don't know why hox genes are active sometimes and inactive at other times. We know the music, we don't know the dance. But, now we know they are both on the dance floor at once.

5 Chemical Markers in the Brain

Introduction to Brainmarks

Don Jacobs

Points on a compass—north, south, east, and west—give travelers the direction of a journey by pointing the way across two sets of 180 degree opposite points (N versus S, E versus W) and all continua in between. Next, the indispensable road map not only shows travelers direction but also the *exact path* to follow from start to finish. It is easy to see the futility of going the wrong direction or taking the wrong path.

Metaphorically, in neuropsychology—psychology at the tissue level—the starting point for all behavior, thinking, and emotion (except the reflexes of the spinal cord) is the brain. Coincidentally, the brain has a chemical compass of its own that "marks" behavior that leads to the pathways we follow in life. The powerful neurotransmitters of the brain *create and run on chemical pathways* that "mark" by the brain in often irreversible ways. Two conditions are influential in deciding which paths the brain eventually "marks" for behavior—the first is emotion, responsible for lighting chemical pathways—and second are "thinking maps" of cognition known as *neurocognitive mapping* that develop directly from an individual's mindset of experiences and developmental influences in family, peer, and social milieus.

THE DANE BRAIN: 'The Jazzers'

Catecholamines—dopamine (DA) and norepinephrine (NE)—together DANE—comprise the principle chemical "jazzers" that lie beneath behavior, moods, and feeling. When we feel 'jazzed' *what lies beneath* is liberated DA (and perhaps NE) at synapse.

Without adequate DANE, individuals display *anhedonic* and *dystonic* behavior—a confusing mix of feeling wretched, sluggish, uninterested, unfocused, and perhaps depressed. The result feels like a one-dimensional blasé life. With adequate levels of DANE, the same individual would be driven to pleasure and reward, displaying plenty of vigor, libido, motivation, and focus, accentuated by goals and "gusto." Suddenly, life is three-dimensional, interesting, appetizing, fun, and worth living.

DA is manufactured in the *substantia nigra* of the midbrain and with wide broadcast throughout the brain with special delivery to the *caudate nucleus* of the basal ganglia buried deep within the cerebrum with targets in the cerebral cortex, particularly the frontal lobes and prefrontal cortex.

Dopaminergic neurons ("ergic" means "working as") extend from the *ventral tegmentum* of the midbrain to the nucleus accumbens and the limbic system's emotionality and reward centers, while another DA bundle proceeds to deep connections within the *frontal cortex* where thoughts merge with emotions resulting in prescient strategizing (given the prefrontal cortex is not damaged).

The medial forebrain bundle of the hypothalamus is rich in DA receptors and has been labeled the "pleasure pathways" of the brain. With DANE liberation across undamaged cortices, sensory, motor, and hormonal activities reflect jazzed, goal-directed behavior.

Norepinephrine (NE)

The noradrenergic neurotransmitter *norepinephrine* (NE) produces alerting, attention-focusing behavior, facilitates learning, memory, and awareness, as well as the well-known "fight-or-flight" behavior within PNS adrenaline (epinephrine) secreted in the periphery as a hormone from the *adrenal medulla* priming motor and sensory behavior. NE is widely distributed in the brain to induce a general level of arousal and excitability. Half of NE activation stems from the *locus ceruleus* and *lateral tegmentum* within the midbrain.

Are we supposed to feel pleasure? Are we supposed to feel jazzed, focused, and interested in what we embrace? Yes, due to the life-affirming characteristics of DANE the chemistry of *what lies beneath* behavior and thinking feels exciting and stimulating. We feel really privileged that we have a seat reserved at the banquet of life.

The Initiators: Testosterone (TEST) & Phenylethylamine (PEA)

As one of two primary chemical "initiators" of a range of behavior (prominently lying beneath *libido or* the sex drive), *testosterone* is the steroid hormone of aggression that acts like a neurotransmitter with binding sites in the CNS. Both males and females have receptors for testosterone. Males experience the rush of testosterone at puberty, as do females with less fanfare (but the chemical effects are still emotionally powerful combined with estrogen). A high level of testosterone in males, although *not directly a cause,* is nonetheless positively correlated with violence and high crime rates. When testosterone—the ubiquitious hormone of libido and aggression–is high with serotonin low, violent, aggressiveness can be predicted, especially when exacerbated by alcohol or other drugs. Without cognitive or chemical restraint, unbridled testosterone can lead to chaotic behavior, aggression, and violence criminality. In males, testosterone peaks in the early twenties and falls gradually by age thirty to much lower levels by age fifty.

In mature females, estrogen, which has masked testosterone for over forty years, subsides to make way for testosterone. At about forty-to-fifty years of age females suddenly become as driven as their husbands or partners used to be.

Phenylethylamine (PEA)

To neuropsychologists, the "romantic rush" is indeed chemical and attributed to a new twist on the "PEA-brain" with the powerful neurotransmitter *phenylethylamine.* Amphetamine-like in chemical structure, PEA lies beneath the physical attributes, demeanor, and body language we find personally attractive in others giving us a "romantic rush" similar to methamphetamine except the chemistry is already within the brain. In fact, when PEA peaks in the brain we become attracted to the person responsible for peaking it. It may be the face, body build, eyes, or hair—but no one else makes us feel the same way.

In the normal brain when PEA is peaked in an individual, he or she is suddenly interested and focused on the person that was responsible for peaking it! Neuropsychologists who are also criminal profilers suspect PEA lies behind why one victim is chosen and not anther. For example, serial killer Ted Bundy chose females with long brown hair split down the middle, physical traits that obviously peaked his PEA. What else could drive behavior with such specificity?

5-HT & GABA: 'The Calmers'

Serotonin (5-HT) is an *inhibitor of activity,* therefore, its effect upon behavior and mood operates as a "calmer." The opposite of NE, 5-HT mitigates aggression, play, and sexual behavior leading to rest, sleep, and recuperation as well as sleep cycling and mood states. Most of the brain's serotonergic neurons are located in the ancient *raphe* (ridge) of the *brainstem.* Besides activity inhibition, normal to elevated levels of 5-HT accompanied by normal levels of DANE, result in the well-known "4-Cs" of liberated serotonin at synapse—calm, cool, collected, and confident. Low levels are associated with low self-esteem, depression, chemical dependency, and eating disorders. From monkeys to humans, mammalian leaders always show elevated 5-HT. However, there is a downside to insufficient 5-HT levels. Male behavior can turn violent with low serotonin accompanied by high testosterone, especially when *deviant cognitive maps* reflect a lack of impulse control, a prefrontal cortex function.

For some males, the ingestion of alcohol exacerbates low serotonin resulting in aggressive or criminality behavior. The formula for chronic depression is consistently low levels of DANE and 5-HT exacerbated or initiated by aberrant cognitive mapping—negative thinking exacerbated by dysfunctional milieus.

The pineal gland contains ample concentrations of serotonin for conversion to *melatonin,* the hormone precursor to sleep (along with GABA). So called "designer drugs" in wide use in pop culture destroy serotonin-producing neurons such as *methylene dioxyamphetamine* (MDA, or "The Love Drug") and MDMA *methylenedioxymethylamphetamine* (or Ecstasy).

Gamma Aminobutyric Acid (GABA)

Like the monoamine serotonin (5-HT), the amino acid GABA is a restrainer of brain activity and is distributed widely in the brain. Hypervigilant behaviors—such as arousal, aggression, anxiety, and excitation—are reduced with GABA liberation. Anticipating sleep, GABA combines with melatonin to prepare for rest, sleep, and rejuvenation. Found primarily in the hippocampus, amygdala, and hypothalamus, GABA is the *major inhibitory neurotransmitter in the CNS* with brain-wide influence. When GABA registers low, the feeling of anxiety in mood is noted. *Anxiolytics* (anti-anxiety medication) such as alprazolam (Xanax®) promote GABAergic transmission at GABA-benzodiazepine clefts in the hypothalamus and mitigate the feeling of stress and anxiety. With "calmers" 5-HT/GABA in balance with "jazzers" DANE, we might have just crossed into Nirvana of peace, love, and happiness.

Acetylcholine (Ach): 'The Restrainer'

In my own thinking over the past few years of presenting neurotransmitter activity as simple as possible to beginning students of psychology and forensic science, the neurochemical "restrainer" category became differentiated (or evolved) into a separate category from the "calmers," primarily the monoamine 5-HT and the amino acid GABA. Produced primarily in the basal forebrain, the ubiquitous CNS neurotransmitter and PNS agent Ach certainly qualifies as a neurochemical "calmer" since it lies behind parasympathetic neurons, conservation of energy, and other distinctly non-impulsive human characteristics. Cholinergic innervation of two strategic brain centers—the *hippocampus* via the medial septal nuclei and the *neocortex* via the basal nucleus of Meynert, leave little doubt of the influence of Ach connections to the prefrontal and frontal cortex, the lobes of "second-thought," restraint, and reasoning in light of consequence.

Obviously, neural damage to the prefrontal cortex, frontal lobes, or hippocampus (the center for learning and memory) would have devastating consequences on behavior intended to restrain or mitigate inappropriate or criminal behavior. A neural brake would no longer be operative allowing impulsivity and "acting out" behavior full reign.

In the behavioral continuum starting with normalcy and continuing along a wide range of pure psychopathologies, psychopathy, criminality, psychosis, or a combination thereof, it's the chemical "jazzers" versus "calmers," "initiators" versus "restrainers" that dictate the directionality of the compass that denotes human mental, emotional, and cognitive characteristics within the cortices of the brain. We are perceived "normal" if and only if the brain compass "points" toward behavior, thinking, and feeling perceived by the standards or ethics of a given society to be "normal" or "appropriate."

By the same token, we are perceived to be "abnormal" or "dysfunctional" or "criminal" if our neurochemical levels are imbalanced to the "point" that is considered "abnormal" or "inappropriate" or psychopathic by societal standards. As we embrace the triune brain paradigm, it is essential to understand how regions are "marked" as conduits of chemical pathways pass across and through these regions.

6 Dopamine: The Pleasure Palace of Mind

PO Box 'The Brain'

Don Jacobs

For *exogenous* drugs of any kind—drugs originating outside the body—to be efficacious (psychoactive) in the brain, and intoxicating, and potentially addictive, they must find affinity to NATURAL CHEMICAL PATHWAYS in the brain. The Chemical Players presented those choice pathways in the brain. If they did not exist, drugs would not 'work' in the brain. Drugs do not contain highs; rather, they trigger highs in chemical pathways *endogenous*—originating inside the body—to the brain. In this essay, we will outline the major psychoactive chemical pathways, systems, and connections in some detail, including the process of intoxication and addiction.

The Dopamine System (DAS)

DAS Pleasure Pathways

When the cortex has received and processed a sensory stimulus indicating a reward, it sends a signal announcing this reward to a particular part of the midbrain *per se*—the *ventral tegmental area* (VTA) whose activity increases. The VTA then releases dopamine not only into the nucleus accumbens, but also into the septum, the amygdala, and the prefrontal cortex by way of the mesocortical DAS pathway. The nucleus accumbens then activates the individual's motor functions, while the prefrontal cortex focuses his or her attention on the pleasurable feelings appertaining to the drug. These regions are connected by what is called the *pleasure or reward bundle.* In neuroanatomical terms, this bundle is part of the medial forebrain bundle (MFB) of the hypothalamus, whose activation leads to the repetition of the gratifying action to strengthen associated pathways in the brain.

First described by James Olds and Peter Milner in the early 1960s, the MFB is a bundle of axons that originate in the reticular formation, crosses the ventral tegmental area, passes through the lateral hypothalamus, and continues into the nucleus accumbens as well as the amygdala, the septum, and the prefrontal cortex. The MFB is composed of ascending and descending pathways, including most of the pathways that use *monoamines*—dopamine and norepinephrine—as neurotransmitters. The mesolimbic and mesocortical dopaminergic systems comprise its main components resulting in the one-two punch of intoxication and addiction.

Most illegal drugs become addictive due a chemical process known as BLOCKING REUPTAKE, which produces a garish liberation of dopamine and the intoxicating rush of euphoric pleasure that follows. At the neuron level, this is accomplished by liberating DA into the synaptic cleft for generous binding to receptors. This is the pathway of intoxication, where the frequency, recency, and amplitude of dopamine 'hits' determine addiction.

Four DA Systems

For starters, there are actually *four dopamine systems* in the brain, TWO of which are directly implicated in intoxication and addiction. The two DA systems NOT implicated are the Tuberoinfundibular DA Pathway and the Nigrostriatal DA Pathway.

In the Nigrostriatal pathway, the *substantia nigra* of the midbrain is connected to the *striatum,* a component of the basal ganglia and directly implicated in movement. Loss of dopamine in the substantia nigra is one of the defining pathological features of *Parkinson's Disease,* a disease not manifested until eighty to ninety percent of dopamine function has been compromised. This DA loop is also implicated in *tardive dyskinesia,* the irreversible neurological side-effect of sustained neuroleptics (antipsychotic) drug use, where neuroleptics block D_2 receptors in multiple pathways in the brain.

In the tuberoinfundibular pathway, dopaminergic neurons in the region of the mediobasal hypothalamus—the 'tuberal region'—regulate the secretion of *prolactin* from the anterior pituitary gland. Prolactin is the peptide hormone primarily associated with lactation; in breastfeeding, suckling stimulates the production of prolactin, which fills the breast with milk. Oxytocin, a similar hormone, triggers milk let-down.

Now, to understand the two DA Systems which are implicated directly in addiction.

The Mesolimbic DAS Pathway (MLDAS)

The powerful Mesolimbic Dopamine Pathway (MLDAS) is a major chemical system for producing pleasure, emotions, and euphoria often linked to eroticism and sexual desire (along with testosterone). The mesolimbic DA loop connects the following associated regions:

- the nucleus accumbens of the limbic system
- the ventral tegmentum area (VTA) of the midbrain

The medial forebrain bundle (MFB) of the hypothalamus connects the nucleus accumbens to the VTA so these connections are dopamine-rich. Just beside the ventral tegmental area is another part of the midbrain—the substantia nigra—that contains a great deal of dopamine. Neurons in the substantia nigra project their axons into the corpus striatum, a region associated with the control of movement. Essentially, all of the dopamine that modulates brain activity comes from the ventral tegmental area (VTA) and the substantia nigra. It is a certainty that The MLDAS pathway is one of the MAJOR REWARD SYSTEMS of the brain, not just for euphoric mood states, but also for INCENTIVE SALIENCE. *Incentive salience* refers to stimuli associated with 'drug-taking behavior', acting as a *conditioned stimulus* (CS) to seek out and re-experience the drug's effects (CSs have long been known, since the time of Ivan Pavlov and his salivating dogs, which should ring a bell for all psychology majors). *Incentive salience reinforces craving for the addict's favored drug.* It works this way: If an addiction has been extinguished, but an ex-addict is presented with cues or stimuli associated with drugs used in the past, then a craving to re-experience the drug occurs, exemplifying 'incentive salience.'

Cocaine, amphetamines, and METH congregate their effects within the central areas of the reward circuit—the VTA and the nucleus accumbens (NAcc). These areas contain especially high concentrations of dopaminergic synapses, which are the preferred target of these drugs. Cocaine's effects on other structures such as the caudate nucleus may explain certain secondary effects of this drug, such as increased stereotyped behaviors such as nail biting or scratching. Regular consumption of cocaine also reduces metabolic activity in several other parts of the brain, which may cause various cognitive deficits.

The Mesocortical DAS Pathway

The Mesocortical Dopamine Pathway (MCDAS) explains how the entire brain is HIJACKED and taken hostage by addiction as the MLDAS loop implicates the frontal lobes, specifically interior regions of the prefrontal cortex (PFC)—the *cognitive* regions of the mammalian brain in the Mesocortical DAS Pathway. The PFC is the final tollbooth for RECONSIDERATION of inappropriate behavior. Addictive substances ranging from endorphins and analgesics—heroin and the opioids—to the euphorostimulants—cocaine and ecstasy—to alcohol and marijuana effectively numb prefrontal regions by the action of this pathway. Truly, in this way, the PFC is the final nail in the coffin of addiction as the MCDAS *connects the VTA of the midbrain to the cerebral cortex and the frontal lobes.*

The Dopamine Transporter (DAT)

In the normal chemical process of the brain, dopamine is released by a neuron into the synapse, where it can bind with dopaminergic receptors on neighboring neurons. Normally, dopamine is then recycled back into the transmitting neuron by a specialized protein called the *dopamine transporter* (also known as the dopamine active transporter—DAT). If cocaine is present, it attaches to this transporter and blocks the normal recycling process, resulting in an excessive buildup of dopamine in the synapse thereby contributing to the pleasurable effects of the drug.

If TWO dopamine systems and pathways of connection were not enough to provide a wide boulevard to addiction, there exists an affiliated system known as The DANE System—The Dopamine-Norepinephrine System—'stimulant control central' the brain.

The Dopamine-Norepinephrine System

Stimulant Central: The DANE System

The DNRIs—dopamine-norepinephrine reuptake inhibitors—explain how euphorostimulant and stimulant drugs can affect one or BOTH of two systems: (1) sympathetic neurons in the sympathetic branch of the peripheral nervous system (PNS) and/or (2) neurons in the central nervous system

(CNS). Some of the most addictive substances known to mankind—methamphetamine, caffeine, and nicotine, for example—find affinity as agonists in the CNS.

Caffeine, Nicotine, and Stimulants

CAFFEINE, found in the common beverages of most societies—coffee, soft drinks, and energy drinks—as well as NICOTINE, found in tobacco smoke, are among some of the world's most commonly used stimulants. Stimulants display significant abuse potential and comprise some of the most carefully controlled substances in America such as ephedrine, amphetamines, cocaine, methylphenaidate (Ritalin®), MDMA (ecstasy), and by prescription only—Desoxyn®, and Dexedrine®.

Amphetamines elevate mood, increase heart rate, respiration, blood pressure, and produce a feeling of euphoria. The sheer availability of amphetamines makes them a prime target of abuse.

In the clinical picture, stimulants are used therapeutically to:

- enhance or maintain alertness
- counteract fatigue in situations where sleep must be suppressed (such as while operating vehicles)
- counteract sleep disorders such as narcolepsy
- suppress appetite and promote weight loss
- enhance the ability to concentrate with a diagnosis of Attention Deficit Disorder (ADD) and ADHD
- boost endurance and productivity
- occasional use as anti-depression meds

How They Work

Stimulant compounds inhibit the reuptake of the catecholamine neurotransmitters *norepinephrine and dopamine*—into storage vesicles of axons. The most

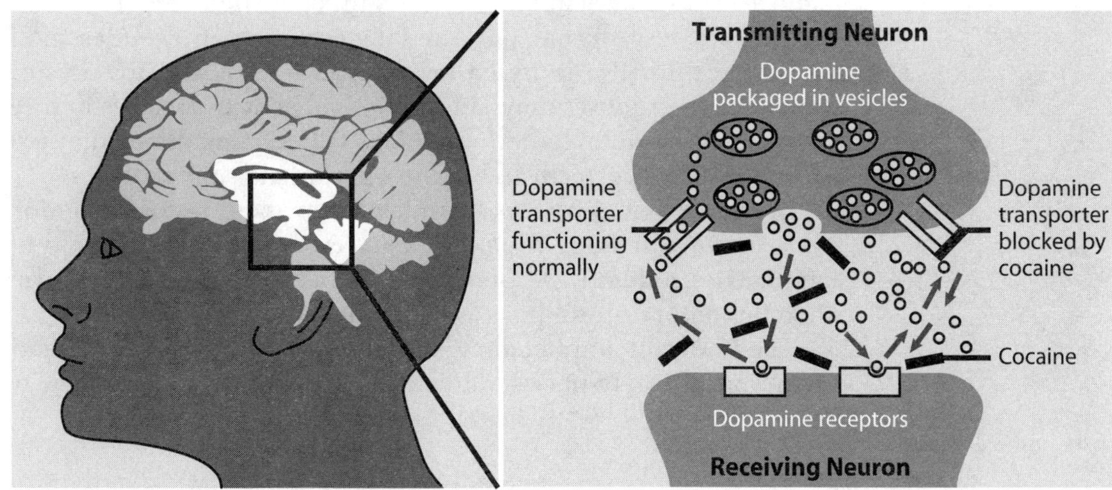

Image 4.1 Blocking Reuptake & Liberation of Dopamine at Synapse.

abundant catecholamines in the brain are epinephrine (adrenaline), norepinephrine (NE) and dopamine(DA)—both produced by phenylalanine and tyrosine. By inhibiting reuptake into the molecular 'pumps,' this action garishly populates DA/NE in the synapse, resulting in their efficacious euphoric and stimulating effects. The most popular and well-known DNRI is the antidepressant Wellbutrin.

The Riot Squad of The DAS: Testosterone & Phenylethylamine (PEA)

As one of two primary chemical 'initiators' of thinking, imagery, and behavior, prominently lying beneath the sex drive (*libido),* the steroid hormone *testosterone* is tied to *aggression,* acting in the role of a neurotransmitter with binding sites in the CNS. Both male and female brains have receptors for testosterone. It is well established that testosterone is aphrodisiacal for BOTH sexes, perhaps enhanced by the release of dopamine and norepinephrine. Women need much less testosterone to sustain their sex drive, perhaps because *estradiol* and testosterone synergize so powerfully. In females, estrogen dramatically increases the density of testosterone receptors in the genital area; testosterone governs the erectile function of this tissue and of the nipples.

Males experience the rush of testosterone at puberty, as do females, although with less fanfare, although the chemical effects are still enormously powerful when combined with estrogen. A high level of testosterone in males, although perhaps *not directly a cause* is nonetheless positively correlated with violence and high crime rates, *especially in the presence of low serotonin levels.* When testosterone registers high in spinal fluid with low 5-HT, aggressiveness and proneness to violence is almost a certainty, especially when exacerbated by alcohol or other drugs.

This condition is apparently mitigated in Alpha males who have elevated levels but are not criminals; instead, they may be 'thieves of the heart.' Driven to succeed in the business world, he may move undetected between varieties of lovers, even though happily married. The perennial pattern of Alpha males benefiting from one major chemical system of his brain (the soothing comfort of the endorphin pathway in a supportive marriage) PLUS the dopaminergic high (from a series of affairs), dictates his cherished lifestyle.

In males, testosterone peaks in late teens to early twenties and falls gradually by age thirty-five to much lower levels by age fifty. It's no surprise while testosterone is robust, a myriad of criminal activity surfaces in juvenile delinquency and in adult 'career' criminals. Is it a coincidence that testosterone's decline matches perfectly the decline in violent crime statistics?

In mature females, *estrogen,* which has masked testosterone for over forty years, subsides, making room for testosterone. At about forty to fifty years of age, females suddenly become as driven as their husbands or partners were when this natural steroid was a 'raging hormone.'

The powerful amphetamine-like chemical that lies behind addiction to a person or person-to-person addiction appears as PEA—the new twist on the PEA-Brain.

Phenylethylamine (PEA)

To neuropsychologists, the 'romantic rush' is indeed chemical and provides powerful effects from *phenylethylamine* (PEA), a biogenic compound or amine ($C_8H_{11}N$) with pharmacological properties similar to those of *amphetamine.* PEA occurs naturally as a neurotransmitter in the brain. PEA lies behind fantasy, and a 'romantic rush' initiated by cues from a person's face, body build, eyes, hair, or scent. This appears to be a powerful chemical in mate selection. PEA lies beneath the physical attributes, demeanor, body language, sights and smells we find personally attractive and sexually appealing in others resulting in the giddy orbit of infatuation.

Interestingly, chocolate is full of phenylethylamine which, as we have seen, is a chemical cousin of amphetamine. Chocolate also contains sugar and *anandamide*—a compound that binds to the same receptors in the brain as marijuana. Hence, we experience the slight feeling of elation and pleasure upon eating a chocolate 'candy kiss.' Could this be why chocolate is the #1 choice of lovers on Valentine's Day?

Cannabinoid Receptors

Cannabinoid receptors are part of the largest known family of receptors—the G protein-linked receptors—which are known to modulate and prolong chemical signals, in contrast to ion-channel linked receptors, known for rapidity in post-synaptic activity. The cannabinoid CB1 receptor is one of the most numerous G protein-linked receptors in the nervous system. Cannabinoid receptors were originally discovered as being sensitive to delta-9-tetrahydrocannabinol (THC)—the primary psychoactive cannabinoid found in cannabis. Interestingly, the discovery of anandamide came from research into CB1 and CB2, as it was inevitable that a naturally occurring (endogenous) chemical would be found to affect these receptors.

Anandamide has been shown to be involved in working memory and in the regulation of the appetitive drive (eating), motivation, and pleasure. Both anandamide and exogenous cannabinoids enhance food intake in animals and humans—hence the attack of the 'munchies' following THC intoxication. In addition, anandamide injected directly into nucleus accumbens enhances the pleasurable responses of rats to a rewarding sucrose taste. It is also important in the implantation of the early stage embryo in its blastocyst form into the uterus; therefore, exogenous cannabinoids might interfere with early stages of human pregnancy.

In the normal brain with surging PEA, we seek to interact as close as possible to the person responsible for spiking this energizing chemical. Forensic neuropsychologists suspect PEA lies behind *victimology*—why one victim is chosen and not another. For example, serial killer Ted Bundy chose female victims with long brown hair, parted down the middle, physical traits that resembled his former girlfriend, Stephanie Brooks, who jilted him in college. What else but powerful brain chemistry could drive behavior with such ferocity and specificity?

It is easy to see how The Dopamine System and The DANE System are enhanced by the ubiquitous androgenic hormone testosterone, anandamide, and PEA. Interestingly, the word 'anandamide' comes from the Sanskrit word meaning 'internal bliss.' Are we especially 'geared' BY NATURE to experience pleasure?

The active ingredient in cannabis is THC, which concentrates chiefly in the VTA and the NAcc, but also in the hippocampus, the caudate nucleus, and the cerebellum. THC's effects on the hippocampus might explain the memory problems that can develop with the use of cannabis, while its effects on the cerebellum might explain the loss of coordination and balance experienced by people who indulge in this 'internal bliss' drug.

7 Chimera

A Mythical Beast & DNA

Ashleigh Portales

The mythological genesis of the Chimera can be traced to the ancient Sumerian civilization originating in the mid-fifth millennium BC. With the emergence of cuneiform (the oldest form of writing) as well as by the spoken word, the legend of this monstrous deity traveled down through the ages from one society to the next, morphing along the way through Babylonian and Assyrian times to become what we know today as the Greek myth related in the writings of Homer and Hesiod. According to the former, the Chimera (pronounced kih-MEE-ra, and alternately spelled Chimaera) was "in the fore part a lion, in the hinder a serpent, and in the middle a goat." She was the child of the gods; her father was the great Typhon and her mother the half-serpent Echidna. She had three brothers, Cerberus, Hydra, and Orthrus, the three-headed hound that guarded the gates of Hell, a nine-headed aquatic serpent, and a two-headed dog, respectively. Chimera herself was something to be seen, or rather imagined. The three-headed beast breathed fire from the mouth of the lion's head, making her a considerable force to be reckoned with. According to myth, she rampaged throughout the land wreaking havoc on the whole of civilization until she met with Bellerophon, the hero of divine origin. Flying in valiantly on the winged horse, Pegasus, Bellerophon dodged the flames of the beast that were so hot they melted the lead tips of his arrowheads. When he thrust this weapon into the creature's throat, her internal inferno melted the lead and sealed her entrails, killing her.

The image of the Chimera can be found on numerous vases, mirrors, and coins as well as in many statues that have been excavated throughout the years. The various images of the creature are quite remarkable in their similarity, revealing that the artisans of the day remained forever faithful to the essence of their subject. Most renderings depict the three separate heads in clear distinction, with that of the goat sprouting from somewhere in the middle region. In posture, the creature's body is eternally arched as if for the attack, the lion's head pointed upward with flames spewing forth.

The most famous of these artifacts is the statue unearthed near Arezzo, Italy in 1553, labeled by the city's chancellor as *Chimerae Bellerofontis simu-*

lacrum, and known today as the *Chimera of Arezzo.* Modern scholars speculate that such an image is the personification of the many evils men perceived to be contained within women. According to the author, Valerius, writing in first century AD, "You do not know that woman is the Chimaera, but it is good that you should know it; for that monster was of three forms; its face was that of a radiant and noble lion, it had the filthy belly of a goat, and it was armed with the virulent tail of a viper." Kramer and Sprenger interpreted this comment in their fifteenth century work, *Malleus Maleficarum,* as meaning "that a woman is beautiful to look upon, contaminating to the touch, and deadly to keep."

Chimera: The Stranger Within

Today, thankfully, modern medical science has dragged us out of the chauvinistic quagmire that was the land of the ancients to give a more accurate and believable definition: a chimera is an organism with at least *two genetically different types of cells.* While at first this concept sounds just as mythical as its age-old counterpart, science has proven it to be quite true. There are several ways in which an individual may become a chimera. The truest avenue to chimerism is *tetragametic chimerism,* when one individual is formed from four gametes: two eggs and two sperm. In this case, two developing fraternal twin embryos fuse into one very early in the stage of embryonic proliferation. As the resulting single fetus develops it is healthy but contains *two separate sets of DNA,* his/her own and that of the absorbed sibling. Resultantly, various tissues of the body are affected by the phenomenon, such that an individual's blood could have one DNA type and the liver, or some other organ, another. Should the original two embryos have been of opposite sexes, the *ambiguous gender* of the child will be evident at birth as hermaphroditic, and thus easily detected as chimera.

However, should the chimera be carried only in the blood of an individual, it is not likely to be uncovered until blood testing, usually for the purpose of organ transplantation. More common than the actual fusion of two fertilized embryos is the occurrence of *fraternal twins sharing the same placenta and thereby the same blood supply.* In this way blood flows freely between the two genetically separate individuals, transferring cells from one to the other and vice versa. About eight percent of all fraternal twins have this type of chimeric blood.

A larger percentage of the population are *microchimeras,* carrying much smaller amounts of contrasting blood cells due either to blood cells passed from mother to fetus via the placenta, or leftover from blood transfusion.

These various types of chimera have occurred with increased frequency since the advent of *invitro fertilization* (IVF) aiding women to get pregnant. The insertion of multiple fertilized embryos into the uterus to ensure a higher rate of success has resulted in *twenty-five percent more twin pregnancies than normal,* also increasing the chance for chimera.

Chimerism: When DNA Can Lie

As blood typing has become more common, the number of documented cases of chimera has increased. One such individual, who shall be referred to as

"Jane," was a fifty-two year old married mother of three whose entire family underwent blood and tissue typing tests as she was in need of a kidney. Rather than telling her that they had found the perfect donor, doctors informed the woman that two of her three sons were not biologically hers, though they both contained their father's DNA. Jane insisted that she had conceived all of her children naturally with her husband and had given birth to them, facts that would normally ensure she was indeed the biological mother. For two years a team of baffled physicians worked to solve this mystery. They centered their investigation on a set of genes known as the *HLA complex, a group of immune proteins that help to distinguish the body's own tissues from foreign material.* Each person inherits two blocks, or haplotypes, of HLA, one from the mother and one from the father. The medical team identified Jane's haplotypes as "1" and "3" and those of her husband as "5" and "6." When matched with their sons' tests, all three were found to share a haplotype with their father while only one of the boys shared one with his mother. The other two had a previously unidentified haplotype marked as "2." When this haplotype was found in other members of Jane's family, doctors knew that the boys were related to her and decided to test DNA from some of Jane's other body tissues, including her thyroid gland, mouth, and hair, because her blood DNA did not match that of her sons DNA.

When the team received the results they were dumbfounded. While some of Jane's tissues carried haplotypes "1" and "3," others carried types "2" and "4." Jane was the result of the merging of two separately fertilized fraternal twin girls, a tetragametic chimera.

Though the twin embryos that fused to become Jane had been of the same sex, those of a British boy were not. His case was documented in the *New England Journal of Medicine,* vol. 338, p. 166. The boy was born at the Western General Hospital of Edinburgh, where doctors noticed that he had normal right testes and semi-developed left testes. At fifteen months of age the boy underwent surgery to remove what doctors believed to be "an abnormal gonad and vas deferens." However, further pathological investigation of the removed structures revealed that they were actually an ovary and attached fallopian tube. Studies showed that some of the child's body cells were female (XX) while others were male (XY). In all likelihood, two fertilized embryos, one male and one female, were implanted in the mother's uterus by IVF and then fused, developing into a single fetus.

Additionally, the same doctor who led the investigation into the case of Jane is currently studying the claims of a woman who is suing her previous partner, whom tests have proven is the father of her child. Yet those same tests showed that the woman was not actually the child's mother, despite the fact that she gave birth to him. The possibilities of a *Jerry Springer* episode aside, the doctor speculated that this could be another case of maternal chimerism. Through this situation, the possibility arises that, should the father of a child be a chimera, he could be ruled out as the biological father through traditional paternity testing when he was in fact the father.

The legal complications in such cases are obvious, but they become larger and even more disturbing in the light of *serious violent crimes.* In a recent case, a woman was raped by a man whom she subsequently identified as the perpetrator. Much to the dismay of the victim and police alike, blood DNA tests revealed that he had not been her attacker. The investigative perseverance of all parties involved revealed that the man matched to the crime in every way but his DNA. Another test was done, this time using a sample of his hair, and matched perfectly to evidence collected off the victim and at the scene. Though conviction was difficult, the prosecution revealed the phenomenon that the perpetrator was in fact a chimera and the man was found guilty of his crime.

The Chimeric Criminal

While chimerism may be rare, the phenomenon poses a significant threat to law enforcement and the apprehension of *criminals who leave behind the DNA of one individual at the crime scene and give another to forensic investigators for testing.* If even one violent offender is set free to unleash his wrath upon the many unsuspecting victims that live within society, the validity of the methods of modern forensic testing comes under question in terms of thoroughness. In cases where the guilt of a suspect is almost certain, it could very possibly be beneficial to go that extra mile. The mere existence of chimerism has an extremely high shock value, but once the initial period of disbelief has passed, one is forced to examine how heavily he or she relies on DNA evidence and whether or not it should always be interpreted as having an infallible face value. Any test result is only as good as the information with which it began, but what if that information was not entirely correct? To catch modern criminals we must use modern weapons, but what is the weapon to catch the man who holds another's identity captive within himself?

8 The Critical Difference

Between Psychopathic Personality & Antisocial Personality Disorder

Don Jacobs

> "Psychopaths have a peculiar, striking *affect disorder*—superficial pleasantness, facile lying, the capacity to kill in cold blood. In some cases, cold blood best captures what is most characteristic about the psychopath."
>
> Jose Sanmartin, *Violence & Psychopathy*/Adrian Raine Ed.

The Notorious "Cluster B" Personality Disorders

According to the DSM IV, TR (2002), a personality disorder is an "(1) enduring pattern of (2) inner experience and behavior that (3) deviates markedly from the expectations of the individual's culture, is (4) pervasive and inflexible, has an (5) onset in adolescence or early adulthood, is (6) stable over time, and (7) leads to distress or impairment." This is a direct quote from the pages of DSM-IV-TR. The parenthetical numbers (1–7) are mine and require comment in light of psychopathy relative to personality disorders (PD). First a comparison of definitions:

Psychopathic personality is defined as an (1) emotionally and behaviorally disordered state characterized by (2) clear perception of reality except for the individual's (3) social and moral obligations, and often by the pursuit of

(4) immediate personal gratification in (5) criminal acts, drug addiction, or (6) sexual perversion (Webster's collegiate 10th ed.).

Aligning with psychopathy conceptualized by Hare *et al* psychopathy is like PDs an (1) enduring pattern of (2) inner experience and behavior ("inner" because it is "felt" and "behavior" because characteristics or features can be objectively observed.

Milder psychopathy does not fit (3) "deviates markedly" but *severe* and *extreme severe* characteristics certainly do. All markers of psychopathy across the continua are "pervasive" as in "diffused through all parts of personality," but if "inflexible" refers to prevalent in overall personality functioning, the psychopath is different relative to *adaptability*—he or she is markedly adaptable, not inflexible, to almost any milieu.

Onset is indeed very early in psychopathy and is stable over time matching both (5) & (6). (7) "Leads to distress or impairment" is perhaps where the psychopath and PD differ the most. The PD diagnosed patient experiences or feels what Freud termed ego-dystonic—an inner experience of emotional discomfort, dysthymia, or anhedonia. These feelings are reversed in psychopathy where the person's (again using Freud's term) felt inner experiences are ego-syntonic—appropriate, stimulating (including hypersexuality), or highly anticipated to the psychopath.

On the other hand, PD individuals (especially the moderate to severe variety) certainly match (1) "emotionally and behaviorally disordered state" but do not match up with "clear perception of reality" especially with borderline and psychotic states in pure psychopathologies and with some cases with severe paranoid PD and borderline PD. Antisocial PD individuals share similarities with psychopaths relative to detriments to (3) "social & moral obligations" and (4) and (5) "immediate personal gratification, criminal acts, and drug addiction." However, PD individuals (even severe antisocials) do not match up with the sexual deviance of extreme severe psychopaths (ESP).

According to differential diagnoses while psychopaths and PDs share some of the same characteristics as we shown above, in *some important features* they are *markedly different* especially relative to:

1. Cultural expectations and socioeconomic status. Psychopathy is actually "nourished" by a hedonistic pop culture and the Western notion of success via "self-interest" and philosophically by social Darwinism and "survival of the fittest." Unlike antisocial PD, psychopathy occurs across all socioeconomic dimensions.
2. Adaptability and longevity. Psychopaths are highly adaptable and remain psychopathic for longer a timeframe than antisocial PD.
3. Ego-syntonic denial of psychopathy supplants ego-dystonic emotional distress of antisocial personality.
4. Degree of sexual deviance (far and away the most salient difference between DSM antisocials and all forms of psychopathy from the hypersexuality of hubristic psychopathy (Jacobs' term for mild to moderate versions) and pronounced deviance in extreme severe psychopathy.

Therefore with side-by-side comparison of DSM pure psychopathologies and PD, psychopathy is not just a special incidence of a PD subcategory to the

DSM. Clearly, psychopathy, especially severe psychopathy, stands alone as a diagnostic criterion regardless of the obvious oversight by DSM editors.

After a short discussion of "Cluster A" and "Cluster C" PDs, we will present in the exact language of the DSM-IV-TR the notorious "Cluster B" PDs often observed in violent criminality. The DSM PDs are grouped into "clusters" based primarily on descriptive similarities. "Cluster A" PDs include Paranoid, Schizoid, & Schizotypal PDs:

1. *Paranoid PD* is a pattern of distrust and suspiciousness such that other's motives are interpreted as malevolent.
2. *Schizoid PD* is a pattern of detachment from social relationships and a restricted range of emotional expression.
3. *Schizotypal PD* is a pattern of acute discomfort in close relationships, cognitive or perceptual distortions, and eccentricities of behavior.

While psychopaths may have **features** of paranoid PD (distrust and suspiciousness) and **patterns** of schizotypal PD, they seldom display schizoid PD characteristics due to their engaging, charismatic personality, and by their verbal "gift of gab."

"Cluster C" PDs include the Avoidant PD, Dependent PD, and Obsessive-Compulsive PD:

1. *Avoidant PD* is a pattern of social inhibition, feelings of inadequacy, and hypersensitivity to negative evaluation.
2. *Dependent PD* is a pattern of submissive and clinging behavior related to an excessive need to be taken care of.
3. *Obsessive-Compulsive PD* is a pattern of preoccupation with orderliness, perfectionism, and control.

While psychopaths may have *features or patterns* of obsessive-compulsive PD, they seldom display characteristics of avoidant PD or dependent PD, again due to their "gift of gab," **extraversion,** and engaging personalities. Moving to "Cluster B" PD, the notorious clusters that most often display both features and patterns of psychopathy across the entire continua. As a reminder, the exact language will be used in order to convey the exact meaning of DSM diagnostic features followed by a brief discussion of the PD in relation to psychopathy.

301. 7 Antisocial PD

The essential feature of Antisocial PD is a pervasive pattern of disregard for, and violation of the rights of others that begins in childhood or early adolescence and continues into adulthood. (2002)

Psychopaths may have features of more than one of the "Cluster B" PDs such as histrionic PD, borderline PD, or narcissistic PDs. Yet, in the seven diagnostic criteria required for diagnosis of antisocial PD the words "sexual deviance" or "hypersexuality" so prevalent in psychopathy are missing. The lame word "pleasure" in the same sentence with "personal profit or pleasure" is woefully inadequate.

9 Behind the Monsters' Eyes—

The Role of the Orbitofrontal Cortex in Sexually Psychopathic Serial Crime

Ashleigh Portales 2006

"The truth is that men are not gentle, friendly creatures wishing for love, who simply defend themselves if they are attacked, but that a powerful measure of *desire for aggression* has to be reckoned as part of their instinctual endowment. The result is that their neighbor is to them not only a possible helper or sexual object, but also a temptation to them to gratify their aggressiveness on him, to exploit his capacity to work without recompense, to use him sexually without consent, to seize his possessions, to humiliate him, to cause him pain, to torture and kill him. *Homo homini lupus* (man as wolf); *who has the courage to dispute it in the face of all the evidence in his own life and in history?*"

Sigmund Freud

Precision Science News (2006) defines the orbitofrontal cortex as:

"a small area of the brain that is located just behind the eyes. It is involved in cognitive and affective functions such as assessing emotional significance of events, anticipating rewards and punishments, adaptation to changes in rule contingencies, and inhibiting inappropriate behaviors."

In the dynamics of the divisions of the prefrontal cortex, the *orbitofrontal is the decision-maker,* deciding whether or not to carry out the plans formulated in the adjacent dorsolateral prefrontal cortex. As stated by Jacobs and Mackenzie, (2006): "individuals who have had damage to [this] area can observe social situations, but fail to respond to these situations in an appropriate manner." Given this information, it is easy to see the implications of this brain area's involvement in violent predatory psychopathy from the initial development of the psychopath, to the planning stages of the crime, through the process of victim selection, in the midst of the act of murder, and well into the conduct displayed and emotions expressed in the aftermath of the crime.

The Making of a Psychopath

In courtrooms and clinics across the globe, the debate rages on over nature versus nurture: whether psychopaths are born with genetically inherent tendencies or made by socialization in milieu into what they are. The most logical answer is 'yes' to both. While a person may be born more prone than another to develop a psychopathic personality, the way in which they are raised and their experiences in the world trigger those already 'primed switches' within their genetic code to produce what popular culture knows as a serial killer. This assertion is duly supported by research in the area of the orbitofrontal cortex. In his landmark neuroimaging study of forty-one murderers, Dr. Adrian Raine of USC found that, in an overwhelming majority, the prefrontal lobes of the murderers were underdeveloped and functioning far below normal levels required for adequate social behaviors. While the reason for the malfunctions in their brains were various (head trauma, physical abuse, emotional neglect, antisocial parenting, etc), the result was obvious: stagnated prefrontal development and/or function, for whatever reason, showed a direct correlation to violent criminality (Raine, et. al., 1994).

McDonald's Homicidal Triad is a well known list of three major 'red flags' of psychopathy shown in early adolescence: enuresis (bed wetting) at an inappropriate age, cruelty to children and/or animals, and pyromania. While these impulsive actions may originate deep within the brainstem and midbrain limbic system (MLS), the fact that the brain regions are interconnected one with the other implies dysfunction in the area charged with keeping such impulses at bay—the prefrontal cortex, specifically the orbitofrontal. Studies have confirmed enuresis as a direct side effect of decreased function or injury to the orbitofrontal cortex. In the 1920s, 30s, and 50s, multiple researchers documented the commonality for patients with known orbitofrontal injuries to freely urinate or even defecate on themselves, not only while sleeping, but also while watching television, eating in restaurants, or conversing with friends. More recently, researchers at the University of Pennsylvania used fMRI technology to image the brains of six people with and six people without good bladder control while filling and withdrawing liquid from their bladders.

Those with good bladder control exhibited increased activity within the orbitofrontal cortex while those without it showed little activity in the same area. However, those with poor bladder control did register activity in other parts of their brain, evidence to support the theory that more primitive parts of the brain (for example, the R-complex—'The Reptilian Brain' or brainstem) may take over when the intended 'controller' is inhibited. The other two criteria of McDonald's Triad are repeated cruelty to children and/or animals and pyromania, both of which suggest a blatant disregard for the consequences of one's actions. This is often displayed by serial murderers who will return to the scene of the crime despite the knowledge that it is being watched, or who continue in their murder spree even though they are under surveillance, believing their intelligence to be greater than that of the law enforcement officers who are after them. In their grandiose sense of self, they think they can get away with anything. But is there more to it than simple arrogance? Research suggests that there is: they have never been properly conditioned to behavior suppression for fear of punishment. According to Sabbatini (1998),

> "normal humans learn very early in life to avoid antisocial behavior because they are punished for it and because they have the brain circuits to associate fear of punishment (feeling emotion) to behavior suppression . . . When there is no punishment, or when the person is unable to be conditioned by fear, due to a lesion in the orbitofrontal cortex, for example, or due to lowered neural activity in this area, then it develops an antisocial personality."

Due to the effects of antisocial or negligent parenting, they have developed the inability to register the severity of the consequences that their actions could bring and, as a result, their actions are governed by whatever whim they feel at the time.

To quote New York University neurophysiologist, Dr. Elkhonon Goldberg:

"Orbitofrontal damage robs people of the ability to anticipate the consequences of their actions."

Planning the Crimes and Selecting the Victims

Extensive research since 1990 has shown that damage to the prefrontal cortex in general results in poor planning and judgment skills, which most would agree applies to just about anyone who seriously breaks the law. But could the word 'poor,' when compounded by various other risk factors, be substituted with 'deviant' planning? Look at the sexual predator. With the planning mechanisms of his (or her) brain set to antisocial tendencies, a violent sexual crime is meticulously designed, step-by-step, in preparation for the day it will be carried over from fantasy into reality. Running on the tracks of aberrant *neurocognitive maps* laid down by a steady diet of hardcore pornography and other sexually explicit and degrading materials, they go with what they know and seek to recreate things they have seen that, by extreme action, elevate their substandard brain chemistry to somewhat normal, or even elevated, levels. Once the perfect plan has been laid, the perfect victim must be found. It has been suggested (Jacobs, 2003) that the selection of a specific victim hinges on whether or not that person peaks the phenylethylamine (PEA) of the predator. PEA is the neurotransmitter that creates the 'romantic rush' we feel upon initial attraction to an individual. In the deviant mind of a sexual psychopath, the characteristics he desires in a victim trigger PEA and help him decide who 'The One' is. Once he decides, there is no turning back.

Evidence suggests that the orbitofrontal cortex plays a large role in the selection of that victim. It has been documented that patients with damages to this area become very stimulus-bound and, once their interest is peaked by a certain object, they cannot shift focus to anything else. This behavior is congruent with the sexual predator that will stop at nothing but fulfillment of his fantasy once he has set his eye on the prize (his desired victim). Only that victim will do. In fact, neural imaging studies back this up, finding that the orbitofrontal cortex was activated in subjects who were presented with several desirable food choices and asked to choose the one most desirable to them. The orbitofrontal cortex seemed to be weighing the prospective incentive value of the stimuli in order to choose the one that would produce the most satisfaction in the subject. This is just how sexual predators select one victim from a world of many by assessing how well each potential one will fulfill his fantastic desires (Arana, Parkinson, Hinton, Holland, Owen, & Roberts, 2003, Tremblay & Schultz, 1999). The results of this study are supported by a 2004 study on rats with orbitofrontal cortex lesions. These rats were able to resist less desirable but immediate stimuli when they knew that a more favorable alternative with a greater reward would come if they denied their impulses just a little while longer. Even when punishments were assigned to choosing the greater reward, the rats with the OFC lesions were unable to resist choosing it (Winstanley, Theobald, Cardinal, & Robbins, 2004), suggesting that a psychopath would be unable to resist his specific victimological stimuli regardless of the consequences associated with his choice. It is the orbitofrontal cortex that keeps the predator focused on his victim, according to the dynamic filtering theory proposed by Rule, Shimamura, and Knight (2002):

> "The OFC initiates control via reciprocal efferent projections that are used to maintain task-relevant activations and inhibit irrelevant or inappropriate neural activity."

In a sense, the predator develops tunnel vision, able only to see his chosen course of action and the victim he has chosen to act it out upon.

> "The problem is in shifting sets and in inhibiting the recurrence of a previous response when the next action is initiated. Once a behavior occurs it tends to persist and contaminate the performance of unrelated actions" (Joseph, 2000).

Thanks to his orbitofrontal cortex, which is already predisposed to sexually deviant stimuli, all other distractions are inhibited from reaching his attention until he completes that which he has set out to do.

The Thrill of the Kill

For the sexual predator, the actual commission of murder is the magical point where fantasy meets reality. Like all other aspects of his crime, this too can be scientifically linked to dysfunction in the orbitofrontal cortex. As reported in *The American Journal of Psychiatry,* patients with orbitofrontal lesions displayed behaviors that were more impulsive and inappropriate. (Berlin, Rolls, & Iversen, 2005) There is no behavior more socially inappropriate than the sexual torture, mutilation, and murder of an innocent individual. Also in this study, those with orbitofrontal lesions reported a faster perception of time (overestimated time) than did the healthy comparison subjects. Many serial killers have reported an unclear perception of the passing time as they committed their murders, almost as if they blacked out or were in a trance. One question that arises in the study of serial predators is, "Do they choose to kill, or are they compelled to do so?" As in the question of nature versus nurture, the answer appears to be both. While they at first choose the actions that lead them to murder, the overwhelming influence of brain chemistry drives them toward behaviors that allow such chemicals to cascade at synapse and create ultimate feelings of pleasure and euphoria, however short-lived they may be. Current theories propose that, because the brain chemicals that produce pleasure register so low in sexual psychopaths, they must overcompensate in their actions in order to get the feeling they desire. Jacobs (2005) has referred to this phenomenon as *The Dull Hypothesis.* Yet, once the chemical surge has subsided, they must carry out the actions again and again to maintain the feeling they so desperately seek. This is supported by the dynamic filtering theory of Rule, Shimamura, and Knight (2002), which states that: "a failure to modulate (i.e., select and inhibit) neural activity associated with emotional events leads to less refined or less context-bound associations. That is, a filtering deficit leads to greater interference (increased noise) and thus reduces contextual memory for emotional events."

In short, the psychopath's dysfunctional orbitofrontal cortex fails to integrate emotion to memory and thus, he must repeat the stimulus (murder) serially to relive the feeling he desires. Subsequent findings reported by Joseph (2000) further elaborate on the behaviors associated with the malfunctioning of the orbitofrontal cortex:

> "Rather than a loss of emotion there is a loss of emotional control and the subject becomes disinherited, hyperactive, euphoric, extroverted, labile, over talkative, and develops perseverate tendencies . . . Patients are frequently described as markedly irresponsible, antisocial, lacking in tact or concern, having difficulty planning ahead or foreseeing consequences, and suffer from generalized disinhibition . . . There can result tendencies toward impulsive actions, to laugh inappropriately and make trivial jokes, or to behave in a demanding or transiently aggressive manner. Proneness to criminal behavior, promiscuity, grandiosity, and paranoia have also been observed."

The implications are clear, as Joseph's list of observed results reads like an excerpt from the Hare Psychopathy Checklist.

In addition to the inappropriate affect noted by Joseph, emotional flattening, also known as *blunt affect,* which is an outward expression of an inner "blunted" brain function, has also been noted. Interestingly, the orbitofrontal cortex may have a connection to yet another, more uncanny aspect of sexualized serial crime. In many cases, the offender will cannibalize parts of his victims either in an effort to keep some sort of a trophy with him or to completely control and devour the symbolic will and resistance of the victim. As reported by Joseph (2000), orbitofrontal damage can have adverse affects on the appetite. Some patients experienced insatiable cravings for all sorts of food, stuffing themselves even to the point of death, while others experienced the need to eat "non-nutritive objects." It is not a large jump to see the correlation between these results and the need of a predator to eat parts of his victims, things which the average person would never desire to consume. Maybe, given the knowledge that the killer's prefrontal cortices are functioning abnormally, these bizarre behaviors become logical.

Aftermath: No Regret and No Recovery

In a 2004 study by Camille, Coricelli, Sallet, Pradat-Diehl, Duhamel, & Sirigu, regret is defined as: "a cognitively mediated emotion triggered by our capacity to reason counterfactually . . . Regret is an emotion strongly associated with a feeling of responsibility."

Sexual psychopaths do not experience this common human emotion because they do not feel that what they did was their fault; the victim had it coming. In the Camille study, subjects who had orbitofrontal lesions did not report feeling regret over their choices regardless of the outcome. When they did report experiencing emotion, it was with considerably less contrast than that seen in the normal test subjects. This effect of orbitofrontal dysfunction is mirrored by The Dull Hypothesis (Jacobs, 2005), which accounts for understimulation in the brain of a psychopath by a *genetic predisposition for a low autonomic arousal threshold.* Psychopaths simply do not experience emotion the way normal individuals do, and lack the capacity to perceive how their actions affect the lives of others. Their own stimulation is their only concern.

Due to the fact that the brain does not regenerate, by the time the psychopath kills his first victim, he is too old (brainwise) to develop the brain regions in which he is lacking. Therefore, there is no chance that he will ever begin to feel the emotions of which he is currently unable. There is no hope of recovery or rehabilitation. It is impossible to be rehabilitated to something one was never habilitated to in the first place. However, because they are psychopathic and therefore inherently crafty conmen, they have the ability to exhibit what is defined in the prison system as 'good behavior' in order to obtain an earlier release so they can resume the only activity that brings them the perverse pleasure they crave: murder. As quoted in Joseph (2000):

> "if a patient or orbitally lesioned animal is purposefully distracted, e.g. via a novel stimulus, this pattern of perseverative attention in momentarily halted and the ability to shift response and attention is briefly regained."

However, evidence suggests that once the strict pattern of purposeful distraction is gone the focus returns again to the initially desired stimuli. This is perfectly applicable to prison life. When placed in a highly structured environment with no available source of stimuli, a psychopath often functions as model prisoner, deceiving authorities into believing he is a changed man who is sorry for what he has done. Yet once he is released into society, he will return as quickly as possible to the stimulation his brain was always conditioned to: serial sexual homicide. Because brain anatomy and chemistry is at the core of behavior, the rapacious mind will always seek to prey on live victims.

In the current age of neuroscience, it would be analogous to an ostrich sticking its head in the sand to deny the role of the brain and its anatomy in behavior, especially behavior that is sexually psychopathic in nature. Research has shown that right behind the eyes of the psychopath, which have ironically been romanticized as the windows to the soul, lays the orbitofrontal cortex, a prime agent in the development of psychopathy and the exhibition of its related behaviors. Evidenced by the research of Dr. Jay Giedd of the National Institute of Health Clinical Center in Bethesda, MD, the prefrontal regions of the brain are the last to develop, and extensive evidence exists to show that the brains of sexual predators, for any of a myriad of reasons, are deeply flawed in the most crucial stage in neurological development. According to Canadian scientist Dominique LaPierre:

> "Both the psychopath and the orbitofrontal or ventromedial frontal patient show an exaggerated preoccupation with sexual matters, acting in a promiscuous and impersonal maladaptive way. Both are remarkable for their lack of social and ethical judgment. Both neglect long-term consequences of their actions, choosing immediate gratification over careful planning."

What lies behind the eyes of such a monster is the key to his psychopathy.

10 Charlie & Caril Ann on a Rampage

Teenage Killers

Ashleigh Portales

Time Span of Murders:
December 1, 1957 to January 29, 1958
Offenses Prior to Killings:
Charles: Several physical assaults & Robbery
Caril: None
Victimology:
Completely randomized victims—anyone unfortunate enough to cross paths with the young spree killers

Quoting Charles Starkweather:

"Mr. Bartlett was moving around, so I tried to stab him in the throat, but the knife wouldn't go in, and I just hit the top part of it with my hand, and it went in."

Charles Starkweather

Charlie Starkweather was born on November 24, 1938 in Lincoln, Nebraska, the third of seven children born to Guy and Helen Starkweather. Though the couple lacked formal education and was poor, they were hard workers; their children never went without food or shelter despite the Depression. The family's resiliency to survive was largely due to Charlie's mother since his father was afflicted with several ailments, and worked only sporadically as his failing health permitted.

School proved to be a double-dip of misery for Starkweather. He was the natural target for bullies as he came from a poor family, he had a speech impediment, he was small in stature and his legs were bowed, and he did not perform well in the classroom. His academic woes could have been corrected but his severe myopia went undetected until age fifteen. In retrospect, his condition prevented him from reading the largest letter on the eye chart, much less the classroom blackboard. The only subject Starkweather excelled in was gym class. Despite his physical disabilities he displayed excellent coordination and physical strength, making him an ace at gymnastics. For Charlie, this was his only source of pride.

Lurking beneath in the ancient, vengeful, reptilian center deep in his young, developing brain, savagery was slowly emerging with his festering rage for being different. He would soon repay those who had treated him so badly. Predictably, fighting and violence soon became his way of life. According to Starkweather biographer William Allen:

"He blamed all of his fights on being made fun of as a child. Sometimes his battles were brief outbursts of violence, but other times they

> were frenzied and prolonged, not ending until they were broken up or his opponent lay senseless. He earned a reputation for being one of the meanest, toughest kids in Lincoln."

One of his many fights oddly resulted in a friendship with his former foe, Bob Von Busch. His description of Charlie illustrates the psychopathy and predatory behavior, both already well developed, that his friend displayed in his lifelong quest to prove his superiority. Quoting Von Busch:

> "He could be the kindest person you've ever seen. He'd do anything for you if he liked you. He was a hell of a lot of fun to be around, too. Everything was just one big joke to him. But he had this other side. He could be mean as hell, cruel. If he saw some poor guy on the street who was bigger than he was, better looking, or better dressed, he'd try to take the poor bastard down to his size."

As were many teenagers of the period, Von Busch and Starkweather were fanatical about movie star and teen idol James Dean. The movie, *Rebel Without a Cause* (1957), brought Dean to the silver screen in a classic portrayal of the savage brain versus parental attempts to mitigate the dangerous scenarios the midbrain-limbic system feeds upon. Finally, Charlie found his alter ego, someone whom he could identify with and soon he began to imitate Dean by adopting his mannerisms, clothes, and his hairstyle. This persona by proxy soon showed several flaws as Dean's looks, talent, and intelligence were missing in Starkweather. However, Starkweather was able to perfectly align himself with the isolation that his rebellious idol portrayed. Another biographer, Jack Sargeant, contends that Starkweather was:

> "acutely sensitive, not just to the taunts of his fellow students but also to his family's low social position and poverty. For [him], poverty was a trap, he could map its confines, and trace its borders, but Charles could see no escape for himself; he believed that his very life was rigidly controlled: he saw that he would not be able to flee the bludgeoning poverty which had characterized his working class childhood but instead would be condemned to repeat it, eventually finding himself a manual job, a wife, having children and then simply dying."

Nihilism, the feeling of 'doom and gloom' that nothing is ever really going to get better or change for the good, is often the predictable philosophy of those who live in social isolation within the headlock of the savage brain. In many ways, the young Starkweather was living up to this philosophy every day of his miserable existence. Dropping out of school at the age of sixteen, he bounced from one menial job to the next, ending up on the loading docks of the *Western Newspaper Union* warehouse, where his foreman would later describe him as "the dumbest man we had." Disappointed in himself and his life, Charlie was looking for a way out, and he thought he had found it when, in 1956 he met the just turned thirteen year old Caril Ann Fugate.

Caril Ann Fugate

Five years younger and barely more than a child, Caril was immediately attracted to Charlie. She came from poverty below even the level of the Starkweather family. A visit to her house revealed the absolute squalor in which her mother and stepfather were raising Caril and her infant half-sister. Living and family conditions aside, Fugate was a perfect match for Charlie in temper and rebelliousness as well as her lack of intellect. Most of all she was young and impressionable. Predictably, she hungered for fun, adventure, and danger. Fugate, like Starkweather was searching for a way out of her miserable life.

Starkweather perceived Fugate as 'a genius' and treated her like gold, easily seducing her in the process. Fugate thought of him as a God—her savior who came with the promise of salvation from the sewage pool of which she was drowning. The two fell hard and deep into lust, the front burner of the MLS, burning much brighter than cognitive regions of 'second thoughts' reality. Apparently, both saw each other as their ONE CHANCE at happiness.

Predictably, the parents stepped into the parlor and had 'second thoughts', especially Caril's parents who strongly disapproved of their daughter's liaison with such a troublemaker. They believed Starkweather was too old for her and feared he was not a person of substance. Despite their objections, Charlie quit his job at the loading dock and took a position as a garbage collector so he would be finished with work by the time Caril got out of school. After a fight with his father over his relationship with Caril, he moved into a nearby rooming house where his friend Bob lived with his new bride, Caril's older sister, Barbara. But his menial pay (only $42 a week) would not allow him to continue at this residence; his landlady soon locked him out. On his daily trash route Charlie passed the houses of the affluent members of society, a constant reminder of everything he and Caril so deeply desired but could not afford. Quoting Jack Sargeant:

> "He saw what he was being excluded from . . . While heaving heavy, stinking sacks of trash for a minimum wage Starkweather came to the realization that, for him, there was one real leveler of class, one way in which he would find himself equal with the rest of society which had oppressed, dominated and alienated him, a method by which he would find retribution: dead people are all on the same level."

Fantasy Life

Caril became the center of Starkweather's universe; he was soon telling others that the two were engaged to be married and that she was pregnant with his child, a rumor that crushed Caril's parents. Charlie felt the whole world was against his relationship with Caril, a belief that would play a role in his first murder.

On the last day of November, 1957, the couple stopped at a gas station where Caril saw a toy stuffed dog that she wanted. Charlie did not have enough money to buy even this small token of his affection and became enraged when the clerk would not let him buy it on credit. This humiliation was more than his already unstable mind could take. At 3 A.M. the next morning Charlie, armed with the 12-gauge shotgun he had recently pilfered, headed back to the gas station to even the score. There he found the same clerk who had degraded him in front of Caril the day before: twenty-one year old Robert Colvert, a newlywed with a baby on the way. Charlie entered the station and purchased a pack of cigarettes before driving away, only to return a few minutes later for a pack of gum. He got in his car and appeared to drive away, but parked nearby to change into a disguise: a red bandana to cover the majority of his face and a hunter's hat to hide his red hair. With his shotgun in one hand and a canvas bag in the other he returned to the store and pointed the gun at Colvert, demanding he open the safe and fill the bag with money. Colvert did not know the safe's combination, so Charlie got little more than the $100 in the cash register. Angered once again by his failures, he pushed Colvert toward his car at gunpoint and ordered him to drive. When they stopped, he ordered Colvert out of the car and shot him. Unsatisfied, he sealed the deal with one more shot, this time at point blank range to the head.

The story caught like wildfire in the press, forcing Charlie to paint his car a different color to avoid detection. Though it was well known that most of the money stolen from the register was in coin he used it nonetheless to buy himself some new clothes, evidence of the fact that the killing had left him feeling invincible (a natural chemical consequence of the MLS).

Believing a transient had committed the murder; police did not bother investigating Starkweather, a fact that further contributed to the belief that law enforcement could not touch him. Now the score was Charlie: 1, the World: 0. He felt entitled. With each murder he felt more grandiose by his actions. And he was just getting started.

Charlie admitted to Caril that he had robbed the station, but his futile attempts to deny the murder did not fool her. Yet, she seemed in awe of his power and bravery, a common *abreaction* of midbrain—emotionally disconnection from cognition resulting in cognitive adoration with denial of the emotional impact of murder, for example. Sadly, she interpreted his actions as a way to get what ever she wanted whenever she wanted it. In a view terribly corrupted by youth, she saw her 'knight in shining armor,' and she vowed to follow him wherever he went. Along with Charlie, she had sealed her fate.

In Search of Paradise

When the natural high the murder created via the coursing chemicals in Charlie's brain wore off, he was faced with a reality that was far less than ideal. He had been fired from his latest in a long line of menial jobs, he had been kicked out of the boarding house where he was living, and both sets of parents were staunchly against their relationship, a fact that was not helped when Caril put on weight, assuring (although falsely) her folks that she was indeed pregnant with Starkweather's child. Once again in the mindset of the star-crossed 'lovers' the whole world stood in opposition to them. But this time, he knew how to take things into his own hands.

On Tuesday, January 28, 1958, Charlie drove to the 'trash heap' Caril and her family called home, taking along yet another rifle to help direct the conversation. Caril's mother, Veleda Bartlett, answered the door. Devoid of specific details, it is presumed that an argument ensued over the Bartlett's disapproval of his relationship with Caril. What is known is that Charlie left the house with Caril still at school, and phoned Fugate's stepfather Marion's place of employment to inform them that he was sick and would be missing work a couple of days. When he returned to the house, he had gathered Caril from school and she stood by his side. When the dust had settled, her entire family lay dead: her mother had been shot in the face and with blunt-force trauma to the head from a rifle butt, her stepfather was shot and stabbed, and her baby sister, only two-and-one-half years old, was also stabbed and beaten with the butt of the rifle.

Even more gruesome than the murders themselves was what transpired in the aftermath. Starkweather, most likely with the aid of Fugate, drug her mother's body to the outhouse and stuffed it down the toilet. The baby was put in a trashcan and placed in the outhouse. Marion Bartlett's body was separated from the rest of his family and thrown on the floor of the chicken coop as discarded 'trash.' Then the young lovers, free from parental interference, cleaned up the mess and wasted away the evening drinking Pepsi and eating potato chips. For the next week they holed up in the house, just yards away from the rotting corpses of Caril's family, with Charlie venturing out only for groceries. During that time, a sign Caril placed on the door that read, "Stay a way Every Body is sick with the Flue," turned visitors away. But Caril's brother-in-law and Charlie's friend, Bob Von Busch, were not so easily swayed. When Von Busch was persistently denied access to the house he involved police who assured him that Caril was credible and there was no cause for alarm. Unsatisfied with such a cursory explanation, Von Busch continued to raise questions until, with the help of his brother, he found the bodies in the outhouse and chicken coop. A bulletin was issued over police radio to be on the lookout for Charles Starkweather and Caril Ann Fugate, but by that time they were already on the run.

Life on the Run

The couple fled to the home of seventy-two year old August Meyer, a friend of the Starkweather family, who lived on a ranch approximately twenty miles outside of Lincoln. On January 27, they pulled into his drive and quickly got stuck in the mud, an ordeal that ended with a bullet in Meyer's head. Charlie wrapped his old friend in a blanket and hid his body in the barn before returning to the house where he and Caril took the old man's money and guns, ate his food, and slept in his bed. The next morning, a neighbor helped them get their car out of the mud and they returned to Meyer's farm by a different road. When they were ready to leave they got stuck in the mud again and abandoned Charlie's old Ford, taking only their weapons and leaving on foot.

With their shotguns hidden from view, they flagged down another teenage couple: seventeen year old Robert Jensen and sixteen year old Carol King. As they leaped into the car the driver found himself being robbed at gunpoint. Jensen was forced to drive Starkweather and Fugate to an abandoned storm cellar near Meyer's farm.

Starkweather marched Jenson to the cellar and then shot Jensen six times in the head. Carol was also shot once in the head, and her body was left with her jeans and panties around her ankles, revealing several stab wounds to the abdominal and pubic areas. Later while awaiting trial, Starkweather attributed this aspect of the murders to Caril, who had become jealous of the other girl's attractiveness. There is a strong possibility that Starkweather mutilated the girl's body out of his own frustration at being unable to perform sexually. The two spree killers escaped in Jensen's car and, in a stroke of incredible stupidity, drove back to Lincoln, past Caril's home, curious to see if the bodies they left had been discovered. The police cars surrounding the residence confirmed their doubts and they drove on, stopping on the affluent side of town where they fell asleep in the car they had stolen.

It was not long before the bodies of Meyer, Jensen, and King were discovered and a major manhunt ensued. Suddenly, the world that had once rejected Starkweather now focused its full attention on him. In keeping with the celebrity that he now believed he was, Charlie turned his attention to the more affluent part of Lincoln, the same streets he had worked in his days as a garbage collector. From these, he chose the home of C. Lauer Ward, the forty-seven year old president of the *Capital Bridge and Capital Steel* companies. That morning, however, Ward himself was not home, but his wife, Clara, their housekeeper, Lillian Fenci, and their dogs were at the residence. While Caril waited in the car, Charlie brazenly walked right up to the door and rang the bell. When Fenci answered she found the barrel of a shotgun thrust in her face. Charlie ordered her to lock up the larger of the two dogs in the basement and, realizing that she was hard of hearing, wrote her a note instructing her to continue preparing breakfast for Mrs. Ward, who found the intruder waiting when she entered the kitchen. He then fetched Caril from the car and, after having some coffee, she fell asleep in the library. Charlie then ordered Mrs. Ward to make him some pancakes, and then waffles, two requests with which she quickly complied. This did great things for Charlie's already inflated ego to have one of the richest women in town waiting on him. Yet he turned on his servant quickly when she drew her own gun, throwing a knife, which landed in her back before stabbing her numerous times in the neck and chest. As he dragged her body into the bedroom her poodle, Suzy, began to bark at him, so he silenced her with a blow from the rifle butt, breaking her neck.

He then sat down to write a warped attempt at a confession or apology or justification, which he addressed to *"the law only:"*

> *"I and Caril are sorry for what has happen, cause I have hurt every body cause of it and so has caril. But I'n saying one thing every body than cane out there was luckie there not dead even caril's sister."*

The killer couple then loaded the Ward's black 1956 Packard with food and anything that looked valuable from the house. At 5:30, the *Lincoln Journal* arrived at the Ward home, and Charlie was ecstatic at the headlines. "Hey Caril," he yelled. "Get a load of this! We're stars! Made the front page of the Journal!" He finally had the recognition he had always felt he deserved. Soon, Mr. Ward arrived home and was promptly shot by Starkweather. He and Caril then tied the maid to a bed and stabbed her to death.

Later, while awaiting trial, he would deflect the blame from himself to Caril as the one responsible for the murders. The bodies were discovered the next day.

The Hunt

This time, Charlie had made a mistake. Ward was a good friend of the Nebraska governor, who quickly called out the National Guard and requested the aid of the FBI when he heard of the savage attack. Yet once again the killers returned to familiar territory, driving past Caril's house just one more time. Seemingly, they wanted to return to the place where their life together had truly begun. But it was not to be; a car was in the drive and all the lights were on in the house, so they continued driving for the rest of the night, crossing the border into Wyoming the next morning, January 29th. In search of a new, less conspicuous set of wheels, they settled on the vehicle of Merle Collison, a traveling shoe salesman from Montana who had parked his Buick on the side of the highway to get some sleep. Charlie decided to take the car, shooting Collison several times in the head, neck, arm, and leg. With the salesman's body in the front seat and Caril in the back, Charlie started the car but was unable to release the emergency brake. When another motorist who assumed they had car trouble stopped to help, he was greeted with the barrel of Charlie's gun and the orders to, "Raise your hands. Help me release the emergency brake or I'll kill you!" It was then that the man saw the dead body lying on the seat and instinctively began to struggle with Starkweather for the gun. During the battle, Wyoming deputy sheriff William Romer pulled his patrol car over to see what was going on.

Realizing that the end had come, Caril turned on Charlie and jumped from the back seat of the car, running toward Romer and shouting, "Take me to the police! He's killed a man!" Starkweather ran back to the car and sped away; a police chase ensued at speeds of over one hundred miles per hour. And then, just as quickly as he had begun, Starkweather stopped in the middle of the highway. Police ordered him out of the car and told him to put his hands up. Instead, Charlie decided to tuck in his shirt, perfecting his image in the event that he might encounter the media. A warning shot quickly coaxed him into compliance, but only because he thought he had been shot and was bleeding to death. He had actually only been hit in the ear by a shard of the window that had been shattered with the officer's shot. In the words of Douglas, Wyoming Police Chief Robert Ainslie, "That's the kind of yellow sonofabitch he is."

Fifteen Minutes of Fame

With his fantasy of James Dean's rebel-like appeal, Starkweather caught the attention of America. At first, he remained Caril's advocate, saying, "Don't be rough on the girl. She didn't have a thing to do with it." However, he quickly turned on her when he learned that she was claiming to have been an unwilling hostage, blaming him for all of the murders and mutilations. "She could have escaped at any time she wanted. I left her alone lots of times. Sometimes I would go in and get hamburgers she would be sitting in the car with all of the guns. There would have been nothing to stop her from running away." Just as it had for Caril, the teen killer's appeal ended for the nation as well.

After being extradited to Nebraska to stand trial, where he declared himself sane despite the fact that his attorneys entered a plea of "innocent by reason of insanity," it took a jury less than twenty-four hours to send him to the electric chair, a sentence that was carried out on June 25, 1959.

Caril, on the other hand, was able to manipulate a jury just as she had Charlie and the officers who had apprehended them. They took pity on the fourteen year old murderess and sentenced her to life in prison. She was sent to the Nebraska Center for Women where she served only eighteen years until her release in 1976. Today she lives in Michigan where she works as a medical assistant. Caril never married and refuses to discuss the case.

11 'The Relational Psychopathy Risk Scale' (RPRS)

Don Jacobs

Measurable Tendencies of Possibly Embracing Psychopathic Personality
About as Unempirical As You Can Get . . .

Answer the following question with numerals 1, 2, or 3. Mark (1) Prefer caution or Disagree; (2) Probably True some of the time; (3) Definitely True, most of the time.

Make your choice a RESPONSE as soon as you read the statement.

1. I find attractive people of both sexes intriguing. I would not mind at all if they engaged me in conversation. After all, it's just talk. 3 2 1
2. I don't always know why I am so attracted to some people, but when I am, I find it hard not to show it. 3 2 1
3. It really doesn't bother me that most people stretch the truth. It's even human nature to tell little 'white lies.' So, what? Most people do it on occasion. 3 2 1
4. If I really like someone I don't really care what he or she had done in the past. I know the past cannot be changed. If we strike up a relationship, it only matters what we do together in the present. 3 2 1
5. I would give a person I was really attracted to (or had been in a relationship with) *one more chance* if they cheated on me, or told me a boldface lie. I think everyone deserves a second chance. 3 2 1
6. If I fell in love with someone, I would help him or her out if they were short of cash, it's really no big deal who pays most of the time; it all balances out in the end. 3 2 1
7. I am a sucker for a smile, sparkling white teeth, and great body. 3 2 1
8. I love to flirt. 3 2 1
9. I love attention. 3 2 1
10. If I truly loved someone I could change him or her for the better. 3 2 1
11. I think most people regret doing things that hurt people. If they don't show it, they just have more strength of character than most people. 3 2 1
12. Regardless of what my family or friends say, if I really like someone, I really don't care what other people think about him or her. 3 2 1
13. I seldom 'give up' on people I'm really attracted to. 3 2 1
14. I believe most people really feel bad when they hurt someone they love. 3 2 1
15. I really think some people really deserve each other as 'misery loves company.' 3 2 1
16. I would breakup in a second with someone I felt was not good for me. 3 2 1
17. Individuals who keep their emotions in check at all times must have strong resolve; they are to be admired. 3 2 1
18. We settle for what we settle for. 3 2 1
19. I will not settle for someone with less values or ethics than what I have. 3 2 1

20. I believe most people mean well by what they do, even though it may not turn out that way. 3 2 1

______ **Total Score**

Scoring Summary

50 TO 60: HIGH vulnerability to Psychopathic Characteristics
39 T0 49: MODERATE vulnerability
BELOW 31: LOW vulnerability
(subtract 6 for adjusted score of 24 due to social desirability answers)

Interpretation

Question 1
The first 'con' of the psychopath is PHYSICAL engagement. Being natural charmers and disarmers, they rely on their physical appeal (aestheticism) to get close to prey. A score of 2–3 reflects a perfect target. *Maturity* brings with it a 'caution mode' to even the most casual of acquaintances. Best Responses: 1 or 2.
Question 2
Permission is granted the psychopath when prey give permission to 'come on stronger' with positive feedback which further engages the advances of the predatory, reflected in a score of 2–3. An emotionally mature individual is careful not to show 'too much too soon' as observed in a skillful poker player. Best Responses: 1 or 2.
Question 3
The second 'con' of the psychopath is willingness of prey to accept 'half-truths' (and close relative to moral compromise), a highly appealing trait to psychopaths who are natural born liars; they will eventually open the floodgates to compulsive lying very soon in a relationship. A score of 2–3 indicates a dangerous weakness to be exploited by the crafty and glib psychopath. An emotionally mature person has observed what lies have caused in the past and makes every effort to tell the truth. Best Response: 1
Question 4
Willingness to ignore one's past performance and mistakes is a highly sought after commodity in psychopathy as a score of 2–3 indicates. The past is often the best indicator of present behavior as all mature individuals know first hand. Best Response: 1.
Question 5
The third 'con' of the psychopath is the ability to masquerade domination and control with *faux* care and concern so an endless parade of chances are given by prey indicated by a 2–3 score. Best Response: 1
Question 6
A fourth 'con' of the psychopath is the ability to line their pockets with YOUR MONEY. A score of 2–3 indicates your willingness to be your predator's

banker. The key here is habits and patterns that become highly predictable. Best Response: 1

Question 7

Internal validity check tied to PHYSICAL APPEARANCE questions 1, 2, & 5. A high score of 2–3 indicates a physical match in the predatory-prey SUPERFICIAL APPEARANCE sweepstakes. Best Responses: 1 or 2

Questions 8 & 9

The psychopath is 'betting on extremes' in targeted prey such as extremely low self-esteem, extreme superficiality, extreme hypersexuality, extreme obsession, extremely addictive personality, etc. Internal validity checks to 1, 2, 5, & 7. Best Responses: 1 or 2.

Question 10

Indicates willingness to 'hang in there' once the abuse starts; scores of 2–3 on the internal validity checks for questions 1, 2, 5, 7, 8, & 9 reinforce 'sexual chemistry' so vital to the 'con' of psychopathy. Best Responses: 1 or 2

Question 11

The salient trait of psychopathy is a *reptilian remorselessness* that masquerades as 'strength of character'; perfect prey score 2 or 3 on the question. Best Response: 1

Question 12

The persuasive abilities of psychopaths to create 'tunnel vision' in prey so input from friends or family is always diminished, eventually to the point of total rejection by prey of family and friends. Best Response: 1

Question 13

Internal validity checks to questions 5 & 12. Best Responses: 1 or 2

Question 14

Internal validity check to question 11. Psychopaths are incapable of love. They are users and abusers. Best Responses: 1 or 2

Question 15

The 'nail' of coffin of psychopathic domination over prey: this mindset reflects the conviction of prey that 'others just settle for mediocrity when they really want what I have.' The psychopath has been so utterly convincing and domineering that almost no one can save prey from the abuse that sure to come. Best Response: 1

Question 16

Denial is the whirlpool that keeps dragging prey back into the funnel of psychopathy. Many of the prey's friends and family have a 180 degree difference in opinion of the character of the predator. Best Score is a 3 but for a far different reason.

Question 17

The most salient, tell-tale sign of psychopathy is also the most difficult to camouflage: the prevalence of *emotional disconnects,* especially sadness over hurting others and concern for the physical and psychological welfare of others. They are masters at this type of subterfuge. Best Responses: 1 or 2

Question 18

Prey who are in denial (they always are) read this sentence as confirmation that they are absolutely with the 'love of their lives' while friends and family see the real truth that prey have lost the ability to perceive. Best Response: 3 for seeing through truth not denial.

Question 19
An internal validity check for denial apparent in questions 16 & 18. Best Score: 3 for truth not denial.
Question 20
The fifth 'con' of the psychopath is the exploitation of the naiveté (or immaturity) in their prey allowing rapidity of results: The more naïve or immature the better. Best Response: 1

UNIT III

Brainmarks of Personality

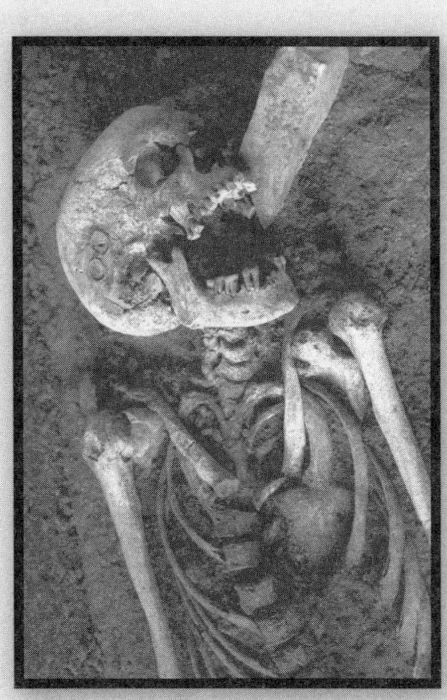

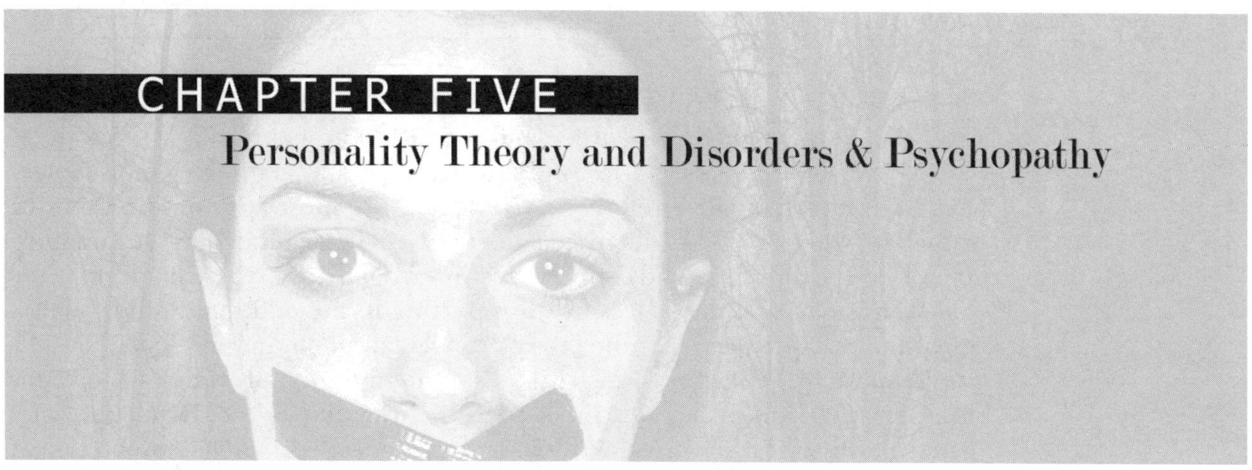

CHAPTER FIVE
Personality Theory and Disorders & Psychopathy

Personality: Total Package of Self

Individuals are known to each other by virtue of personality—the total package of self—that we present to the others comprising, but not limited to physical appearance and demeanor, ways we communicate and impart our feelings, individualistic modes of expression, communication, and relatively predictable patterns of behavior, thinking, and moods.

Personality theorists have often been referred to as *maverick theorists* due to their *courageous speculation* and willingness to defend their theories in the face of a variety of criticism. Perhaps, Freud best exemplifies this archetypal maverick as he weathered a storm of protest from colleagues when he wrote about his theory of personality around 1900. Other theorists, some who had abandoned Freud's analytic school featuring unconscious conflicts, to form their own school, followed in the next fifty years protesting Freud's *biological determinism* visualized as the tyrannical id—"I WANT IT NOW"—contrasted by ego—"How do I get what I want?"—and superego—"Should I get what I want?"—and other theoretical concepts surrounding so-called repression and hypothesized defenses of ego.

Since personality is such an all-encompassing component of everyday experiences it's no wonder why theorists were so eager to study its etiology.

We start our survey of personality theories with the grandmaster of metaphor himself, Sigmund Freud. Freud believed that personality was largely driven by biology (*biological determinism*), where early developmental influences from milieu, while significant, were no match for the overpowering influences of the instincts of sex and aggression, powered by id—the biological component of personality.

It is important to understand that Freud's garish influence in the early decades of psychology were more attributable to the man himself, not his "science" which was non-existent. Freud reads more like a gothic novelist than a meticulous fact-based scientist. Much of Freud's pseudo-science of psychoanalysis embedded in metaphor—ego, superego, id, etc—passed into historical archives on the date of Freud's death in 1939.

Here's an example of his Conflict Theory played out with his metaphorical "characters:" Superego (or conscience)—developed due to parental influences and societal roles and rules. Superego opposes id in a lifetime of psychological struggle (conflict theory) between biological libido, expressed as: "I want it NOW!" versus "Should I do it?" which is expressed as cognitive ambivalence. Ego, a kind of cognitive referee, must decide the most appropriate thing to do—"How do I do it?" As noted, id's demands are exaggerated by exclamation (!), while ego and superego counter with interrogatives (?), hence our *perpetual conflict* as smart mammals. It is worth repeating in the end that Freud's elaborate theories are woven into the fabric of speculation but never science.

Trumping Freud, Carl Jung, Freud's contemporary, provided a more *esoteric* perspective in personality theory. His theories, like Freud, have been relegated to historical archives as philosophical tidbits of Jung's psychological "spiritualism."

Early nineteenth and twentieth century theorists who were influenced by the new disciplines of sociology, anthropology, and family dynamics include the following:

- Alfred Adler
- Karen Horney (Orr-nay)
- Harry Stack Sullivan
- Erik Erikson
- Carl Rogers
- Abraham Maslow
- William Sheldon

Personality disorders will conclude the discussion of personality. Neurochemical disorders, dysfunctional family milieus, and personal experiences contribute to the clusters of personality disorders' diagnostic criteria. The following disorders will be defined and discussed:

Odd, Eccentric
Cluster 'A' characteristics
- Paranoid
- Schizoid
- Schizotypal

Dramatic, Emotional
Cluster 'B' characteristics
- Antisocial
- Borderline
- Histrionic
- Narcissistic

Anxious, Fearful
Cluster 'C' characteristics
- Avoidant
- Dependent
- Obsessive-Compulsive
- Passive-Aggressive (in newer classifications, no longer considered a personality disorder)

Personality Captivates

While it's true everyone loves a good story, it is usually the *vivid personalities* we tend to remember the most—the characters in literature and movies. Award-winning actors such as old-timers Dustin Hoffman and Jack Nicholson, and more recently Adrian Brody have created memorable characters from Ratso Rizzo in *Midnight Cowboy* and Raymond Babbitt in *Rainman,* to Jack Torrance in Stephen King's *The Shining* and the obsessive-compulsive writer Melvin in *As Good As It Gets,* and Brody who is the youngest actor to win the Best Actor Oscar®.

Female and ethnic characterizations have recently expanded in the same way as Hollywood continues to embrace interesting personalities. For instance, in 2003, Charlize Theron won the best acting award for playing serial killer Aileen Wournos in *Monster.*

While good stories captivate us, even more memorable are the personalities we love and sometimes loathe. Most recognized actors have similar resumes with colorful personality portrayals. Movies, plays, short stories, and novels create characters we often wish we knew or are extremely glad we don't.

Independent Films: Personality Profiles

The rise of independent films, the new *zeitgeist* of contemporary personality exposition, often portrays less than savory (also known as *noire* personalities) individuals who are often seedy, quirky, or downright scary. Or, they break our hearts with their humanity.

For the most part, we line up to buy tickets to see just how revolting or inspiring the portrayals come across. Teen 'slasher' movies feature good-looking, not quite twenty-something actors, who fend off psychos with an eclectic collection of personality deficits right out of the current DSM psychiatric manual. While the stories are entertaining, it's still the characterizations rich in personality that rivet our attention. Some portrayals are so riveting that *typecasting* occurs which describe actors who are cemented in a particular role. No matter what roles they eventually come to play, the actor William Hurt, for example, will always be remembered as the seedy lawyer, Ned Racine, and actress Kathleen Turner will always be remembered as the seductive manipulator, Mattie Tyler Walker in *Body Heat.*

Personality often defies precise description, yet it is an individual's *unique qualities* that draw us to them like a magnet, or in others, repel us. Sports personalities are abundant and range from the shocking (Dennis Rodman) to mundane, (nameless wannabes warming the bench). Want shock value? After all, serial killer trading cards have been around for some time now.

Persona and Print

Carl Jung, a Freudian contemporary, first described *persona* as a powerful archetype inherited into our *collective unconscious* psyche—Jung's most esoteric concept. To Jung, an *archetype* is an emotionally-charged idea or concept that is pre-formed, requiring little or no learning, akin to an instinct; instinctively, *we just know it.* Our collective unconscious is defined as our *gene pool inheritance* that predisposes all members of a race or of a given ancestry to share *common experiences* such as the Native American archetype of *kindred spirits.* Again, pure speculation, not science.

Persona is defined as one of a number of social masks individuals use in the social presentation of self. By definition, personas have strong contextual bias as different personas are tied to any number of social contexts. In contrast, *print,* a term often used in relationship counseling, refers to who we perceive ourselves to be in our deepest heart of hearts. In this regard, personality print is like a fingerprint, an identifier that no one else has.

The following is a partial list of situations in which we expect to see persona, and we should not be shocked when we do. Personas are aspects of personality that we perceive to be advantageous to show to others.

The following contexts cue predictable personality characteristics.

Context	Persona Characteristics
First dates	Lots of smiles and laughs, light and airy conversation; nothing too serious
Job interview	Best foot forward; you can do absolutely everything requested in the interview
Political persona	Regardless of events swirling around, your public persona is clear—you're in control
Boxer's persona	Meeting in the middle of the ring, an intimidating stare to gain an advantage
Reaction Formation	By acting the opposite way we feel, we show calm instead of fury, the classic persona matches the response, "Am I angry?—heavens, no!"

Personality Theory

While personality theory comprises one of the most interesting and fascinating aspects of general psychology, it still remains highly theoretical, not hard science regardless of what you hear from e-Harmony.com® or any free personality profile promising a harmonious match!

Even though it is a psychological specialty, it is almost entirely an *academic* specialty. However, this does not diminish the importance of personality theory. Exploring personality, relative to characteristics, etiology, and development remains one of the top priorities of college freshmen taking psychology courses.

In this section, *five classic theories of personality* will be presented as well as how they have weathered pop culture. Naturally, each theory has a different perspective on a variety of philosophical, biological, behavioral, and societal influences. At the conclusion, students are free to decide the credence of each theory in the context of their experience. When this is accomplished, students will understand how each theory contributes to the gestalt (or holistic) theory of personality.

In general, we can become *eclectic by* picking and choose the best and brightest aspects of several competing theories; in doing so, this approach fits better with our own experiences.

Personality & Stable Patterns

Traditionally, a well-known definition of personality is a collection of *stable patterns of behavior and thinking that characterize individuals most of their lives.* The focus on *stable patterns* does not preclude the *dynamic* or changeable aspects of personality (such as refining or redefining self). However, most theorists see *patterns or traits that are stable and distinguishable* as the most significant.

Personality as *Self*

Personality is a synonym for self. With personality we can often predict how a person will act in a given situation, but as we all know a person can act *capriciously* (out of character) at times, in ways that are unpredictable. Especially in adolescence, as a teen's behavior may be anything but predictable or follow any pattern. Our friends may say, "I've never seen that side of you."

Additionally, as we mature and grow—intellectually and emotionally with experience—we are not expected to be the same persons we were at an earlier age. If fact, as we get older, a memory of ourselves as children or teenagers is just that—a memory—the person we no longer recognize.

Also, we cannot begin to explain every little nuance of personality. This provides a perfect segue into the theorist who focused his entire theory on *hidden aspects of personality and the conflicts that often go unnoticed in conscious awareness.* Appropriately, we begin our survey of traditional personality theories with the most controversial *as well as most comprehensive personality theorist* in the history of psychology, Dr. Sigmund Freud, M.D.

Sigmund Freud: Why He Won't Go Away

What is the esteemed pediatric neurologist, Sigmund Freud, doing in psychology in the first place? No doubt, every general introductory psychology text lists him as a charter member of the Founders Club. Today, Freud continues to be a pop culture icon on par with Darwin and Einstein. How many individuals can be universally identified by their last name only—DiVinci, Einstein, and Freud?

To further sort out his monumental influence in the early days of psychology, we must remember Freud was a nineteenth century medically trained neurologist who eventually specialized in mental and nervous disorders. His interest in mind (the *unconscious mind*) did not occur until he was more than forty years of age. Somehow, Freud's psychoanalytic theory, derived from a medical perspective, has been continually "psychologized" as a *hybrid psychological perspective.*

Collectively, Freud's theory and others that likewise stress *unconscious dynamics* can be categorized into the current nomenclature of *psychodynamic* perspectives. Until Freud, no one in North America was remotely interested in studying unconscious conflicts. How can this subject of inquiry be objectively studied?

It was only after Freud's 1909 speech at Clark University in Worcester, Massachusetts (the occasion of receiving an honorary Doctor of Laws) that American colleagues began their *intellectual courtship* with Freud. In both Europe and North America, his focus on *unconscious determinants* of behavior cast a long shadow over the conservative medical community. In the Clark University speech, Freud's theory became a worthy competitor to behaviorism—the *zeitgeist* of North American psychology. At this time, according to Freud's biographer, Peter Gay (1990), "Freud's ideas remained the *property of the few, and a scandal to most.*" The Real Story has found nothing in Freud to contradict Gay. It is our position that Freud's study of the *Id by Odd is more appropriate.*

Trained Introspection and Psychoanalysis

Sigmund Freud stands alone among personality theorists in the following ways:

- his *ad infinitum* theoretical comprehensiveness
- his focus on deeper dynamics related to complexity of unconscious dimensions
- his reliance upon *trained introspection* and dream analysis produced the drain plug of science going down the drain

Trained Introspection

Essentially, *trained introspection* is a carefully analyzed self-report given by the *analysand* (the patient in the nomenclature of psychoanalysis). The charge to analysands is to begin exploring their own psyches introspectively and unabashedly. Of course, such an introspective report cannot escape being *highly subjective,* therefore, *highly unreliable as a scientific tool.* "Show me" demands science. In contrast, Freud replied "believe me, even though you can't see it. Just because it's hidden does not diminish its importance."

Freud's theory of psychoanalysis has two central components:

- A comprehensive *theory of personality*—a design of mind replete with descriptive metaphors—id, ego, superego, unconscious conflicts, and ego defenses against anxiety.
- A *treatment methodology* based upon his theory of personality and development.

Originally, Freud conceptualized a theory to understand and treat *hysteria,* the most prevalent psychological disorder of Freud's day. Patients with hysteria suffered from *bodily symptoms,* such as blindness, fainting spells, numbness in the arms or legs, nausea, and vomiting. Yet, physicians could find no discernable underlying medical cause. Absent physiological etiology, Freud postulated unconscious mind and repression of experiences to be the real culprit.

Conversion Disorder

Today, hysteria is known as *conversion* disorder, characterized by approximately the same description as observed in Freud's day, that is, physical symptoms, without physiological cause (etiology) as the root source.

Perhaps the stress of time compression to solve modern problems has contributed to conversion disorder being displaced today by depression and anxiety as the most prevalent psychological disorders. Social philosophers refer to our present day as *post-modern,* where *skepticism and cynicism* have produced a considerable mindset typified by uncertainty regarding the future. In this climate, it would not be surprising that depression and anxiety would be prevalent.

Unconscious Process

In psychodynamic perspectives, dynamic refers to the emphasis placed upon unconscious processes. Whatever one chooses to call Freud's theory (as some describe his perspective 'The study of Id by the odd'), it remains unique and prescient—ahead of its time. Often revolutionary, practically scandalous, and always controversial, psychoanalysis has yet to be hustled out of mainstream psychology texts.

Psychology professors continue to present Freud in their lectures and authors continue to include Freud as a virtual centerpiece in their books, especially in pop culture with *au courant* bylines such as "Don't Stick the Fork in Freud Yet . . . He's not Done." The reason is deceptively obvious. Ask this question: how many times do we do things that we simply cannot explain away with conventional wisdom? Why did we do it after we swore an oath we would never be caught dead doing it? We may be ashamed or completely stupefied by our actions. Apparently, we simply "lost our mind."

To Freud, the culprit was our more complex and powerful mind—the unconscious mind! Perhaps for this reason alone, psychology continues to embrace Freud.

Metaphors of Mind

No doubt, Freud's status as a pop culture icon is due to in large part to his theory of development and personality, both rich in metaphorical reference (although poor in empirical data). He states that personality is forged out of conflict between three psychological components which compete for dominance as *energy systems:*

- *Id.* Metaphorical for our biological inheritance—the instincts of sex and aggression. Linguistically, id would say: "I WANT IT NOW!"

- *Superego.* Metaphorical for parent-taught system of social learning, rewards and punishments in societal milieu. If superego could talk it would ask: "Should I do it?"
- *Ego.* Metaphorical for cognitive decision-making and self-restraint. If ego could talk it would ask: "How do I do it?"

It is helpful to note in Freudian analysis, mind and personality are practically interchangeable. When you're reading about one, you're reading about the other. From his nineteenth century *zeitgeist,* today Freud would be intrigued by behavioral neuroscience (neuropsychology); he profoundly believed that someday *behavior would be explained largely in terms of biology.* That day has arrived.

Master of Metaphors

The use of *metaphor* became a handy device used to communicate hypothetical aspects of personality to largely uneducated audiences. It's also handy when a theorist cannot prove what he is contending. Freud was not privy to the sophisticated high resolution neuroimaging brain tools we have today such as PET, CAT, SPECT and fMRI scanners. Freud's theory was constructed from his own personal insights bolstered by testimonials from his emotionally crippled patients.

Students should know that in no way does Freud's theory derive from healthy patient testimonials. Psychoanalysis was constructed *from psychopathology to treat psychopathology.*

To be fair, it must be noted that Freud was not alone in using metaphor. Today, most modern cognitive psychologists use metaphor and simile by insisting the brain and memory work like computers. In fact, the list of theoretical metaphors is rather lengthy with most any specialty in psychology.

Conflicted

Freud's *metaphors of mind*—id, ego, and superego—provide the battle ground for personality (mind) formation. To Freud, personality is an often times polished façade hiding the real picture—psychological *conflict* in unconscious mind. Personality is an ongoing psychological skirmish highlighted by maneuvering and unconscious scheming of the three combatants in an all-out mental tug of war—one against the other.

In this day-to-day bickering, the outcome of libido and feeling (id), versus appeals to conscience (superego), versus thinking and deciding the best course (ego) are often compromised. Foreshadowing post-modern sentimentality, Freud believed the best we could hope for was to be slightly neurotic. In his view, smart mammals were deeply conflicted and prone to unhappiness. But, were we prone to sexually psychopathic serial crime?

The Iceberg Metaphor

Metaphorically, Freud likened the mind to an iceberg where the larger more formative part—the unconscious aspects of mind—are cocooned neatly away beneath conscious awareness. The result is to extract a fair amount of psychological damage through *repression*—the principle defense mechanism of ego's endless battle to ward off anxiety from superego and id.

The smaller visible tip of the iceberg projecting up from the water represents conscious mind, characterized by the purely conscious (ego) and conscience (superego). However, ego and superego often pale in comparison to the power of the undercurrents of libido (id). Like the Titanic disaster, the small peak of the iceberg on the surface is only a glance at the disaster coming underwater. What then is the powerful unconscious 'submerged' part of personality?

Id: Pleasure-Principled

Id represents our biological inheritance. Freud's training as a nineteenth century physician presented the view of the *body as an energy system* with corresponding generators. Id functions as the genetic energy required for the running of ego and superego. Id is largely instinctual with biological urges to seek *sensuous pleasure* as well as sexual pleasure regardless of the outcome. Might id be the fuse to violent criminal behavior?

The Pleasure Principle

Even today, there exists an ongoing debate between Freudian scholars on the intensity of the sexual connotations of id. This global, sensuous instinct of id operates by virtue of the so-called pleasure principle. The *pleasure principle* states that sensuous pleasure is highly sought and may be the underlying perspective in behavior. Therefore, id seeks pleasure and avoids physical pain (unless pain becomes associated with sexual pleasure in a perversion known as *sadism*).

Taken to the apex of sexual energy, Freud hypothesized that id as *libido* compels us (consciously and unconscious) to seek sex as the life wish. The logical conclusion of the life wish is procreation, which assures our species will continue. Theoretically, by satisfying libido, mortido—the death wish or violent aggression—is prevented from gaining leverage over libido. Apparently, Freud visualized *libido-mortido* to oppose each other. Freud speculated that some individuals are extremely libido-driven (hypersexual) while others are mortido-driven or, perhaps, in milder versions *action addicts* who engage in dangerous activities as though they have a 'death wish.' Bullet bikers who drive one hundred miles an hour weaving recklessly through traffic most likely have a death wish.

When id is the driving force of personality, Freud described the person as *id-dominated.* From a personality perspective, this person is impulsive, selfish, egotistical, and self-serving, basically adhering to a life patterned after the pleasure principle. Freud observed that the pleasure principle might prove to be the ultimate downfall of humanity. Id provides Freudians with the most profound conundrum (puzzle) of all. What do we really want from life and what motivates us to get it? Being self-absorbed sounds like the beginning of a psychopathic personality.

Superego versus Ego

Before we go into more detail, we will provide an overview of the three components of personality. First, Freud described id as a "seething cauldron of excitement," as the original energy system of personality. Anthropomorphically, if id had a face, it would shout, "I WANT IT NOW!" showing id-driven impulsivity toward self-gratification and the psychic blueprint for the development of the other components—ego and superego. Perhaps psychopathy.

Around age two, with language acquisition well underway, ego and superego start to emerge, not unlike the special ascendancy of epigenetic development. The *psychosocial component* (superego) rises to oppose id. Superego springs from parental discipline where rewards and punishments teach children right from wrong. Superego is roughly equivalent to what we call *conscience.* So, conflict is underway. A superego-less person would qualify as a psychopath.

Finally, *ego* emerges as the *cognitive decision maker*—how things can be accomplished with as little conflict as possible. Ego represents the cognitive pretzel of being a *smart mammal.* Appropriate *social behavior* is the priority of ego.

To Freud, personality is akin to *Shakespearean tragedy.* How can an impending doom be avoided? The answer is complex. It can be partially accomplished by traveling down the dimly lit royal road to the unconscious—sleep, dreams, and nightmares, our next stop.

Interpretation of Dreams

Theoretically, id is the seat of unconscious influences affecting personality in mysterious and profound ways. With the publication of Freud's landmark book, *The Interpretation of Dreams* (1900), Freud outlined three ways unconscious experiences could be accessed by experienced clinicians:

- Dream interpretation
- Free association (unbridled expression of thoughts)
- Significance of 'slips of the tongue'

Dream Interpretation

To Freud, dreams exist as the *royal road to the unconscious.* Through dream analysis, Freud believed we could decipher the complex *symbolic nature* hidden in the *latent content* (realm of symbolism) in dreams. Repression automatically blotted out conscious awareness of the threatening content of dreams—sexual or aggressive—denied disclosure in the *manifest content* (literally, what happened in the dream). In Freudian analysis of dreams, the interpreter merely listens to the manifest content, but analyzes *what lies beneath* (that is, the latent content) which was theorized to be riddled with symbolism.

Through *symbolism* of latent content, Freud believed dreams actually *preserved sleep* by tricking the dreamer into a realization of the dream as non-threatening. Freud believed dreams always contain sexual content albeit in disguised form. As you might expect, dream interpretation is not an empirical science, but again, highly theoretical.

Due to repression, *overwhelming anxiety* free-floated into ego and conscious awareness. Theoretically, repression distorts dreams by wrapping them in a disguised package that can be tolerated by the dreamer's ego. Like blood pressure, repression is always there, acting automatically to blot out sexual or aggressive themes that would otherwise threaten to awaken the dreamer.

Today, so-called *recovered memory therapy* is an example of Freudian theory revisited, where individuals are helped by therapists to recall painful events. These painful events may have been repressed from consciousness perhaps for years.

Recovered memory therapy is based on the belief that anxiety-producing experiences are extinguished from awareness much like amnesia. In this way, repression does not disappear emotionally; instead repression lingers like a dormant virus. Often, the person feels an ominous emptiness or dread—a nagging reminder that something is not right. Repression is theorized to be the ultimate defense against feelings or experiences that are too sexual or too aggressive to process cognitively. The amnesia-like qualities of repression allow ego and superego to appear (almost) normal on the surface. In Freud's perspective nothing is what it appears to be due to the power of unconscious mind.

An exceptionally bright student remarked one day in summer school—"there's simply nothing normal with Freud." She was completely correct.

Relative to recovered memory therapy, what if these allegations of traumatic sexual encounters (or sexual abuses) were simply fabricated and untrue? You can see the irreparable damage that could be done when therapists are not knowledgeable enough to guard against such a likely eventuality.

Free Association

Another method used by Freud to probe unconscious mind was actually a technique developed by a colleague. *Catharsis—the* 'talking cure'—involved allowing the *analysand* (psychoanalytic patient) to talk in an unrestricted and unbroken "stream of consciousness," without regard to filtering inappropriate or scandalous thoughts.

Trained introspection, as it came to be known, was a technique that often required months to perfect. That repression could 'bubble to the surface of awareness' was the goal of simply talking about previous experiences in this unfettered way, leading eventually to *sudden insight,* giving rise to emotional catharsis—the psychological purging of pent-up repression.

Freud had abandoned hypnosis earlier in his career in favor of trained introspection. He developed this f*ree flow of uncensored thoughts* through the years to be one of the principal therapeutic tools of his theory.

Slips of the Tongue

Freud's third method of accessing unconscious, repressed experiences was through the acknowledgement of *slips of the tongue.* During therapy, when the analysand said one thing, but retracted it immediately, by the infusion of humor, or by cognitive reversal, Freud found *word slips* highly significant. He believed the slip was actually what his patient intended to say, and thus proved to be an important insight into unconscious repression such as, when the analysand said, "You know, I really hated my mother for how he raised me!" In one statement, the analyst discovered two slips. Hastily, the patient corrected the reference of hated to loved, and he to she. However, it was too late as the analysand's underlying meaning can be analyzed as *familial ambivalence* characterized by strong negative feelings toward his mother.

Family Values & Superego

Our first caretakers, usually our biological parents, become the architects of a second metaphorical component of personality known as *superego,* comprising the learning experiences in family milieu where the child is rewarded, punished, indulged, or ignored.

In a global sense, superego is synonymous with social structure and organization. A society (like an individual) cannot survive without order, guidelines, and laws; otherwise living in chaos would result. *Id-as-chaos* (where sexuality and aggression go unbridled) contrasted to *superego-as-disciplinarian* represents one of the great dichotomies of human behavior. In modern neuropsychology, the contrasting agendas of the fore front cortex versus the midbrain limbic systems are analogous.

Superego as Conscience

As the child matures and begins to understand the parenting as influential on his or her behavior, superego represents the child's conscience—the parental presence *in absentia.* Obviously, conflicts are destined to arise between the child's desires and parental restraint regarding appropriateness. For example, the first time toddlers reach for an off-limits object, parents may swat his hand which symbolically marks the beginning of superego development. Lines have been drawn. Limits have been set. Repercussions and consequences enter the picture. The child (as id-dominated) has been challenged by (superego-training) parents.

Self-serving children, sooner or later, must contend with rules and regulations in the form of behavioral expectations from parents. If superego could talk it would immediately ask questions: "Should I do it?" "How will I feel later?" "What are the likely consequences?" While id is biological and inherited, superego is learned in family and peer milieu and exists as a measuring rod for discipline and appropriateness, heavily laden with morals and ethics. Playing by the rules with an eye to consequences characterize superego.

Ego: Cognitive Decision-Maker

Another personality (mind) metaphor, ego, is used to describe the *cognitive decision-maker.* Developmentally, the emergence of ego appears to coincide with language acquisition and superego internalization with behavioral habits. Ego is forged from an awareness of self as a separate entity and the necessity to make difficult decisions. Ego seeks appropriate behavior as long as id is not demanding or too overwhelming and or superego (conscience) does not protest too much. Ego's rational course of action may cause a collision with the values of superego.

As id is driven by the pleasure principle, ego functions according to the *reality principle,* or social appropriateness. In this way ego has at least a one-third chance of winning the battle over pleasure-seeking demands of id and the judgmental standards of superego.

Defensive Maneuvering

Freud postulated that ego must confront feelings of anxiety due to the constant pressure and conflict from id and superego. So-called *defense mechanisms of ego* are required to refine or redefine situations brought on by living day-to-day. Defensive maneuvering has three operational strategies as they deny, distort, or falsify reality; they operate unconsciously. Although there are many more, the following nine ego defenses are most often observed:

Repression: In Freud's view his cornerstone psychoanalytic principles where threatening feelings of a sexual or aggressive nature are denied access into consciousness through a kind of 'automatic amnesia.' Freud believed this principle was his greatest discovery and it became the focus of analysis in therapy. Repression must be dislodged from its unconscious 'holding pattern' and 'bubble to the surface' into conscious awareness for processing.

Resistance: The term used to describe the tight grip of a repressed experience within unconscious mind. Upon release, *abreaction* is the *emotional catharsis* (purging) felt as the corrupting repression is understood through *cognitive insight* by ego in analysis, thereby, healing old psychological wounds.

Rationalization: A cognitive-like aspect of ego that allows the rationalizer to justify his behavior against the 'audit' of superego and libidinal demands of id. Suspicious feelings and guilt must be squelched. Freud postulated that rationalizations were somewhat helpful in muting anxiety by finding excuses.

Intellectualization: Closely associated with rationalization, intellectualization consists of (1) repressing the emotional component of an experience and (2) restating it in an *abstract intellectual analysis.*

Denial: Absolute refusal to acknowledge a painful or threatening reality to the point that the person comes to believe the painful or threatening reality is false or not that bad.

Projection: A cognitive defense that rebuffs anxiety by projecting on another person his feelings or attitudes as though they were projected upon a screen for viewing. The result produces a deflection of uncomfortable feelings or attitudes that were once threatening. "Blaming others" is the focus of projection. In doing so, anxiety is lessened in two ways: First, from superego pronouncements—"you shouldn't feel that way," and from ego—"everybody deserves the benefit of the doubt."

Identification: Identification is the reverse of projection. The defense of projection as we have noted allows "projectors" to rid themselves of threatening feelings by attributing them to someone else. In contrast, identification mitigates anxiety by modeling or adopting another's ideas or philosophies to share the limelight and avoid feeling incompetent. We can observe identification at work when parents' unfulfilled dreams are rekindled through their children.

Reaction Formation: Showing the opposite feeling than the one felt is the main aspect of reaction formation. A person is emotionally distraught, yet in observable behavior, the person appears calm, cool, and collected. This defense protects the person with an emotional insulation to hide behind and the cognitive resolve that everything is OK, but it's a deception.

Displacement: Shifting repressed feelings regarding an original threatening object to a less threatening substitute, such as kicking the family pet or pounding the door over the anger caused by someone else.

Regression: Reverting to childlike behavior under stressful conditions, such as "sticking out your tongue" to show your displeasure at another person or simply throwing a temper-tantrum.

The Conflicted & Deceptive Life of Albert DeSalvo

In conclusion, mental life and personality, according to the psychoanalytic perspective, consists primarily of a *Conflict Theory*—conflicts experienced by our id-nature, id conflicted (or opposed) by our thoughts, feelings of doing the right thing to avoid feeling guilty—the sum total of our *socialization experiences* embodied in ego and superego. Dreams, emotions, habits, and behavior are expressed suddenly or gradually as a result of unconscious mind's conflicted nature.

Freudian theory has special *focus on early childhood experiences* as determining the development of adolescent and young adult behavior. A famous quote underscores Freud's deterministic perspective with an emphasis upon past development regulating present behavior, "The child is the father of the man." It required the development of neuroscans of the brain to show how early developmental influences and conflicts within family could be toxic to the brain.

The conflicted life of serial killer Albert DeSalvo is presented in Predator Profile 5-1.

PREDATOR PROFILE 5-1

Albert De Salvo "The Boston Strangler"

Don Jacobs

Alternate Media Monikers

"The Measuring Man"
"The Green Man"

The world knows serial killer Albert De Salvo as "The Boston Strangler." Born to Frank and Charlotte De Salvo in Boston, Massachusetts, Albert grew up in Chelsea, an overcrowded and impoverished suburb of working class Boston. Here his father Frank worked by day as a laborer and a plumber. At night, as an alcoholic, he physically and mentally abused his wife and children. Sadly, Frank and his wife had two more sons and three daughters.

Before his father left and the dysfunctional couple divorced, De Salvo watched him knock out his mother's teeth and break all the fingers on one of her hands. The abuse was extremely harsh on young Albert. At one point in his childhood, Frank sold Albert and two of his sisters as slaves to a farmer for nine dollars. They escaped their "master" in about a month. At six years of age, Albert's father introduced him to petty crimes such as shoplifting, which led to burglary and breaking and entering.

As Albert grew older, his father brought prostitutes home and forced him to watch their sexual intercourse. By the age of ten, Albert had his first sexual experience. Apparently, the sexual experience at such a young age led to a sexual obsession. Soon, De Salvo became a sex addict. According to reports, De Salvo claimed to have sex with his wife up to six times a day. The DSM, IV, TR (2002) would diagnose him as a sexual sadist, which means he experienced "recurrent intense sexual urges and sexually arousing fantasies involving acts (real, not simulated) in which the psychological or physical suffering including humiliation (including post mortem) of the victim is sexually exciting to the person."

After high school, De Salvo joined the Army and went to Germany. He met and fell in love with a woman whom he brought back to the states with him. While stationed at Fort Dix, his daughter was born. Due to an allegation that he sexually molested a nine year old girl, the Army dishonorably discharged De Salvo. The mother didn't file charges against him, so there was no conviction. After his discharge, he and his wife moved back to Chelsea where he grew up. De Salvo then moved to another suburb of Boston, Malden, where his son was born.

In the late 1950s, De Salvo became a petty criminal by breaking and entering. His wife grew tired of his behavior and soon rejected him. He began to stalk young girls, follow them home, and pretend he represented a modeling agency. "The Measuring Man" became a media moniker because he told them the agency required him to take their physical measurements to see if they qualified as models. He used his charm to seduce some of the girls into sexual encounters. Even though some girls complained to the police, no authorities intervened. In 1961, authorities arrested De Salvo for breaking and entering. He spent eleven months in jail. Then, after his release, he turned to murder.

On his nineteen month murder spree from 1962 until his arrest in 1964, De Salvo killed thirteen victims (the term serial killer did not exist at this time.) These murders produced his lasting alias, "The Boston Strangler." The vast majority of his early victims were older females, but later he preyed on younger victims before police caught him and he went to jail.

Victimology

De Salvo gained entrance into his victims' homes by portraying himself as a repairman or someone of authority, such as a police detective. When he started out, he chose only older females. Authorities believed he killed out of hatred for his mother or for women in general, and De Salvo allegedly said:

> *"I did this not as a sex act but out of hate for her. I don't mean out of hate for her in particular, really, I mean out of hate for a woman."*

At the end of his criminal career, he started preying on younger victims. He spent little time targeting his victims. He sexually molested all in some way and strangled them with an article of their own clothing. How did De Salvo target his victims?

> *"Attractiveness had nothing to do with It . . . when this certain time comes on me, it's a very immediate thing. When I get this feeling, instead of going to work I make an excuse to my boss. I start driving and I start building this image up, and that's why I find myself not knowing where I'm going."*

Signature

When he strangled his victims, he tied a bow around their necks with an article of clothing. Not just any bow, but a big floppy bow. He then placed the body in an obscene position where anyone who entered the apartment after he left would see it lying there with no warning. He penetrated the vagina with a foreign object, such as a broom handle or a bottle. It is unclear whether he did this post mortem or while the victim still lived.

Another charge authorities arrested him for involved a string of sexual assaults that spawned "The Green Man" alias. They believe he committed over three hundred such assaults dressed in green trousers and did not murder one of his victims. These took place over a four state area—Massachusetts, Connecticut, Rhode Island, and New Hampshire.

Insanity Plea Fails

De Salvo's attorney tried to convince the judge and the jury that his client had a mental disorder and was thus unaware of the nature of his crimes. He brought in witnesses to testify as to his mental illness. Regardless, they convicted him and sentenced him to life at Bridgewater Hospital.

Serial killers are crafty and slippery. Along with two other inmates, De Salvo escaped from the hospital. Re-captured, they moved him to a maximum-security prison for the remainder of his sentence. However, six years into his prison term, someone stabbed him through the heart six times in his cell in an apparent prison brawl. Authorities never identified his murderer.

Some of De Salvo's comments show a tinge of remorse for his crimes; however he also bragged about them. Therefore, it is hard to discern their genuineness. Perhaps he just said what people wanted to hear. For example, De Salvo said:

> *"I would go home and watch what I had done on TV. Then I would cry like a baby."*
>
> *"If God didn't mean them to be dead they wouldn't be; it was a case of their time being up."*
>
> *"It* (the murders) *wasn't as dark and scary as it sounds. I had a lot of fun killing somebody. A funny experience."*

Psychodynamic Theories

The Esoteric Carl Jung

Carl Gustov Jung (1875–1961), the world-renowned Swiss psychoanalyst, originally belonging to the Freudian movement, began to disagree with the emphasis Freud placed on sexuality (or sensuous pleasure). This difference proved to be profound and eventually provided the breaking point of the two colleagues. Consequently, the once loyal Jung stopped all correspondence with Freud, and in time, all communication with his former mentor. It was an ugly break that very much distressed Freud who once fainted in Jung's presence at a conference. Jung's central concepts of personality were addressed earlier and will be summarized here.

Collective Unconscious and Archetypes

Jung's central concept of personality (or mind) is centered around the *collective unconscious,* where inborn *archetypes*—emotional images directing behavior—come biologically prepackaged to assist the mind in comprehending the world. Later, learning helped to refine archetypes through immediate experiences. For example, females unconsciously know or comprehend masculine qualities in males due to their archetypal component of gender known as *animus.* Behavioral archetypes of animus (masculine side of females) result in aggressive, hypersexual behavior, or fiercely independent behavior, clearly characteristic of masculine character.

Ostensibly, masculine traits in archetypes are derived from centuries of males competing with each other to comprise animus. Animus qualities can be modified (refined) somewhat depending on the influences from a person's current milieu experiences.

On the other hand, the *anima* archetype stands for feminine qualities of behavior such as emotionality, nurturance, and intimacy residing (by nature) in the collective unconscious. For example, males could have a more empathetic anima-side to their personality as a result of living together with females and raising children. We observe anima-like caring and nurturing in grandfathers perhaps more than we observe it in fathers.

By contrast, females could have an independent and aggressive side to their personalities (animus), due to the experiences of living with and competing with males on a daily basis.

Shadow Knows, Persona Shows

Shadow is the archetype of *animal instincts* that are inherited from the darker side of collective *primordial experiences.* Criminal behavior can be explained as the result of a *strong shadow archetype.* It can be responsible for the appearance in behavior of unpleasant and socially reprehensible thoughts and actions. The shadow-dominated individual habitually acts in dangerous, aggressive, and in antisocial ways. Might this represent Jung's theory of sexually psychopathic serial crime?

In contrast, *persona* is the Jungian archetype reflecting the importance of *socially appropriate behavior.* It is the many 'theatrical masks' that are used for first impression. Socially, persona puts our "best foot forward," intending to make definite *social impressions* on others. Persona develops from the archetypes of the collective unconscious refined by current social milieu and by current societal expectations.

Self-Realization

The ultimate goal of Jungian personality development is *self-realization,* which highlights the forward-going character of development due to development in the direction of stabilization and unity. Growth toward stabilization and unity act as a counter-balance to the biological archetypes of shadow.

Criticism of Jung's Esoterica

One of the major criticisms of Jungian theory (as well as Freudian) is that it is *more philosophical rather than empirical.* Esoteric qualities in Jung's theory, such as archetypes and the mysterious collective unconscious (and shadow), impede empirical verification and acceptance.

In conclusion, both Freud and Jung have serious flaws when exposed to modern day empiricism. They do not translate well in the illumination of today's zeitgeist—neuroimaging of the brain.

Neo-Freudian Alfred Adler

Individual Psychology

Adlerian psychology was a brisk departure from the biological determinism of Freud and esoteric terminology of Jung. Like everyone else, personality theorists have their own demons to tame and struggles often find expression in their theories. Adler was no exception.

Adler (1870–1937) suffered from *feelings of inferiority* most of his life. It comes as no surprise that inferiority feelings are at the center of Adlerian theory. In his view, dependency and social weakness due to the child's small stature during childhood produced inescapable feelings of inferiority. Throughout early life the child, and then the adolescent, is motivated to overcome inferiority feelings by *striving for superiority.*

Style of life is the Adlerian term for the characteristic ways individuals strive to overcome inferiority by becoming superior through application—higher education, sports, politics, business, and ultimately career. Overcoming inferiority is a central theme in life. Breaking free from self-doubt, anxiety, and insecurity is our principal developmental task, according to Adler.

Birth Order

Adler was one of the first psychiatrists to address the impact of birth order and childrearing practices, and the impact on personality. In terms of birth order, he believed:

- *First born* and *only born* tend to be authority-oriented and leaders
- *Middle children* are peer-oriented, rebellious, and sociable

- Babies of the family (*youngest siblings*) tend to be dependent, pampered, and insecure

Cognitive Theory

It was Adler who first set forth the main tenet of *cognitive theory* when he observed that *thinking shapes behavior,* particularly the way thinking defines success. Today, this perspective is known as cognitive-behavioral theory. To most cognitive psychologists, there seems to be an indisputable connection between inferiority complexes and thinking patterns.

In the Adlerian tradition, modern cognitive theorists believe that as long as individuals display self-defeating thinking, yet they remain active and think positively, they will eventually experience a measure of success. Once success is experienced, the feeling is contagious. Thinking about how good the experience feels as a success, leads to behavior that produces more success. In this perspective THINKING WAGS THE TAIL OF BEHAVIOR.

Interpersonal Theorists

Taking advantage of fresh, new insights from social sciences of sociology and anthropology and interpersonal relations, psychoanalysts Karen Horney (1885–1952), Harry Stack Sullivan (1892–1949), and Erik Erikson proposed that personality was forged chiefly by *social relationships in family milieu* and tangential interpersonal relations. The interpersonal theorists provided a strong platform for the development of known offender characteristics within family experiences.

Karen Horney

Dr. Karen Horney (Orr-nay) became dissatisfied with orthodox Freudian psychoanalysis prompting her move from native Hamburg, Germany to New York City in the late 1930s. She founded the *American Institute of Psychoanalysis* during this time and served as its Dean until her death. She believed Freud's approach depended too much on biology and instincts, crassly presented by id. Students of psychology are not surprised that her theories liberated women, in a sense, from Freud's sexist, male-dominated bigotry.

Horney believed that female psychology was based on:

- *a lack of confidence*
- an *overemphasis on the love relationship*

She theorized that basic anxiety corrupted personality integration. *Basic anxiety* comes from disruptions in early mother-child relationships such as overprotection, punishment, and rejection. These disruptions lead most individuals to the formation of a neurotic personality.

Basic Anxiety

According to Horney, an individual with a *neurotic personality* experiences feelings of insecurity and anxiety. In general, anything that disturbs the security of the child in relation to his parents produces some measure of basic anxiety. The child attempts to cope with these negative and hurtful feelings by three strategies that often become permanent fixtures of personality as young adults.

- *Moving away from people.* This strategy produces a *socially detached person* who is emotionally remote.
- *Moving against people.* This strategy produces a hostile and/or dominant personality instilling fear in others.
- *Moving toward people.* This strategy results in a persona of dependency and/or submissiveness.

Neurotic Conditions

Core neurotic needs that lead to the above three headings are listed in the following discussion. The essential difference between a normal and a neurotic personality is one of degree—the ever present behavioral continuum. Students must realize that everyone has similar issues and conflicts, but for some, due to rejection, neglect, or overprotection, exaggerated forms of neurosis (emotional instability) occur.

Horney was an astute observer of human behavior. When she arrived in New York City she began her analysis of American society and the relationships it spawned. In the mid 1940s she wrote a brilliant essay on disturbed human relationships. The essay remains a classic today. Later, she grouped them into the three headings previously discussed. How would a female European psychoanalyst view life and relationships in a big North American city? The list of ten neurotic (irrational needs) follow which produce a neurotic personality. The need for:

1. *Affection or approval* by being all things to all people so that extreme sensitivity to rejection or disapproval is noted
2. *Another person or partner* who will take over one's life, a parasitic need highlighted by abandonment issues often observed in borderline personality disorder
3. *Restricting one's life within narrow borders* resulting in obsession with modesty and predictability
4. *Power seekers* who glorify strength and contempt for weakness
5. Exploiting others
6. *Overindulgence seeking prestige*
7. *Overindulgence in seeking personal admiration*
8. *Overindulgence in seeking personal achievement*
9. *Focus on self-sufficiency* (and independence) where a lone wolf mentality dominates personality where interpersonal disappointment has led to cynicism
10. *Perfectionism* seeking unassailable character traits, the obsessive-compulsive need to cover-up flaws before others can observe them

According to Horney, all neurotic needs can be avoided if a child is raised in a home characterized by trust and love, security and respect, tolerance and warmth.

Harry Stack Sullivan

Harry Stack Sullivan, clinician, scholar, and theoretician created a fresh new viewpoint known as the *Interpersonal Theory of Psychiatry*. Even though he was trained as a psychiatrist within a strong biological perspective, Sullivan's theory reflects his preference for *social psychology*.

In his view, personality is "the relatively enduring pattern of recurrent interpersonal situations which characterize human life" (1953). To Sullivan, personality is essentially a *hypothetical construct* that exists only in relation to social context. It cannot be separated from interpersonal situations. Even a hermit who retreats into oblivion takes with him memories of former social interactions. His former experiences continue to affect his personality.

Sullivan theorized that self-esteem and personality are practically synonymous as personality develops from images we "internalize" through feedback gained from others. He believed that personality disorders come from *disturbed parent-child relationships*. In his view, *family milieu* is a critical determinant of personality.

Cognitive Influences

Cognitive processes in the formation of personality find unique expression in Sullivan's theory. Experiences are the raw materials for cognition, which translates into human awareness. Experiences occur in three cognitive modes:

- Prototaxic experiences
- Parataxic thinking
- Syntaxic logic

Prototaxic Experiences *Prototaxic experiences* occur as a necessary precursor to the other two in the first few months of life as:

- raw sensations
- images
- feelings that are largely disconnected from cognition

They are early perceptions (sensations) that will be cognitively organized as the child matures and acquires language and comprehension.

Parataxic Thinking Parataxic thinking occurs when we see an *apparent causal connection* between experiences but in reality the two experiences have nothing to do with one another. For example, if a dog is drinking water out of his bowl when suddenly his owner throws a bone into the yard, the dog may associate drinking water with being thrown a bone. Subsequently, the dog drinks a lot of water only to be interrupted by his glances for a bone being thrown over the fence.

Behaviorist B. F. Skinner validated parataxic thinking by experiments with laboratory pigeons, preferring the term, *superstitious behavior.* Interestingly, all superstitions appear to be nothing more than examples of parataxic thinking. In the same vein, education attempts to show the value of critical thinking where cognitive and strategic analysis are learned and applied, canceling out *illogical* parataxic thinking.

Syntaxic Logic In contrast, the most advanced mode of thinking is known as syntaxic logic where consensually validated symbols, words, and numbers are communicated and understood. *Syntaxic logic* depends on semantics—the meaning attached to words in context—and paves the way for logic, metaphor, empiricism, and *a posteriori* logic. In turn, this paves the way for the penumbras of semantics, the many shades of meaning and usage required in effective communication as the person matures. Increasing the power of syntaxic logic by way of Word Scholar exercises is a central strategy of this text.

Personifications

Personifications are mental images an individual has of himself or other people—emotionally laden 'pictures' we carry around in our heads. *Personifications* allow us to form scenarios related to feelings and attitudes that grow out of our experiences with others. For example, when other people produce satisfying relationships with us we perceive them favorably. In our experience they can do no wrong.

Apparently, shared personifications are often misleading or false as observed in *stereotypes,* such as the absentminded professor, 'air-headed' blonde, and the 'hard-driving' businessman.

Personifications act as social lubricants and aid in deciphering the perception of others. Experiences can modify or confirm another's perception of our personifications. It's interesting how physical characteristics speak to stereotypes but turn out to be completely false as the beautiful blonde who is also an academic scholar (which they often are), or the athlete who is also a national merit scholar (which they often are).

Erik Erikson

Erik Erikson is regarded as an early architect of *ego psychology* and the role of ego development in forging personality. To Erikson, ego identity is the driving force of all mental life; the resolution of *identity* is the key motivation in all interactions and behavior. In Erikson's theory, *ego identity* is expressed in a series of *psychosocial stages of personality development.* These stages are based on an analysis of major psychosocial challenges—crises—confronting individuals through our life span.

As the individual progresses successfully through each stage, the *positive resolution leads to greater integration of ego identity.* Erikson believes that failure to meet the demands of each stage can lead to *ego disintegration and eventually to mental disorders,* including personality disorders and psychopathy. The early adulthood crisis of Intimacy versus Isolation fits perfectly into the known offender characteristics of the FBI's study of serial homicide. A person who feels isolated from everyone is incapable of intimacy.

ERIKSON'S EIGHT CRISES OF DEVELOPMENT

Stage	Period	Crisis	Resolution
First year	Infancy	Trust vs. Mistrust	Trust
Second year	Early childhood	Autonomy vs. Shame and Doubt	Will Power
3 to 5 years	Play	Initiative vs. Guilt	Purpose
6th year to Puberty	School Age	Industry vs. Inferiority	Competency
12 to18 years	Adolescence	Identity vs. Role Confusion	Sense of Self
20s to 30s	Early Adulthood	Intimacy vs. Isolation	Love
40s to 50s	Middle Age	Generativity vs. Stagnation	Productive
60s and older	Old Age	Integrity vs. Despair	Wisdom

Humanistic Psychology

Philosophical perspectives of human potentials are more descriptive of humanistic theories of personality than identification of structures, developmental stages, or components of personality such as ego, or self. For example, most perspectives in humanistic psychology adhere to the following. Generally, individuals seek:

- Considerable freedom of choice
- To be responsible for actions
- Self-fulfillment a priority in social interactions
- To be essentially good-natured

Finally, psychology has a reason to be good-natured.

Abraham Maslow's Hierarchy of Needs

Abraham Maslow's (1908–1970) famous and often-quoted *hierarchy of needs* explains personality as satisfying needs *by priorities.* For example, by satisfying basic needs first such as food, water, sex, exercise, recreation, and feeling safe, we can progress to higher order *psychological needs* such as love, belonging, and self esteem. When these needs are met, we have the confidence and focus to reach the pinnacle of psychological and personality integration—self-actualization.

A child who comes to school hungry or tired finds it very difficult to pay attention or focus on learning anything. Not able to meet even the most basic of needs of food and rest often present special challenges to already overburdened teachers.

In Maslow's hierarchy, when caretakers do not provide the essentials of food, clothing, and sleep—basic needs of children—they go unsatisfied and personality integration starts to fragment. Until basic needs are met, the person faces emotional roadblocks preventing further progress. Personality and mental growth has hit a "glitch" and if uncorrected can lead to emotional dysfunction and personality disintegration.

Maslow's Hierarchy of Needs

Pinnacle Need
Self-actualization: Development of potential to the fullest extent
Psychological Needs
Esteem needs: Attention and recognition from others, feelings of achievement, competence, and mastery
Love and Belonging Needs
Affiliation with friends and companions, a supportive family, group identification, and intimate relationships
Basic Needs
Safety needs: Need for order, predictability, physical security, and freedom from fear
Physiological needs: food, water, rest, recuperation

To Maslow *personality development builds upon prior needs being satisfied in hierarchical fashion.* Only when basic ones are met can we progress beyond them to meet the challenges of the higher level psychologically-oriented growth needs. As indicated above in Maslow's hierarchy the *most basic needs* are listed on the bottom to the *most psychological needs* at the pinnacle similar to a pyramid. Clearly, due to predatory "toxic" parenting needs are not met.

Carl Rogers and Self-Theory

The renowned humanistic psychologist, Carl Rogers (1902–1987), theorized that self-concept is at the center of personality development. In Rogers' self-theory, personality is a part of our *total subjective experience* referred to as *phenomenological field* providing the raw materials of perception to construct *real self* and *ideal self.* Goals and aspirations comprise the ideal self. When the two selves merge into a similar reflection, Rogers believes we have reached a state of psychological health. According to Rogers, a fully functioning person has the following characteristics:

- open to new experience
- live in the present, real world
- display emotional sensitivity
- self-trust

Like so many of the theorists following Freud, Rogers felt that childhood and family experiences, (*milieu* experiences) are important in the development of personality.

Unconditional Positive Regard and Therapy

Rogers is responsible for the creation of one of the most positive and life affirming terms in all of psychology, *unconditional positive regard,* or 'no strings attached' love by parents. This attitude fosters positive feelings and nurtures a positive self-concept in children, according to Rogers.

Unconditional positive regard satisfies a person's need for belonging and love. Rogers developed a method of therapy that helped patients overcome glitches in self-concept development. Not surprisingly, this approach is based on a *client-centered* focus. Empathy and acceptance are advocated in this humanistic methodology. Not long ago in most graduate school curricula, the humanistic approach was a staple in the education of school counselors and therapists.

Trait Theory: Five-Factor Model

According to trait theory, personality traits can be measured by paper-and-pencil tests, or through insights gained by projective techniques. For example, projective instruments use inkblots or ambiguous drawings or pictures, such as the *Rorschach* (inkblots) and *Thematic Apperception Test* (TAT), a picture-story test, or House-Tree-Person drawing that suggests aspects of self. In recent years, several researchers have presented the so-called *Five-Factor Model* that identifies the five most commonly cited personality factors. They are:

- **Extraversion.** Outgoing, energetic, talkative versus **Introversion**
- **Agreeableness.** Pleasant, kind, trusting versus **Disagreeableness**
- **Conscientiousness.** Organized and prudent versus **Unconscientiousness**
- **Emotional Stability.** Calm, cool, and controlled versus **Emotional Instability**
- **Openness.** Creative, imaginative, cultured versus **Close-mindedness**

As one might expect, there exists considerable controversy over the exact content of the Five-Factor Model. However, it is a starting point as a general trait theory that integrates findings from a variety of studies. Trait theory assumes personality traits are stable, measurable units that can convey patterns of personality characteristics.

Behavioral Repertoire

Behaviorism, the science of observable behavior, defines personality in terms of habits learned through years of conditioning (learning). Personality is comprised of behavior that is rewarded by reinforcements so that it remains in the behavioral repertoire. Some behavior is encouraged not to remain in personality by simply being punished or ignored. *Environmental determinism* is the belief that environmental setting or context offers contingencies that have powerful controlling associations and consequences. In this perspective, it is not mental abilities such as traits or egos that influence personality development but rather the surrounding social context.

Cognitive-Behavioral Perspective and Personality

In a related theory, cognitive-behavioral (also known as social-cognitive learning) perspective merges *learning with cognition and behavior* so that personality is defined in terms of cognitive-behavioral variables. For example, simply observing another's behavior and copying it by *social modeling* is not contingent on association or consequences as observed in classical and operant conditioning. To model or copy behavior and/or personality traits observed in others is purely a cognitive process. This form of learning is merely *observational* where *intrapsychic* components account for personality acquisition.

Albert Bandura, a leading cognitive-behaviorist, believes that three factors define the opportunities that coexist for social modeling—the environmental setting, the behavior itself, and how thinking modifies the interactional process. For example, if we admire the way people express themselves due to some combination of personality traits such as choice of expressions or mannerisms, we may simply copy them to enhance our own personality. This process does not occur in a vacuum. We perceive the setting—where the behavior occurs and under what circumstances, then we copy or mimic the targeted behavior. Lastly, we use our cognitive skills of perception—how close we think we are pulling it off as well as gauging the feedback from others to complete the imitation process.

Personality According to Body Type

If we don't acquire personality strictly from *genetic inheritance* or strictly from *social learning*, or a combination, then where does it come from? Are there other possibilities?

- Could personality develop as a direct result of our physical constitution, our body type?
- Could personality traits develop from our body build?
- Could temperaments result from the same body type?

We now turn to a classic theory of personality development not as a result of instincts, archetypes, or various aspects of social learning, but through *specific body types*.

Constitutional Psychology

Are plus-size persons jolly? Are thin persons morose? Are redheads short-tempered? In important respects, can personality characteristics be directly related to a person's physical makeup—his or her body type? William H. Sheldon, M.D., Ph.D. (1899–1977) makes a convincing case. Sheldon's theory is known as *constitutional psychology;* Sheldon defines it as "the study of the psychological aspects of human behavior as they relate to morphology (body type) and physiology (organic and endocrine function) of the body."

Consequently, behavior, temperament, and personality are due to physical body structure and biological functioning, according to Sheldon. In a classic textbook, Hall and Lindzey's *Theories of Personality* (1970), the following bears mention.

Neuroscience, particularly neuropsychology, continues to show the importance of how the brain is "marked" by neurotransmitter and hormonal chemistry. Two textbooks by Jacobs (2009) including this one and *Brainmarks: How the Brain Marks Pathways in Life,* document the central role of the brain to all modalities of behavior. So, how does Sheldon differ?

Sheldon's Constitutional Psychology

Sheldon and his associates analyzed over four thousand pictures—front, side, and rear of individuals, well documented in the so-called, somatotype performance test as described in his classic book, *Atlas of Men* (1954). His analysis produced three basic body types:

- **Endomorphy, or endomorphic.** This body type is characterized by softness and is spherical in appearance. In the endomorphic body, digestive viscera are highly developed.
- **Mesomorphy, or mesomorphic.** This body type is hard and rectangular with a preponderance of bone and muscle. The mesomorphic body is strong, tough, and resilient to injury such as observed in athletes, soldiers, or outdoor adventurers.
- **Ectomorphy, ectomorphic.** This body type is linear and fragile with a pronounced delicacy of the body—any part of it might appear to break at the slightest touch. This body type is thin and lightly muscled.

Out of the three somatotypes, Sheldon studied thirty-three graduate students and instructors over a year's time to discern personality characteristics. He accomplished the original study through (1) clinical interviews, (2) questionnaires, and (3) observation in the volunteers' everyday professional lives. The results produced twenty-two personality traits showing high correlation to each of the three somatotypes. Out of this group, three components of temperaments were distilled. They are:

- **Viscerotonic.** Endomorphic individuals have a viscerotonic temperament. They love comfort and sociability. Temperamentally, they are *gluttons for food, people, and affection.* The viscerotonic is even-tempered and tolerant of others and observably easy-going. Food acts as a focus, motivator and social buffer.
- **Somatotonic.** Mesomorphic individuals have a somatotonic temperament. This person loves physical adventure and risk-taking with a strong need for vigorous physical activity. The somatotonic is *aggressive, callous toward the feelings of others; over mature in appearance, noisy, courageous, and dominant.* Their body is symmetrical and powerful which projects into an oft times intimidating personality. Vigorous activities, including sexual behavior acts as a focus and motivator.

- **Cerebrotonia.** Ectomorphic individuals have a cerebrotonic temperament. This person is *secretive, self-conscious, youthful in appearance, avoids people, and is happy to be a "home-body" preferring solitude by shunning attention.* Reflection and intellectual pursuits act as a focus and motivator.

Scale for Temperament

Among the original twenty descriptions given the three temperament types, the following nine characteristics are representative. To find your type, give yourself a point value of 1 to 7. A score of 7 represents the closest to what you perceive yourself to be while 1 is the least. Then add your score: the higher your score the more probable the temperament fits you (and your body type).

Continuum of Scores

54 to 63	Severe
27 to 53	Moderate
18 to 26	Mild
17 and below	Miniscule

Viscerotonia

_______ 1. Love of physical comfort
_______ 2. Love of eating
_______ 3. Strong need for affection and approval
_______ 4. Orientation to people (gregarious, extravert)
_______ 5. Tolerance of others
_______ 6. Complacent
_______ 7. Easy communication of feeling
_______ 8. Need of people when troubled
_______ 9. Orientation toward childhood and family
_______ SCORE

Somatotonia

_______ 1. Love of physical adventure
_______ 2. Characteristically, energetic
_______ 3. Loves to dominate, lust for power
_______ 4. Loves to risk, take chances
_______ 5. Bold and direct, courageous
_______ 6. Psychologically callous
_______ 7. Loves to compete, goal-oriented
_______ 8. Indifference to pain, claustrophobic
_______ 9. Need of action when troubled relationships
_______ SCORE

Cerebrotonia

_______ 1. Love of privacy
_______ 2. Hyper-attentiveness
_______ 3. Secretive, emotionally restrained
_______ 4. Inhibited socially

_______ 5. Unpredictable attitude, agoraphobic
_______ 6. Chronic fatigue
_______ 7. Need of solitude when troubled (introverted)
_______ 8. Youthful in appearance and manner
_______ 9. Orientation to later periods of life
_______ SCORE

Structure and Function of Personality

Sheldon's somatotype is the structure that gave expression to traits. For example, the following temperaments are functional with each corresponding body structure:

- Viscerotonic traits characterize the endomorphic body structure
- Somatotonic traits characterize the mesomorphic body structure
- Cerebrotonic traits characterize the ectomorphic body structure

Physique and Mental Disorder

At the extreme end of the affective (emotional) continuum, Sheldon observed the following *psychopathologies in temperament* (with milder forms existing toward the middle of the continuum):

Paranoid episodes of delusions tend to be observed in viscerotonia.
Schizophrenic episodes of extreme withdrawal are observed in somatotonia.
Bipolar (manic-depressive) episodes tend to be observed in cerebrotonia.

Positive Personality Dynamics

Self-Efficacy

Believing in ones abilities is always a desirable trait reflected in Albert Bandura's concept of *self-efficacy* expectancies. Exhibiting positive self-efficacy often produces a positive outcome, much like a placebo—"if you believe you can, you will do it!" Lack of self-efficacy can quickly become a habit. In this way, it becomes a vicious cycle of expecting failure, then actually failing, and anticipating more failure. Recent studies show that self-efficacy can be a factor in diagnoses of such psychological disorders as anxiety and depression.

Internal and External LOC

The social-learning theorist, Julian Rotter (1950) proposed a theory of behavior and personality in relation to one's expectations, or *locus of control (LOC)*. Where is the focus of one's expectations? Internal or external?

Internal LOCs (looking inward) are characterized by:

- exercising self-control
- known to be responsible
- hard work eventually leads to success
- trust plans and strategies

External LOCs believe that forces outside of themselves such as luck or fate control their destiny. These individuals often display traits of personality that are far different from the strategies of the internal LOCs.

Psychological Hardiness

Kobasa (1982) proposed that *psychological hardiness* may be a personality trait or characteristic that can help some individuals be more resilient to stress. Studies show that these individuals often live healthier, more productive lives. To Kobasa, psychologically resilient individuals display:

- more control over their lives (personal and/or career), so that direction, purpose, and outcome become intrinsically motivating
- feel more challenged by their goals
- are more committed to what they do

Eclectic Personality Theory

Personality theorists often combine the best and brightest of several competing theories in order to provide a more comprehensive perspective. In this way, an *eclectic perspective* can be tailor-made to more effectively deal with a client's presenting symptoms and psychosocial history when used by clinicians. What factors—brain, biology, parenting, or peer relations—can be hypothesized to account for the violent sexually psychopathic serial crimes of Joel Rifkin?

PREDATOR PROFILE 5-2

Joel David Rifkin

Ashleigh Portales

Time Span of Crimes
March 1989 to June 28, 1993
Offenses Prior to Serial Murder
Soliciting sex from an undercover policewoman (August 1987)

Quoting Rifkin
"I will not go away as a monster, but as a tragedy."

Preferred Prey

Rifkin killed drug-addicted prostitutes who were in their twenties and thirties, not because they were prostitutes, but more likely because they were the only kind of women he could attract and ultimately control.

Society's Outcast

Rifkin was born January 20, 1959 into the Baby Boomer generation in much the same way he would spend most of his life: unloved and unwanted. His mother, who bore him as the *illegitimate child* of a college romance, *gave him up for adoption*. By Valentine's Day, the infant had gone to Bernard and Jeanne Rifkin, a childless couple living in upstate New York who christened him Joel David Rifkin. The Rifkins adopted a daughter three years later, and the *seemingly* happy family moved to East Meadow, Long Island.

Rifkin began the first grade in 1965, the birth date of GEN Xers. From the very beginning, Joel found school a nightmare. He was an immediate *outsider* and a *favorite target for bullies* who called him "The Turtle." The name came from his slow walk and slouch, which accentuated his neck and head so that it resembled a turtle's head thrust from beneath its shell. He suffered *constant physical assaults* and *had his head pounded into concrete* on more than one occasion. At other times, bullies forcefully stripped him of his clothing in the hallway and left him partially nude. At still other times, they stole his books and lunch.

Joel's *failure to socialize with peers* carried over into the classroom where he flunked his courses due to *undiagnosed learning disorder (dyslexia)*, despite a tested IQ of 120. To add insult to injury, Joel's father was a member of the East Meadow School Board and saw his son as little more than a major *disappointment*. He often ranted at the boy, "Why can't you do anything to please me?" Though Joel's mother shared her love of gardening and photography with him, she was the model of *incompetent parenting* (emotional apathy) and completely oblivious to the hell her son suffered in school. Later she stated that she just "thought of him as a loner."

By the time Joel reached high school, things only had gotten worse. One of Rifkin's many tormentors recalled him as "an abuse unit . . . who was subtly obnoxious . . . his presence annoyed you." In a desperate attempt to fit in, Joel joined the track team where his only reward was the moniker "lard ass." Teammates routinely hid his clothes and shoved his head into the toilet. Determined to win over a few friends, Rifkin invited them to his home to drink and watch TV, but as one classmate later stated, "No one wanted to associate with him . . . we used him, to be blunt about it . . . he was easy to make fun of." Abandoning the track team, Joel turned his talent for photography into a position on the yearbook staff where his camera subsequently got stolen. Still he slaved to produce the yearbook, but, despite his efforts, the staff excluded him from the year-end wrap party.

Rifkin failed even more with romance. Rifkin's track teammates destroyed one date by holding Joel prisoner in the gym and pelting him with eggs. On another occasion, Rifkin and his date arrived at a local pizza parlor, but school bullies chased them out and pursued them until the couple reached safety inside a public library.

Rifkin's parents gave him a car in his senior year. While it did nothing to boost his popularity, it did allow young Joel to cruise the streets at night, picking up prostitutes in the nearby town of Hempstead and later in Manhattan. According to Robert Madinich, author of *From the Mouth of the Monster,* "Joel's fantasies included 'some bondage' and 'some rape' plus 'a gladiator type thing with two girls that would fight to the death.' He raped and stabbed women in some daydreams, but his fantasy victims were silent, 'just passive about it.'" He fixated on strangling prostitutes after a 1972 viewing of Alfred Hitchcock's *Frenzy,* loosely based on London's "Jack the Ripper" homicides of 1888.

Rifkin graduated at the bottom of his class in 1977 and looked forward to the prospect of life after high school, which he assured himself could only get better. Unfortunately, he was wrong. He proved an *academic failure* again at Nassau Community College on Long Island, where he completed only one course his first year due to constantly skipping class. Rifkin then transferred to the state university at Brockport, where similar behavior as well as *rejection by his first real girlfriend* resulted in his *dropping out.* From there, Joel drifted back home to his parents and subsequently failed at a second attempt at Nassau Community College. He again attended only sporadically and earned only twelve total credits before dropping out of school for good in 1984.

Rifkin's life continued its downward spiral from there as he bounced from job to job. According to Court TV's "Crime Library," "poor hygiene, chronic absenteeism, and general ineptitude" blocked the road to advancement. His employer at a local music store described Joel as "a total piece of work—this guy couldn't even count to ten."' Rifkin aspired to be a famous writer, but his bleak verse did not suit publishers. While he actively pursued his interests in photography and horticulture, neither produced a paycheck, and Rifkin soon found himself back at home with his parents. In Joel's own words, "I couldn't put two nickels together," and most of what he did earn went to prostitutes. "The whole focus of my life was on the streets." Even on the streets among the lowest of the low, he proved unfit. On at least a dozen occasions, hookers or pimps robbed him; he fell for the same trick twice from the same girl who made off with his money before having sex with him.

Successive failures spelled *depression* for Rifkin. The depression deepened, as did his *commitment to sexual deviance,* when his father committed suicide in 1987 to escape the pain of prostate cancer. In August that same year, Rifkin got arrested in Hempstead after soliciting sex from an undercover policewoman. He escaped with only a fine that, though substantial, suggested to his *damaged brain* that he could act against society with minimal repercussions.

For a brief period in his life in 1988, Rifkin experienced success in his horticultural pursuits. After graduating with straight A's from the State College of Technology in Farmingdale, New York, he received an internship at the renowned Planting Fields Arboretum in Oyster Bay, New York. His *inadequacy with women* once again proved his downfall though. Infatuated with a co-worker who did not reciprocate his feelings, Rifkin created an *elaborate fantasy affair.* Her blatant rejection of his numerous advances finally pushed him over the edge. Payback time had come.

The Outcast Fights Back

Rifkin took matters into his own hands for the first time in March 1989 at the age of thirty. The first murder coincided with the second anniversary of his father's death and Rifkin later noted that most of his killings connected in some way to a number of personal anniversaries.

During this particular time, his mother was traveling out of the state, and Joel was home alone. While cruising Manhattan for prostitutes, he picked up a woman he remembered only as "Susie." She asked Rifkin to take her out to buy drugs after having sex because she was a hard-core crack addict. However, Rifkin began beating her instead, later recalling, "I just lost control. I stopped when I got tired." But Rifkin had "gotten tired" before his victim died. When he attempted to move Susie, she retaliated by severely biting into one of his fingers. Angered, Rifkin began strangling her. This would become the killer's *modus operandi.* In Rifkin's words, he would "just grab and hold on, basically till my hands got tired." When the girl finally died, Rifkin put her body in a trash bag and cleaned up the mess he had made of his mother's living room before falling asleep. When he awoke, he took the body in the bag down to the basement where the washer and dryer became an autopsy slab. He dismembered the girl with an X-acto knife, an event that Rifkin recalled was "reduced to biology class."

Fearing the body still could be identified he then removed the girl's fingertips and pulled out her teeth with pliers. He placed her severed head in an old paint can and the other parts in various garbage bags before loading them into the back of his mother's truck. Driving across the state line to New Jersey, Rifkin deposited the head and legs in the woods before returning to Manhattan, where he threw the arms and torso into the East River. He believed that his actions would erase the possibility of detection, and maybe even the crime altogether, but a golfer playing a course that backed up to the New Jersey woods proved him wrong by stumbling upon the paint can containing Susie's head.

Rifkin *closely followed media coverage* of the discovery and suffered an anxiety attack when he learned the girl had tested HIV positive. He felt pleased though when police efforts produced no positive identity. "Susie's" identity and her murder remained unsolved until Rifkin's 1993 confession.

More than a year passed before Rifkin's mother left town again, leaving him free to kill again. He chose prostitute Julie Blackbird because of what he called her "pseudo-Madonna look," and took her back to his mother's house where they spent the night. Upon awakening the next morning, he beat Blackbird with a table leg and then strangled her. This time, however, he determined that the body remained buried, so he took a trip to the store and returned with cement and a large mortar pan before once again dismembering his victim.

He modified his MO by weighting several buckets with the concrete and placing Blackbird's head, arms, legs, and torso in separate ones. He then loaded the buckets into his car and drove to Manhattan where he threw those with the head and torso into the East River and dumped the arms and legs into a Brooklyn barge canal. Rifkin's attempt at concealing his crime worked; no one ever found Blackbird's body. In fact, authorities only discovered her murder upon Rifkin's confession and through details written in a diary found in his room.

In April 1991, Rifkin went into the landscaping business for himself and rented a space at a nearby nursery to store his equipment. Though the business proved a failure, the storage area proved the perfect place to keep bodies before permanently disposing of them. However, he would not store his next victim there. He wrapped thirty-one year old Barbara Jacobs in plastic, folded her into a cardboard box, and dumped her into the Hudson River. Rifkin did not dismember her body because he was "put off" by the thought. Though someone found her body, she went unidentified until Rifkin's confession two years after her death.

A few months after Jacobs, Rifkin picked up twenty-two year old Mary Ellen DeLuca and drove her to a seedy motel. Rifkin says that she rushed through their sexual encounter, complaining all the while. He contends that he asked her if she wanted to die and complied with her affirmative response by strangling her. He alleges she "did nothing, just accepted it." Her murder was "one of the weird ones." Because of the public location of her murder, Rifkin feared simply dragging her body across the parking lot to the car. Reenacting a scene from Alfred Hitchcock's *Frenzy,* he bought a cheap steamer trunk, transported his latest victim to a rest stop in Orange County, and dumped her body there. When discovered on October 1st, the body's advanced state of decomposition foiled attempts at identification. Authorities buried DeLuca in an unmarked grave, like Jacobs before her, until they learned her identify from Rifkin's confession.

Rifkin used the same trunk for his next victim, thirty one year old Yun Lee. He strangled her after she made fun of his inability to attain an erection after he picked her up. Her ex-husband's identification saved her from an anonymous burial.

"Number six" remained unidentified, as Rifkin could not recall her name. After strangling her during oral sex, he stuffed her body in a fifty-five gallon oil drum and rolled the drum into the East River. Pleased with the success it gave him, Rifkin purchased three more oil drums to serve as the coffins of victims seven through nine. Number seven was twenty-eight year old manic depressive Lorraine Orvieto. Rifkin killed her in December, 1991 and dumped her into Coney Island Creek, where no one discovered her until July, 1992.

One week after the Orvieto murder, on January 2, 1992, thirty-one year old Mary Ann Holloman met her fate at Rifkin's hands. He dumped her inside her designated oil drum in Coney Island Creek, where someone discovered her two days before Orvieto. However, Holloman's dental records identified her, and authorities returned her to her family for burial.

The last of Rifkin's oil drums went to a prostitute he remembered only for her tattoos. Rifkin dismembered her body before disposing of her in Brooklyn's Newton Creek. Her discovery would not take place until May, 1992, and she would remain unidentified.

Rifkin picked up victim number ten, twenty-five year old Iris Sanchez, on Mother's Day of 1992. After strangling her during sex, he hid her body under a rotting mattress close to a runway at JFK airport. Police discovered her remains in June, 1993 using a map Rifkin had drawn during his confession.

Soon after Sanchez's murder, on Memorial Day weekend, Rifkin killed thirty-three year old Anna Lopez and dumped her body along I-84 where a motorist discovered her the next day.

Rifkin had not brought one of his girls home in almost a year. After strangling twenty-one year old Violet O'Neil, he mutilated her corpse in his mother's bathtub, wrapped the various pieces in plastic, and deposited them in several Manhattan waterways. Her arms and legs appeared in an abandoned suitcase. Thirty-one year old Mary Catherine Williams was a former homecoming queen, college cheerleader, and ex-wife of a pro football player. After failing as an actress in New York, she had turned to drugs and prostitution, which led her to Joel Rifkin on October 2, 1992. He dumped her body in Yorktown where its discovery occurred two months later. Authorities buried her as a Jane Doe. Rifkin's confession also identified her.

"The toughest one to kill," twenty-three year old Jenny Soto, broke all ten fingernails in her fight to live, a fight she lost to Rifkin on November 16, 1992. The following day her body appeared in the Harlem River, which authorities identified by her fingerprints taken on a recent arrest.

The first victim of 1993 was twenty-eight year old Leah Evans. Rifkin strangled her after she started to cry when he refused to give her privacy while undressing. He drove to a wooded area in Long Island and buried her in a shallow grave where hikers found her unearthed hand on May 9th.

Rifkin claimed two more victims before capture, one in April and one in June. He strangled twenty-eight year old Lauren Marquez before sex, yet her body, discovered after Rifkin's arrest, showed signs of battery. Rifkin could not remember beating her.

Tiffany Bresciani was Rifkin's last. He picked her up in the early morning hours of June 24th, his fourth hooker in half as many days. After strangling her, he stowed her body in the backseat of his mother's truck and drove home, stopping to purchase rope and tarp. He wrapped her in it and moved her to the truck bed. Immediately upon his arrival home, his mother took her keys from her son and went shopping, never knowing there was a dead prostitute in the back. After his mother retired for the evening, Rifkin relocated Bresciani to a wheelbarrow in the garage where she remained for three days. He then loaded her back into the truck and drove fifteen miles to Melville Public Airport, where he intended to dump the body. This time, however, luck was not on the side of Joel Rifkin.

The Outcast Cast Away

At 3:15 A.M. on June 28, 1993, two New York State Troopers noticed a Mazda pickup truck with no rear license plate driving along Long Island's Southern State Parkway. The troopers activated their lights in an attempt to make a routine traffic stop. When the driver did not pull over, they ordered him to halt over the loudspeaker. Instead, the truck sped away down the next off-ramp, and the chase was on. The troopers called for backup, and soon five patrol cars joined the ninety mile-per-hour pursuit, which ended when the Mazda's driver missed a turn and crashed into a telephone poll at 3:36 A.M. The driver did not resist when police removed him from the vehicle. They found an X-acto knife in his pocket upon frisking him. The man's driver's license identified him as thirty-four year old Joel David Rifkin of East Meadow, Long Island. The thick layer of Noxzema smeared across the man's upper lip puzzled officers, as well as the reason he fled to avoid the minor citation for having no rear license plate. They got their answers when, searching for the source of a rather pungent odor, they pulled back a tarp covering the truck bed and discovered the decomposing body of a naked woman. Confronted with the discovery, Rifkin replied, "She was a prostitute. I picked her up on Allen Street in Manhattan. I had sex with her, then things went bad and I strangled her. Do you think I need a lawyer?"

Back at headquarters, Rifkin embarked upon an eight-hour confession, identifying Bresciani as "number seventeen." Though police never recorded any part of the confession, they took a written transcript as well as various sketches of body locations and victim lists Rifkin made. Throughout the entire proceeding, Rifkin acted emotionally detached and smiled when describing the most grisly of details, displaying characteristic *blunt and inappropriate affects.* He responded to questions about how he felt during the murders by indicating he *felt no hesitation or remorse.* "I'm not sure I felt anything, it's just something that happened." He stated that, in the moment of murder, he was "behaving mechanically, autopilot." When he omitted Williams' name from one of his lists, it confused police, who falsely elevated the death toll to eighteen. Rifkin found this humorous, saying that "the clumsy cops had counted Williams twice."

Police obtained a search warrant for Rifkin's home and served it around 8:00 P.M. that evening. Jeanne Rifkin had no idea what was going on until she saw news reports of her son's arrest on TV. The search of her residence, specifically Joel's room, yielded at least two hundred and twenty-eight "trophies" he had taken from his victims, including clothing, jewelry, IDs, and makeup. Investigators also confiscated literature on the Green River Killer and Arthur Shawcross.

Rifkin went on trial for the murder of Tiffany Bresciani on April 20, 1994. He pled not guilty by reason of insanity and, judging by the way he slept and snored through the prosecution's arguments, he expected to get off. However, prosecution psychiatrist Dr. Park Dietz, who had testified against Arthur Shawcross, Jeffrey Dahmer, and John Hinckley, found that Rifkin was "sick but not insane. He knew exactly what he was doing, and he did it." The jury agreed and convicted Rifkin of murder and reckless endangerment (for the car chase), sentencing him to twenty-five years to life.

In subsequent trials for the murders of Evans, Marquez, Sanchez, Orvieto, Halloman, and "Jane Doe," Rifkin received one hundred and eighty-three years. The New York Supreme Court rejected Rifkin's appeal in 2002, and he currently is serving two hundred and three years to life in the Clinton Correctional Facility, isolated high in the Adirondack Mountains. He will become eligible for parole in 2197.

Aftermath

Rifkin has spent most of prison life in solitary confinement, labeled "involuntary protective custody," where he is confined to his cell twenty-three hours of every day. Here he has amused himself with civil suits filed by the victims' families. When the Orvieto family sued him for wrongful death, he responded with a handwritten note labeling his victim an AIDS carrier who "may be responsible for the eventual deaths of numerous individuals" and suggesting her surviving relatives shared responsibility for "what might have been."

Rifkin sells his paintings from prison to compensate his victims' families. While most are scenes of flowers or wildlife, one work, entitled *Guardian's Failure,* depicts an angel weeping over a bare foot clad with a coroner's toe tag.

Perhaps the greatest irony occurred in August, 1999 when Rifkin revealed to the public his plans for "Oholah House," a shelter for prostitutes seeking rehab, counseling, medical care, and job training. Rifkin claimed the project was so named because "Oholah" was both the Hebrew word for "sanctuary" and the name of an Old Testament prostitute who suffered a gruesome end. (The spelling of the latter name actually is "Aholah.") In Rifkin's own words, this gesture was "a way of paying back a debt, I guess." This attempt at deception did not fool prosecutor Fred Klein. He had the plan promptly dismissed on the grounds that Rifkin had included in the plans a "Motivation Room" where residents of the house would be "scared straight" by viewing photographs of prostitutes murdered while tricking. His reasoning for such callous treatment: "These girls think, 'I can't be touched.' Well, seventeen girls thought that, and now they're dead."

Introduction to Personality Disorders

The DSM—Diagnostic and Statistical Manual of Mental Disorders

Personality disorders (PD) are classified by the DSM (Diagnostic and Statistical Manual of Mental Disorders, Fourth Edition, Text Revision, 2002), as traits that are:

- inflexible and maladaptive
- display significant functional impairment and/or
- characterized by subjective distress

Individuals diagnosed with PDs do not function well in relationships and often experience difficulty remaining employed, and may experience chronic anxiety and/or depression as manifestations of *subjective distress.* Usually, individuals with *severe personality disorders* do not recognize the source of dysfunction coming from themselves; they do not judge their own personality traits as unacceptable or part of the problem in an instance of what Freud labeled an *ego-syntonic* personality.

Ego-syntonic goes to the very heart of the seriousness of personality disorders characterizing the afflicted individual as *blind to their own deficits;* it's other people who are jealous, or have hidden agendas—they are the trouble makers. Ego-syntonic traits set the stage for the disordered person to use denial, rationalization, and other defense mechanisms to justify being victimized.

Showing the deep-seated nature of PDs, Freud preferred the term *character disorders* (instead of personality disorders). In fact, character disorders may be more descriptive of how pervasive they are in *personality disintegration* in severe cases.

Spectrum of Severity

Personality disorders are recognized by adolescence or earlier and can be characteristic of adult behavior most of a person's life. They are indexed in the DSM (2002) across a *continuum of severity* mild, moderate, or severe.

In practice, most individuals with *mild* PDs may be barely diagnosable (unless they are psych majors!); *moderate* disorders are more pronounced and typical of the specific diagnosis, but usually do not require hospitalization. In *severe* cases, some type of psychiatric or psychological intervention is usually required such as psychiatric hospitalization or medication protocols.

When no psychiatric intervention is available (usually due to lack of insurance or a responsible party initiating inpatient care), legal remedies often are employed. This accounts for the individuals from all walks of life that 'fall through the cracks' of MHMR (the Texas state mental health agency: Mental Health Mental Retardation) who are arrested, detained in an acute psych unit, or imprisoned.

Personality Disorder Etiology

Although theories exist with biological and/or social learning explanations for PDs, there are no clear cut answers regarding etiology (causation). A widely used allegorical anecdote may be helpful in visualizing the long-standing, ongoing nature of personality disorders which explains why they are so resilient to treatment. Remember, no one wakes up one morning and decides to have a full-blown personality disorder. Those who have personality disorders have been *becoming* disordered for a long time, perhaps since childhood.

The Scorpion and the Frog Allegory

Once upon a time there was a scorpion named 'Red' who wanted to cross a deep pond. A life of taking by force or intimidation whatever Red desired guaranteed he would find a way assured by his imposing looks and a controlling personality. The menacing stinger curled high atop his segmented tail, let the world know that this was one character not to irritate or cross.

Searching for a ride near water's edge, Red happened upon a playful, carefree frog out for a morning swim. "Hey, Bub," said the scorpion (not knowing this was one of the famous frogs from a once popular beer commercial) in his raspy, mutant-like voice, "Give me a ride?" The frog agreed the moment he laid eyes on the sudden intruder. It is common knowledge that frogs love to accommodate others as co-dependents. They are well-known enablers.

Red crawled upon the back of the frog and off they went. Halfway across the pond the scorpion, without warning buried the stinger deep in the tender back of the frog. Immediately the frog felt the poison's effects and froze without as much as croak. In desperation, the frog let out a familiar refrain. "I was just trying to help . . . your senseless indifference has killed us both!"

"Foolish frog," Red retorted, "you must have noticed my ally Jake the snake slithering just beneath us. He owns me."

"Why are you so heartless?" gasped the frog in a last spasm of death.

"Simple," said the scorpion, as he slipped safely upon the back of the large water moccasin. "*It's my nature* to sting."

"It's our nature" is often the way some theorists refer to personality characteristics. "That's just the way we are," or "don't mind him, it's just his nature to be rude!" are common expressions to describe the powerful aspects of personality characteristics coming from our biological inheritance.

DSM Personality Disorders

Cluster 'A' Personality Disorders: Odd and Eccentric

Paranoid Personality Disorder

Paranoid personality disorder is characterized by *hyper-suspiciousness*. This individual is overly suspicious of everyone and has a difficult time trusting anyone. Obviously, the paranoid individual will have great difficulty sustaining a normal relationship with others in both social relationships and in the workplace.

Sadly, in today's *zeitgeist* it is probably normal for most adults to be mildly paranoid considering the fair amount of potential violence around us. Remember, we are describing the paranoid personality (moderate to severe type) who is obsessively and compulsively paranoid to the point that it significantly interferes with social interactions with subjective distress.

Schizoid Personality Disorder

The schizoid PD is characterized by:

- extreme shyness
- social withdrawal, and
- emotional inhibition leading to detachment from others

Social paralysis is a good way to describe the schizoid personality who finds it emotionally difficult to interact with people. It is not hard to visualize a milieu where children are basically ignored, neglected, or mistreated so that avoiding others is encouraged. From the learning theory perspective, it is not a far stretch to imagine schizoid characteristics due to early learned behavior and, of course, exacerbated by genes.

Schizotypal Personality Disorder

Quoting from the DSM (2002) "the essential feature of *schizotypal PD* is a pervasive pattern of social and interpersonal deficits marked by:

- acute discomfort with, and reduced capacity for, close relationships
- cognitive or perceptual distortions and
- eccentricities of behavior

This PD begins by early adulthood and is present in a variety of social contexts. They display *ideas of reference*—incorrect interpretations of casual incidents and external events as having a particular and unusual meaning. They feel they can see the future and read others thoughts. Their social interactions may appear strained or inappropriate; their mannerisms, looks, or manner of dress stand out as typical characteristics. A distinguishing characteristic is the inability to engage in give-and-take banter of everyday communication

Albert Einstein is often referred to as having features of mild to moderate schizotypal behavior. Often he would come to his university class barefooted in freezing weather. Can you imagine sitting down with Al and having a normal conversation?

JUST YOUTHFUL REBELLION. Bizarre dress, odd, eccentric behavior, body piercing, tattoos, and hairstyles might just mean youthful rebellion, but when combined with the more enduring *traits of personality* might add up to a severe personality disorder. Schizotypal *features* may be manifested as youthful rebellion, while the more severe personality disorder is deeply ingrained, perhaps from childhood, and is a salient personality trait, a relatively permanent fixture of behavior. Do some elements in our pop culture actually encourage mild to moderate schizotypal behavior? If so, give examples of such elements.

Cluster 'B' Personality Disorders: Dramatic and Emotional

Antisocial Personality Disorder

The cluster includes:

- antisocial
- borderline
- histrionic
- narcissistic PDs

Grandiosity and potential for *violence,* is observed in antisocial PD, typified by patterns of irresponsible and antisocial behavior beginning in childhood or early adolescence and continuing into adulthood.

In this severe and potentially homicidal disorder, studies show that individuals, thus afflicted, display early red flags of being far out of bounds relative to normal development. What are the early warning signs? A picture emerges of the *lonely outcast* who is:

- cruel and abusive to animals
- who sets fires in the neighborhood or at school
- constantly fighting, and being expelled (or truant) from school, and/or
- obsessed with pornographic literature

Taking drugs or drinking alcohol exacerbates antisocial behavior. Unfortunately, *dysfunctional family patterns often keep secret these early warning signs.* Also, the budding antisocial has a secretive nature that prevents detection.

According to ex-FBI special agents Robert Ressler (Ressler & Shachtman 1992), John Douglas (1994), and others, antisocials who become serial killers do not possess normal relationships with their mothers or later, normal relationships with females in general. They do not have consensual, nurturing relationships. Also, most serial killers have *multiple addictions* such as hard-core pornography, alcohol and/or other drugs. Recent examples include Jeffrey Dahmer and Ted Bundy.

The serial killer and sexual sadist, Ted Bundy, recounted his early obsession with crime magazines leading to an obsession with violent, hard-core pornography. Additionally, his antisocial behavior was exacerbated by an addiction to alcohol. In all respects, Bundy developed by degrees and stages into a full-blown *serial killer* (*a sexual psychopath,* not antisocial PD even though his behavior was clearly anti-social). Differences—some subtle and others pronouced—between antisocials and psychopaths follow on the next page.

According to researchers trails of red flags litter early home experiences of sexual psychopaths. For example, as a young boy walking home from school, he often picked up 'road kill' along the highway. Later in his basement, Dahmer performed autopsies on the animals and recalled *vivid erotic fantasies* of how a human would look on the inside and what kind of sexual experiments would work to create a 'sex zombie.'

In all psychosocial histories of antisocials (and psychopaths), evidence abounds to suggest that they become deviants *by slow degrees over time.* Absent brain injury or severe chemical imbalance, no one in middle age, for example, suddenly wakes up one morning and decides to become a serial killer or sexual predator.

RED FLAGS. Collectively, the following behaviors characterize the development of antisocial PD:

- compulsive lying and stealing
- vandalism, such as setting fires destroying property
- elopement (frequent running away from home or school)
- physical cruelty to other children and pets are typical childhood signs

Often, one or more deviant behaviors become pronounced in adolescence. In adulthood, the pattern continues with typical behavior such as compulsive failure to honor obligations, compulsive failure to act in responsible ways, chronic unemployment, arrest records, child and spousal abusers, and sexual promiscuity; any combination of the above, which leads to imprisonment or premature death, are salient features.

Once an individual with an antisocial personality disorder crosses the line and becomes violent to others, studies show there is no treatment. Ex-FBI psychological profilers—John Douglas and Robert Ressler—who have collected extensive data on serial killers, agree.

Antisocial personality disorders are reflected as a gender difference where males are primarily the antisocials (as well as psychopaths).

Psychopathy (Psychopathic Personality Disorder)

From high resolution brain scanning to highly valid and reliable psychometric evaluation instruments, such as the *Hare Psychopathy Checklist (HPC),* modern *experimental neuroscience* has shown time and again that both garden variety non-violent psychopaths as well as violent psychopaths are *qualitatively different* from petty criminals and violent antisocials diagnosed with antisocial personality disorder. The clinical terms *sociopathy* and the ICD-10 *dyssocial personality* are qualitatively the same disorders with strong sociological influences, yet remain relatively useless terms in light of neuroscience.

Ushered in by the new discipline of *sociology,* it is no wonder that Partridge (1930) objected to the more medical model-oriented, neuropsych term psychopath. Obviously, he desired to help sociology gain a foothold as a respected science from medicine and psychology.

Psychopaths have to be relatively good looking, glib, and charming to get away with the emotional chaos they bring upon others who cross their path, while most garden-variety criminals (even the violent ones) are distinguished by average looks (or below) and possess few of the charismatic and animal cunning of psychopaths.

Borderline Personality Disorder

A diagnosis of *borderline personality disorder* is characteristic of severe *distortions of self-image, mood, and interpersonal relations.* It is a pervasive *identity disorder* where sexual orientation, values, and one's future career are confused.

Borderline individuals have an extremely *difficult time tolerating loneliness.* They experience marked mood shifts and overall, they are impulsive. A borderline's impulsivity often leads to eating binges, shopping sprees, casual sex, and oppositional defiance. The salient feature of borderline personality disorder is the presence of *severe abandonment issues*. They can become increasingly demanding when abandonment is suspected (recall the Glenn Close character in the movie *Fatal Attraction*). Feeling empty is a common complaint of borderlines. Prostitutes, shoplifters, and false accusers of sexual harassment fit the profile of borderline PD. Gender-wise it's more prominent in females.

Histrionic PD & Narcissistic PD

Histrionic personality disorder and narcissistic personality disorder are related. In terms of gender, a female who is attractive and seductive, attention-seeking, charming without being genuine, is diagnosed with some degree (mild, moderate, or severe) of *histrionic PD*. For example, a histrionic female sees herself as the *perpetual victim* drawn unwillingly into one bad relationship after another. She is often displeased with others and remains superficially interested in most relationships (including friendships) for purposes of *self-advantage*.

She feels empowerment only when she is the center of attention. The context of her victimization always involves the losers she settles for in relationships. Histrionicism can be observed as early as grade school.

Similarly, the narcissistic male—the masculine equivalent of the female histrionic—fantasizes about success, power, ideal love, and sexual conquests. He is often a compulsive procrastinator, unless he perceives getting the job done right away benefits his interests. Friendships are based on self-advantage; promiscuous sex bolsters self-esteem. Often, to narcissistic males, females are perceived as conquests instead of partners or companions and "rated" on a scale from zero to ten, a very adolescent thing to do.

Also, n*arcissists* are very calculating, glib, and manipulating; they are obsessive-compulsive about staying youthful as "sex magnets." They remain hypersensitive to being evaluated by others and have pronounced trouble relating to authority figures.

Narcissists who have become *grandiose* in self-absorption may turn to violence as they feel entitled to do whatever it takes to get what they feel they deserve.

Difficulties in the workplace with both narcissistic and histrionic personalities abound. So much energy and attention is required to satisfy their egos that workplace duties and focus are often lost. They require attention and approval at every turn.

Cluster 'C' Personality Disorders: Anxious and Fearful

Avoidant and Dependent Personality Disorders

The *avoidant personality disorder* is characterized by extreme withdrawal from others by living on the fringes of society. Avoidant is the precise term to characterize this personality disorder as they avoid as much social interaction as possible, being extremely uncomfortable around others. The avoidant is most often observed in males. If these individuals are not self-employed or work on the graveyard shift they are usually homeless or live with parents.

In contrast, the individuals with *dependent personality disorder* are just the opposite of the avoidant as they are clinging-vines to others. Individuals with dependent personality disorder perceives others to be empowered, not themselves. Gender-wise, females tend to be the dependents and co-dependents.

Obsessive-Compulsive Personality Disorder

Individuals who are highly moralistic and judgmental characterize obsessive-compulsive personality disorder, not to be confused with obsessive-compulsive disorder (OCD), an anxiety disorder.

Nothing is ever good enough for the individual with OCPD—obsessive-compulsive personality disorder; they are stingy with emotions and possessions and display a strong need to dominate and control with a high incidence of alcoholism and drug abuse characteristic of this disorder. Ostensibly, drugs are used as self-medication to cope with the reality of imperfection and being challenged by others. Those with the severe type are very difficult to live with due to the chaos and constant micromanaging. The word empathy is not in the vocabulary of the obsessive-compulsive person.

Passive-Aggressive Personality Disorder

Although no longer listed in the DSM (2002) as a Cluster C personality disorder, passive-aggressive *characteristics and features* in personality are worthy of mention. These individuals are:

- stubbornness typified by covert expression of anger through simply forgetting
- passively resist the requests of others
- perpetually late or absent when presence is requested

Additionally, the passive-aggressive personality is often critical of others in authority. This annoying and disruptive personality disorder is diagnosed equally in males and females.

Interactional Aspects

In the final analysis, can we identify any one predictable cause of personality disorders? No. Can we identify any one cause for personality development outcome? No. Due to interactional factors from biology, genetics, social and milieu influences, many factors must be analyzed. By observing psychosocial histories, *dysfunctional family milieus* account for many of the maladaptive behaviors characterized by DSM (2002) personality clusters. Most experts agree that early learning in families is almost always significant. On the other hand, advances in biochemistry and brain imaging have led neuroscientists to speculate on the increased role of biology in the development of personality disorders. However, we know that chemical imbalances can be caused by dysfunctional learning as well as innate biochemistry.

Both biology and learning proponents agree that the existence of any one of the disorders is a developmental scenario. In the interactional perspective, it is impossible to separate biology from social learning and the responsible thing to do is include both in the clinical picture.

Name ______________________________ Date ______________

Forensic Psych Word Scholar V

Define the following words.

1. personality

2. conflict theory of psychoanalysis

3. metaphors of mind

4. defenses of ego

5. Carl Jung

6. Horney & Adler

7. The DSM

8. Cluster B personality disorder

9. Psychopathy

10. Antisocial versus psychopathy

Name ______________________________ Date ______________

Aftermath

Defend in a one page essay why depression or anxiety alone would seldom if ever produce the deception and violence required by sexually psychopathic serial killers.
Why must personality disorder, especially sexual psychopathy, always enter the picture?

UNIT IV

Poisonous Parenting

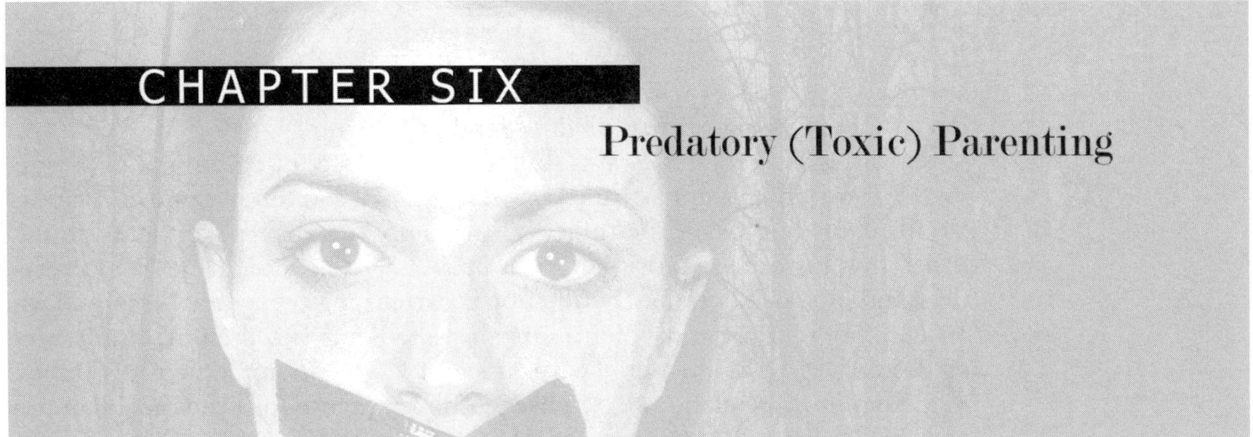

CHAPTER SIX

Predatory (Toxic) Parenting

> And not infrequently parents who are professional people—doctors, lawyers, club women, and philanthropists—who lead lives of strict orderliness and decorum but yet lack love, send children into the world who are as undisciplined and destructive and disorganized as any child from an impoverished and chaotic home.
>
> —M. Scott Peck, M.D. (*The Road Less Traveled,* 1978)

An Analysis of Sexually Psychopathic Serial Crime is ultimately about *causes* told through the *effects* of violent psychopathy perpetrated upon unsuspecting victims. *Toxic, predatory parenting* is a glaring external tonic to all of the predator's internal perversion, anger, and rage. Interviews with incarcerated violent psychopaths provide background information regarding horrific family milieus—*the breeding ground for human predators.* In many ways, the *parents of violent psychopaths are the true criminals of society.*

The faceless, pathetic parents of criminals litter the background in anonymity. Yet, they are responsible to a great extent for raising antisocial misfits. Severely dysfunctional themselves, they seem confused when asked how their son or daughter turned out "so bad." Herein lays a glaring flaw in the perception of predatory parents. As emotionally damaged individuals, they perpetuate deviance by "setting up" pathology in family milieus, ensuring that another generation of dysfunction continues. Rapacious parents—parents who prey upon their young in hateful and abusive ways—are too drug-addled, too emotionally dysfunctional, or too emotionally absent to care for themselves, much less a child who demands clear-headedness, responsibility, and goodheartedness to have a fighting chance at normalcy.

Toxic Parenting

Toxic parenting often arises out of barely surviving horrific parental abuses themselves, and it continues by constructing elaborate defenses against self-hatred, anxiety, and anger. Emotional defenses such as *denial, rationalization,* and *regression* temporarily defuse the anxiety a child feels from knowing "deep down" no one loves or wants him. Sadly, the child interferes with the disfunctional parents' immaturity, addictions, and perversities.

According to Freud, *denial* is the master architect of emotional defenses; operating as thought unconscious *repression* all of a sudden became conscious. Parents deny that they cause their children's problems, just like the alcoholic denies that his excessive drinking is a problem, even though he loses everything and hits rock bottom. Dysfunctional parents may reason: Why should we change if we're not the problem? Obviously, looking into the mirror and seeing themselves as dysfunctional parents is just too emotionally painful. So they use denial or more elaborate defenses. A flimsy excuse for every problem defends *rationalization. Nothing is ever the fault of incompetent and antisocial parents.* The problem is always "other people" and "the kids."

When parents use the defense of *regression,* they show how immature they really are as temper tantrums or assaultive behavior replaces strategic talk and positive reinforcement for the good choices children make.

Growing Up in the Image

Children grow up in the image of their parents. Garden variety psychopaths and other social misfits come from incompetent parenting—from the uninvolved, the passive, the unimaginative, the addicted, and the minimally-skilled parents. *Violent psychopaths and society's serial killers and serial rapists—come from toxic (predatory) parents.*

In our nomenclature, *incompetent parents* are parents who do not, in general, rear criminals but, nonetheless, deeply scar their children emotionally through inadequate emotional expressions of love, caring, and nurturing. They display minimal proactive parental skills and often berate their children with verbal abuses and threats. Garden-variety psychopaths who frequently manifest features of male narcissism and female histrionics, along with borderlines PD, often come from incompetent parents who display ambivalent discipline and emotional uninvolvement, as well as conditional love. Children from these types of homes often grow up with *addiction problems related to chemical abuse and attempt self-medication* to cover their sadness.

When troubled teenagers (who often become sexually promiscuous) have their own children, they perpetuate incompetent parenting by repeating the lessons from their own dysfunctional parenting. Why shouldn't this be the case? Usually, they have not devoted time to learning effective parenting skills from college courses or from local mental health organizations. Why should they? Like all psychopaths with narcissistic or histrionic features, *they have all the answers.* In reality, they have all the wrong answers.

Emotional Nihilism

In our nomenclature, *toxic, predatory parents* rear violent criminals and sexual psychopaths. They go beyond leaving the emotional scars that incompetent parents inflict to producing *irreversible neurological damage* in their children's central nervous systems. This damage stems from inflicting severe physical and sexual abuse to hateful emotional treatment and/or neglect. Emotional **nihilism** characterizes the perception of sexual psychopaths (and their parents) who view people and morality as "nothing." It is no wonder that *they feel destruction is desirable since existence is meaningless.* This has been the consistent message from parents who are emotionally vacant.

The current chapter represents a parental "script" for the development of sons with violent psychopathy, and for daughters who become prostitutes, criminal offenders, and killers. On the flip side, it's about what to avoid when parents seek to raise emotionally normal children. To follow the recipe of a deviant parental script is to transform a sweet and innocent child with all of the promise of excelling in life and finding happiness in normal pursuits into a disgusting, human predator and antisocial misfits. Hateful, predatory parenting results in molding a cold-blooded reptile that hides behind a human face.

Authorities look back after the capture of a serial killer to a history of *oppositional deviant behavior* as teenagers and most likely, to the homicidal triad of enuresis, cruelty to peers and/or pets, and pyromania. Red flags of psychopathy litter his home and school milieus. Daughters raised by antisocial parenting usually turn anger and humiliation *inward* rather than outward. This is due to brain differences related to gender, and other unknown reasons. In the next chapter, *Psychology of Perversion,* Chapter Seven, female sexual psychopaths and adolescent killers who kill kids and adults will be featured.

Childhoods of Violence

In the late 1970s, the FBI's fifty-seven page *Criminal Personality Research Profile* revealed the results of thirty-six incarcerated serial killers talking about childhood influences. For the first time, investigators had insight into the horrific milieus of violent childhoods. In short, serial killers became *conditioned as violent criminals in severely dysfunctional homes characterized by toxic parenting.* In this regard, sins of omission (what parents failed to do) are just as glaring as sins of commission (what parents did). For the present, neuroscientists do not expect to find a single gene for psychopathy or for antisociality.

Applied Predatory Parenting

Richard Trenton Chase "The Vampire of Sacramento"

Parenting milieu characterized by anger, mental illness, drug addiction, physical abuse, and alcoholism.

Chase grew up in a household of anger, mental illness, and alcoholism. His mother was an alcoholic and a drug addict who displayed signs of schizophrenia. His father was a strict disciplinarian who used physical abuse as punishment and as a way to solve problems. At an early age, Chase was a fire-starter.

As Chase grew older, his crimes escalated to more serious offenses, and he showed no remorse for his misconduct. As an adolescent, he was a loner showing signs of psychopathy and displaying deviant fantasies. Acquaintances never perceived him as a "ladies man," and they ridiculed him for being sexually impotent. Often he found dead animals and brought them home. He likely cannibalized them. Once, he injected rabbit blood into his own veins. The procedure made him deathly ill, but somehow he survived. Eventually doctors diagnosed Chase as a paranoid schizophrenic, suffering from somatic (bodily) delusions. He felt his blood would turn to powder without blood transfusions (this comprised his blood fetish).

Chase is one of the few sexual killers who actually had a psychiatric diagnosis. Most serial killers are not psychotic; they know exactly what they're doing.

Jeffrey Lionel Dahmer "The Milwaukee Cannibal"

Parenting milieu characterized by physical and emotional abandonment issues from emotionally distant parents, possible molestation from a neighbor, homosexual fantasies, road-kill autopsies, and alcoholism.

Born ten years after Chase in 1960, Jeffrey Lionel Dahmer grew up with unhappily married parents who spent most of their waking hours in verbal sparring. The family moved often during the early years of his life.

Dahmer had medical problems from birth. He was born with broken legs and had to wear splints until he was three years old. He couldn't walk until he was two. Then he had to have help standing up. At the age of four, he received surgery for a hernia. Dahmer recalled the experience as "embarrassing" and causing considerable pain. He erroneously thought the doctor amputated his penis, and years later he felt this experience emotionally scarred him. As he grew up, he was quiet and kept to himself. At the age of eight, a rumor existed that a neighbor boy molested him, although nothing became proven. However, at this age Dahmer became introverted and behaviorally disruptive.

In school, he made a spectacle of himself in class as the class clown. Often, he ran down the hallways of his school shouting and waving his arms, and peers referred to his behavior as "doing a Dahmer." Teachers considered him a bright child, but he only managed average grades. He never liked school.

As an adolescent, Dahmer began experimenting with the dead carcasses of animals. This macabre behavior coincided with the start of puberty. He brought home road kill and performed autopsies on it. Since his father was a chemist, Dahmer used his knowledge of acids dissolve the tissue away from bones. At this point, he began to fantasize what humans would look like on the inside.

Dahmer discovered along with his puberty that he experienced attraction to men, and only to men. He didn't know how to handle his urges and dared not mention this to his parents, so he began to drink in an effort to self-medicate. The fantasy plagued him of lying next to an unconscious person, someone who could meet his needs but who told no one and asked nothing in return. He became a fierce alcoholic, trying to suppress his thoughts, but eventually, he felt compelled to act out and murder his first victim. Dahmer hid in the woods one day soon after he could no longer control his urges, hoping to disable a male jogger who frequently ran along the path. He took a baseball bat and planned to knock him into unconsciousness and just lie down next to him. Luckily for the jogger, he didn't come by that day, and Dahmer tried to suppress the thoughts again.

The themes of physical and emotional abandonment ran through Dahmer's entire life. This condition strongly suggests that Dahmer not only demonstrated a psychopathic personality but suffered from *borderline personality disorder* as well. The fear of abandonment is such a disorder's central issue. The sufferer becomes emotionally "clingy" and demanding of others. His parents divorced when he turned eighteen. His mother received custody of both Dahmer and his brother; however, they soon abandoned him, leaving young Jeffrey home alone. He called his father, who eventually moved back into the family home, but his father soon abandoned Jeffrey too. Dahmer finally moved in with his grandmother, but she made him move into his own apartment due to his strange behavior. Then the nightmare began as he sought his second victim after an extensive cooling off period.

Albert De Salvo "The Boston Strangler"

Parenting milieu characterized by physical abuse and emotional abandonment producing petty crime and sexual addiction; father forced prostitutes to have sex with his ten year old son. Reportedly, he had sex with his wife up to six times daily; alleged sexual molestation of a young girl.

Before his father left and Albert's dysfunctional parents divorced, De Salvo watched as his father knocked out his mother's teeth and broke all the fingers on one of her hands. The abuse was extremely harsh on young Albert. At one point in his childhood, a farmer paid the elder De Salvo nine dollars for Albert and two of his sisters. They escaped after a month of being the farmer's slaves. When Albert was about six years old, his father coaxed him to begin petty shoplifting, which led to breaking and entering and burglary.

As Albert grew older, his father brought home prostitutes and forced Albert to watch him having sexual intercourse with them. By the age of ten, Albert had his first sexual experience, which led to sexual obsession and sexual addiction. According to reports, De Salvo claimed to have sex with his wife up to six times a day.

Later, the Army discharged De Salvo dishonorably due to an allegation that he sexually molested a nine year old girl. The mother failed to file charges against him, and there was no conviction.

Harvey Glatman "Lonely Hearts Killer"

Parenting milieu produced a sadomasochistic, autoerotic asphyxia addict with sexually perverted impulses.

As a child, Harvey Glatman's own dysfunctional parents perceived him as "a little strange." They caught him at the age of four inflicting pain upon himself and enjoying self-punishment in a sexualized sadomasochistic act. He had tied one end of a string around his penis and tied the other end to a drawer. When he leaned back against the string and pulled it taut, it caused pain.

The string proved a prescient metaphor of things to come, for as an adolescent, he exchanged the string for a rope that he used around his neck in an obsession with autoerotic asphyxia. With the rope around his neck, he looped the free end over a pipe or rafter. He yanked the rope with one hand and masturbated with the other, envisioning the ropes as extensions of his arms. As a serial killer, he used rope on his victims to bind them so he could control and kill them.

Psychiatrists diagnosed him as a psychopathic personality, schizophrenic type, defined as having "sexually perverted impulses."

Richard Ramirez "The Night Stalker"

Mentor was a pot-addicted killer and rapist who filled his mind with vivid images of rape and killing.

Like all serial predators, Richard Ramirez did not have a normal or peaceful childhood. A male relative of Ramirez became a mentor to young Richard; he was a heavy pot smoker and Vietnam veteran who bragged to young Richard about his brutalities and sexual conquest of women during the war. He showed him Polaroid snapshots of the rape and killing of the inhabitants of entire villages. With Ramirez present, his relative shot his wife in the face following an argument.

Arthur Shawcross

Parenting milieu characterized by being sodomized with a broomstick, oral sex fetish, and incest.

Arthur Shawcross's mother was an extremely sick individual. She sodomized him with a broomstick when he was a child and told him to have oral sex with his aunt, who taught him how to do it. He adopted a fetish for oral sex from this experience. Someone caught him as a child having sex with his sister, an allegation his sister denies. In another instance, his girlfriend's brother caught him performing oral sex on her. The brother's condition for keeping quiet was for Shawcross to perform oral sex on him. This was his first homosexual encounter. Shawcross was a chronic bed wetter until age eight.

In his youth, Shawcross was a loner and never really had any friends. As he grew, he became more and more violent and began to break into houses and buildings. He also became more withdrawn from the world. At the age of fourteen, he began to have oral sex with his cousin and sister.

Violence and Lack of "Contact Comfort"

Experimental psychologists Harry Harlow and Margaret Harlow began a series of studies in the 1950s showing the importance of **tactile stimulation** (touch) in normal behavior. These famous "contact comfort" studies extended into the mid-1960s at the University of Wisconsin and provided the first experimental evidence that *inattentive and dysfunctional mothering resulted in abnormal behavior and abnormal brain development.* Moreover, inattentive and loveless mothering "set up" (conditioned) in the offspring the propensity for some *forms of violence* later in life.

One aspect of toxic parenting is inattentive, unaffectionate, and loveless mothering and/or fathering, which produces an emotional detachment from family milieu. As mentioned previously, *milieu* is important social contexts of learning (such as the home and peer groups) that emphasize *emotional connection and emotional development.*

Accordingly, Winnicott (1965) contends there is no such thing as an infant per se, meaning of course, that maternal care given the infant forms a subunit of a milieu, making mother and child emotionally inseparable. Healthy development as the child matures allows this natural bond to be broken; otherwise, the child experiences excessive "separation anxiety" away from the mother.

The Harlow study chose infant rhesus monkeys as subjects because, like human infants, they require long periods of *emotional attachment to caregivers,* observed in *bonding.* The experimenters isolated the infants at birth in solitary cages that prevented touch of any kind, as well as attachment or social bonding with the other monkeys.

Infants *raised in isolation* appeared singularly withdrawn as adults and engaged in *self-mutilating behavior* (pinching and biting themselves). Later, they channeled self-aggression into hostility—"acting out" inappropriately against others. Infant monkeys who became adult mothers were "brutal, indifferent mothers" to offspring. Similarly, male and female monkeys raised in isolation grew up to be "unstable, brutal" parents.

Could these results reveal an early parental blueprint for raising sexual psychopaths?

Surrogate Mothers

In a related experiment, researchers placed a group of newborn infant rhesus monkeys in a cage with **surrogate** "mothers" to test the mothering process further. They constructed one mother of wire mesh and a heating lamp, and provided a bottle with a nipple for nourishment. They covered the other mother in soft terrycloth and gave no further accoutrements (i.e., no lamp or bottle). The infants routinely chose the terrycloth-covered mother under a variety of conditions (such as being frightened by a loud noise). Even when hungry, the infants would cling to the cloth-covered mother while reaching across the wire mother for milk. This experiment verified the importance of *contact comfort* in the *bonding experience* between infants and mothers. Prior to the Harlows' study, the dominant theory of parent-to-child bonding was the *cupboard theory of attachment.* This view holds that infants bond with their mothers because the mother provides nourishment.

After the Harlow results, *touch and cuddling* became significant factors in the understanding of how maternal and social bonding produces healthy, well-adjusted children.

Rocking: The "Nutrient" of Motion

When experimenters replaced the wire surrogate "mother" with a cloth-covered surrogate capable of a rocking motion, the infant monkeys *preferred the sensation of being rocked* to the motionless terrycloth surrogate. Later, as young "children," the rhesus monkeys raised with motionless cloth surrogates showed repetitive rocking movements. The monkeys raised with the surrogate capable of rocking did not display abnormal rocking movements.

HTCR—Holding, Touching, Cuddling, & Rocking

The classic study by Harlow and others has convinced neuropsychologists that *sensory stimulation* before age two of the variety researched by the Harlows is *necessary for normal brain development—holding, touching, cuddling, and rocking, or* (*HTCR*).

Truly, the rocking chair may be the best neuropsych tool ever invented for the development of normal brains! *Sensory enrichment* through *HTCR* leads to changes in the branching of neurons, and in ion conductance, which lie at the heart of normal synaptic connectivity.

The developing brain depends on sensory stimulation to such an extent that some researchers refer to touch as a nutrient.

Shore (1996) contends that "infants' early emotional experiences in relation to the primary caregiver actually influence the production of certain brain chemicals that play a role in the physical development of the cortex, the part of our brain that is responsible for our most sophisticated and complex functions such as thinking, perception, and emotion. When the emotional attachment of the infant to the caregiver is stressful and unsatisfying, the hormones created in the infant's nervous system cause the abnormal development of specific structures and circuits in the cortex of the brain that are responsible for regulating emotional reactions; this abnormal brain development, triggered by negative environmental factors during the critical growth period of birth to two years of age, creates an enduring *susceptibility to various psychological disorders* later in life."

Advanced Attachment Theory

In 1951, British psychoanalyst John Bowlby added a human touch to the Harlow findings in rhesus monkeys when he began a series of studies of homeless children in postwar Europe. He analyzed the mother-infant bonding process that led to his *attachment theory* of bonding. We can analyze the *genetic determination* of social bonding and its centrality to the normal development of self, personality, and behavior by analyzing his studies' further development with the following bonding types:

- **Type A** ***avoidant attachment*** corresponds with *maternal insensitivity to infant's cues.* Infants learn to distrust parental affection as a defense against being overwhelmed by fear or sadness. *They tend to anticipate rejection and become hostile or angry.* They show less resilience in times of distress.
- **Type B** ***bonding*** corresponds to *securely attached.* Infants received optimal and consistent responsiveness from caregivers, and parents routinely comforted them in times of distress. *They displayed considerable positive affect (emotions) and resiliency.*
- **Type C** ***ambivalent/resistant attachment*** with unpredictability of emotional attachment. *Such children become impulsive and helpless.* Although normal, they tend to be clingy and insecure.
- **Type D** ***disorganized attachment*** where parents become *frightening to the child* because of their own traumatic issues. Instead of providing security, parents become elicitors of fear. *Children display anxiety, hostility, and anger.*

According to researchers, Type B—securely attached, and Type C—ambivalent/resistant attachment, fall within what is considered *the normal range without pathological implication.*

While Type A—avoidant attachment—produces *difficult children* who may require professional counseling later in life; Type D is the prototype *for the development of the violent behavior.*

Psychology of Movement

The Cerebellum—"Nutrients" of Holding, Touching, Cuddling, and Rocking

The *cerebellum,* the three-lobed cerebral tissue behind the occipital lobe of the brain, coordinates balance and fine muscle movements. When an inebriated person fails a field sobriety test, it's the cerebellum that nails him! Flunking the test of normal balance and coordination means the effects of alcohol or other drugs anesthetized the cerebellum.

Not surprisingly, the cerebellum is the brain center most targeted when infants and small children experience *sensory enrichment* through holding, touching, cuddling, and rocking. The author believes, somewhat tongue in cheek, that an acid test for whether or not a given three year old child is receiving adequate HTCR is for Mom or Dad to playfully throw him up in the air (not too far!), as most parents do in play. If the child's eyes widen in fear, and he stiffens from head to toe, the implication is that the brain's motion center—the cerebellum—is somewhat dissonant to HTCR (unless this was the first time tossed in the air).

If the child gleefully smiles and says, "Do it again!" we have some **anecdotal** evidence that the cerebellum is becoming finely tuned due to the enhanced development attributable to HTCR. As any parent knows, children playfully thrown in the air and caught love it and never want to stop. The author's experience is that when this occurs outside in the front yard in plain view of other children, a "me next" line soon forms. The same rationale holds true for swinging, sliding, jumping on a trampoline, riding a bike, or circle-riding on the merry-go-round.

Kids love motion—running, jumping, spinning around, and falling down. Apparently, the brain requires motion for normal development. In a practical and beneficial way, youth sports before age two enhance early parent-child interactive play. Non-contact sports such as gymnastics, swimming, tennis, and, to a lesser extent, soccer and volleyball seem especially beneficial. Due to the likelihood of head trauma from vicious blows to the head, the physical violence of football, rugby, and boxing can exacerbate pre-existing neurological damage (currently observed in the violent social behavior of Mike Tyson and the nearly incomprehensible verbiage of former professional boxers Joe Frazier, Muhammad Ali, and Michael Spinks). Some authorities believe the cumulative effects of the "head butt" in soccer and volleyball can lead to blunt head trauma.

According to brain neuroscientist Dr. James Prescott of the *National Institute of Child Health and Human Development:* "when touch and movement receptors and their projections to other brain structures do not receive *sufficient sensory stimulation,* normal development and function [do not occur.]" Dr. Prescott believes that under-stimulation can have devastating effects upon *emotionality* later in life because of the cerebellum's involvement in *complex emotional behavior.*

In relation to complex emotional behavior, we now turn to the Predator Profile of one of the most despised serial killers in history—"Killer Clown," John Wayne Gacy.

PREDATOR PROFILE 6-1

John Wayne Gacy, Jr. *"The Killer Clown"*

Ashleigh Portales

Time Span of Crimes
January 1972 to December 1978
Offenses Prior to Serial Killing
Solicitation
Conspiracy to assault and sodomy of a teenager
Kidnapping
Rape
Quoting Gacy
"I have no remorse. I don't feel sorry for anybody."

Preferred Prey

Gacy preferred teenage boys and hired them as "cheap labor" for his company, PDM Contractors, Inc., to gain easy access.

Recipe for a Monster

John Wayne Gacy, Jr., was born on St. Patrick's Day in 1942 to Irish Catholic parents in Chicago, Illinois. Gacy was close to his mother and two sisters, yet he *never acquired the relationship he desired with his father.* John Wayne Gacy, Sr. was an *alcoholic who verbally and physically abused his wife and son.* He often beat John, Jr. with a leather strap and constantly tormented him with verbal assaults. In a typical middle class existence, Gacy held several part time jobs after school as a paperboy, supermarket bag boy, and grocery stock clerk. He also was a member of the local Boy Scout troop. He was not particularly popular in school, but his teachers liked him.

The neurological damage inflicted by his father became aggravated when Gacy was eleven. A swing struck him in the head while he played by a swing set, resulting in a *blood clot in his brain that doctors didn't discover until he was sixteen.* Gacy suffered a series of blackouts between the initial injury and its diagnosis, which ceased when he received medication to dissolve the blockage. The following year at age seventeen, doctors diagnosed an idiopathic heart condition. Though they never found an exact cause and he never suffered a serious heart attack, John complained frequently throughout his life of chest pains.

In addition to his medical problems, things at school took a downward spiral. Gacy attended four different high schools his senior year and *dropped out* completely when he realized that he would not be graduating in the spring. He drifted to Las Vegas where he found only part time work performing odd jobs as a janitor in a funeral home. Discouraged by the lack of prospects for a man with no high school diploma, Gacy decided to return home and did so three months later. It took that amount of time to earn enough money for the ticket. At home, he enrolled in business school where the *trademarks of a psychopath—his "gift of gab" and charming persuasion,* made him an instant success. Upon graduation he took a job at the Nunn-Bush Shoe Company, which quickly promoted him to manager of a men's clothing outlet in Springfield, Illinois. There he began a long career in community service. He was membership chairman of the Chi Rho Club, a board member of the Catholic Inter-Club Council, commanding captain of the Chicago Civil Defense, an officer of the Holy Name Society, and vice president of the Jaycees, who named him "Man of the Year."

Despite all his hard work, Gacy gained a formidable amount of weight and became hospitalized for his heart condition and a spinal cord injury, as well as for nervous exhaustion on one occasion. Yet he tirelessly continued his work with the Jaycees even after his marriage to Marlynn Myers, with whom he moved to Iowa and had two children. A member of the Waterloo, Iowa, Jaycees, he soon began a campaign for the organization's presidency. Gacy's ability to hyper focus, a salient trait of *ADD,* eventually contributed to his infamous status as one of the country's worst serial killers. Activities consumed his life. One volunteer said, "The Jaycees was his whole life." Friends and neighbors knew him as a "generous, friendly, and hard-working man, devoted to his family and community." However, all that was about to change as Gacy's carefully *constructed façade* started to come crashing down.

A Crack in the Mask

Among the Jaycee members, and subsequently throughout town, rumors spread that Gacy's sexual preference inclined toward young boys. In May 1968, a grand jury indicted Gacy for the sodomy of Mark Miller, a teenager who visited the Gacy home a year before.

Miller testified that Gacy tricked him into being tied up and then brutally raped him. Gacy contended that Miller had participated willingly and accepted payment for his performance. He further postulated that he was "set up" by Jaycee members in opposition to his run for the presidency. Four months later, Gacy found himself in court again. This time he was on trial for hiring eighteen year old Dwight Anderson to beat up his accuser Miller. He had given the teen ten dollars plus an additional three hundred to pay off his car loan if he succeeded. Under the judge's order, Gacy received psychiatric evaluations at various mental health facilities. Psychiatrists concurred about his mental competence. In a prescient diagnosis, however, they found him an *antisocial personality* who in no way would benefit from any known medical treatment. Subsequently, they sentenced Gacy to ten years at the Iowa State Reformatory where he entered prison for the first time at age twenty-six. Soon thereafter, Mrs. Gacy filed for divorce on the grounds that John had violated their marriage vows.

Like many other *organized serial killers,* Gacy manipulated the criminal justice system to work in his favor. He realized that good behavior equaled early parole and acted the role of a *model prisoner.* It paid off. On June 18, 1970, authorities released Gacy after serving only two years of his sentence. From there, the monster behind the mask ventured to Chicago and began to rebuild his dangerous façade.

The Mask Repaired

Gacy began reconstruction by moving in with his mother. Though he struggled with deep depression over the death of his father while he was in prison, he decided it was time he lived on his own, and with the aid of his mother and sisters he acquired a home just outside the Chicago city limits. Gacy flourished in the upscale, family-oriented neighborhood where he threw lavish parties for his neighbors and quickly turned them into friends. Yet his new life could not stifle the old desires. Barely one month after Christmas in 1970 Gacy faced charges of disorderly conduct for picking up a young boy at a bus terminal and forcing him into various sexual acts. Like the *sneaky reptilian predator* that he was, Gacy slipped through the cracks when his accuser failed to appear in court. The court dropped the charges.

Free again, Gacy found another victim in Carole Huff, a recently divorced mother of two children who was emotionally vulnerable, dependent, and pitifully naïve. In spite of Gacy's prior convictions, she believed he would make a good husband and father and married the slick psychopath on June 1, 1972. Together they hosted even greater extravaganzas, at times entertaining more than three hundred guests in one evening.

Though it did not deter them from attending such affairs, neighbors often commented on a horrible stench in the Gacy house. Believing a rat had died beneath the floor, they urged John to fix the problem, but he chalked it up to nothing more than the accumulated moisture in the crawl space under the home.

In 1974, Gacy went into business for himself by beginning Painting, Decorating, and Maintenance, or PDM Contractors, Incorporated, where the majority of his employees were teenage boys. This perfect charade provided a macabre process of selection for victims. Gacy could supply himself with an endless array of potential prey within the confines of a reputable business. His inability to subdue his homosexual and ultimately homicidal urges proved the beginning of the end though. He began drifting away from his wife, preferring repeated masturbation while watching pornography to making love to her. He became constantly unpredictable, bursting out in uncontrollable rages with no apparent provocation. When Carole found homosexual pornography in their home, Gacy confessed to her that he preferred boys to women. She subsequently filed for divorce, which became final on March 2, 1976. His track record in relationships proved a benchmark to his severe psychopathy.

Gacy's political aspirations undeterred, he volunteered his company to clean the Democratic Party Headquarters in Norwood Park. He also began to dress as "Pogo the Clown" and provide entertainment to children at parties and in hospitals. His actions caught the eye of an influential political figure who nominated him to the street lighting committee. Though he seemed on the right track, Gacy strayed soon enough.

Gacy's lack of self-control once again proved his downfall as rumors, inspired by actual events, began circulating about his homosexual activities with young boys. Even those who refused to believe the initial rumors could not deny the gruesome shock that the entire nation received in December, 1978 at the revelation of the real John Wayne Gacy, Jr.

The Monster Unmasked

Lt. Joseph Kozenczak obtained a search warrant for Gacy's residence when an investigation into the disappearance of fifteen year old Robert Piest led him right to the man's doorstep. Lt. Kozenczak learned that Piest had left his job at a local pharmacy to meet with a contractor who had offered him work. The name of that contractor was John Gacy. Though Gacy denied any knowledge of the boy's whereabouts, Lt. Kozenczak ran a background check and uncovered Gacy's criminal past, the fact that produced the warrant. On December 13, 1978, the lieutenant and several officers carried out the search. Among the confiscated items were:

- A box containing marijuana and rolling papers
- Seven erotic Swedish movies
- Aamyl nitrate and Valium
- A switchblade knife
- A stained section of rug
- Color photos of pharmacies and drug stores
- An address book
- A scale
- Books: *Tight Teenagers, The Rights of Gay People, Bike Boy, Pederasty: Sex Between Men and Boys, Twenty-One Abnormal Sex Cases, The American Bi-Centennial Gay Guide, Heads and Tails,* and *The Great Swallow*
- A pair of handcuffs with keys
- A three foot-long 2 × 4 with two holes drilled in each end
- A 6-mm Italian pistol
- Police badges
- eighteen inch rubber dildo
- A hypodermic syringe and needles
- Young men's clothing
- Nylon rope

Among Gacy's three automobiles was a 1979 Oldsmobile Delta 88, whose trunk contained hairs that police later matched to Robert Piest. When police entered the crawl space beneath the house, they believed the smell to be raw sewage, and though the ground was covered with quicklime, it appeared untouched. The police had no grounds to arrest Gacy, but while they awaited forensic testing of the evidence they did place him under twenty-four hour surveillance. Frustrated, they decided to detain him for possession of marijuana and Valium. They did not know that only the day before his arrest he had admitted to a co-worker that he had "killed about thirty people because they were bad and trying to blackmail him."

Gacy finally confessed during interrogation that he had killed someone but said he had done it in self-defense and that he had buried the body under his garage. Police returned to the house but decided to check the crawl space once more before they began digging. This time they noticed a suspicious mound of earth. They found human remains only minutes after digging into it. By the time they completed the excavation they had gutted the entire Gacy home and found *twenty-nine bodies*. They discovered an additional four in the Illinois and Des Plaines Rivers. All but nine of the bodies were identified. The cunning predatory monster that had hidden behind Gacy's façade stunned his friends and neighbors.

Confronted with the finds, Gacy confessed to killing "at least thirty" men, most of whom he had buried beneath his home. He stated that he had laid the bodies so close together and subsequently dumped them in the rivers because he needed more space. He lured his victims into handcuffs before sexually assaulting them then stifled their screams by stuffing their own underwear down their throats. Gacy killed them by pressing a rope or board against their throats, often while raping them. Sometimes he kept the bodies under his bed or in the attic for several hours during the aftermath of the murders, before finally burying them, but he always allowed adequate time to clean up before his wife returned home.

The Monster Caged

Gacy's murder trial began on February 6, 1980 before a jury of five women and seven men. The jury rejected Gacy's *plea of insanity* in favor of the prosecution's assertion of his rational and premeditated actions. The prosecution based this assertion on the findings of several psychologists that Gacy was sane at the time of the murders. Jeffrey Ringall, a man whom Gacy drugged, kidnapped, and raped in May of 1978, was among the witnesses to take the stand. Ringall experienced such emotional duress during his testimony that he began to vomit and cry hysterically, and they had to remove him from the courtroom. Gacy looked on completely devoid of emotion all the while, a condition often observed in severe psychopathy and known as blunt affect. After only two hours of deliberation, the jury found Gacy guilty of thirty-three counts of murder and sentenced him to death. The State of Illinois executed Gacy on May 10, 1994, fourteen years after his trial date.

Displeasure Centers

Neuroscientists, once known as physiological psychologists (and later psychobiologists), study the effects of early environmental influences on behavior, and therefore on the brain. Since the mid 1960s, researchers have known that mere external electrode stimulation of the anterior hypothalamus within the limbic system produces violent behavior. Dr. James Olds, a trailblazer in electrode stimulation of the brain, discovered *pleasure centers* in the **medial forebrain bundle (MFB)** of the hypothalamus. In self-stimulation studies, a rat would level-press up to five thousand times in one hour when the electrode passes through the MFB, a center rich in the neurotransmitter dopamine. Olds also discovered a displeasure center not far from the MFB near the hypothalamus and its associated areas known as the amygdala.

Does lack of motion (and HTCR) in infancy and young childhood contribute to the emotion of displeasure and perhaps acting-out behavior later in childhood? As early as 1975, Dr. Robert Heath of Tulane University showed that the emotional centers of the brain's *limbic system* (specifically the **hippocampus,** amygdala, and septal areas) did interconnect with the cerebellum.

According to Dr. Heath, . . . "violence may result from a *permanent defect in the pleasure centers* (due to) inadequate early mothering. The infant who is deprived of movement and physical closeness will fail to develop the brain pathways that mediate pleasure . . . such people may be suffering at the neuronal level from *stunted growth of the pleasure system.*"

Anhedonia

In Dr. Heath's view, the brain does not "broadcast" the expression of pleasure properly due to a scarcity of cell connections. The immature cerebellum develops abnormally in infants and young children deprived of HTCR. Low activity between the cerebellum and the pleasure centers resulting in fewer connections, may lead to a condition known as **anhedonia,** a decreased ability to feel pleasure. This, in turn, may produce an insatiable desire to experience pleasure. In the absence of pleasure, the person may act out his frustration and anger in an attempt to experience some measure of it.

Apparently, neuronal connections required to experience sustained feelings of pleasure did not become hard-wired in the brain at developmentally significant times. Researchers theorize that the developmental timeframe is before age three.

Might the red flags associated with the developing psychopathy—cruelty to pets, setting fires, fighting, and anger outbursts—occur due to the frustration of not feeling pleasure, as well as *to the low amplitude in or the inactivity of* cerebral neurons?

Interestingly, psychiatric patients at Tulane University Medical School treated with *cerebellar stimulation*—internal stimulation of the cerebellum—reported nearly miraculous results. A pacemaker technique of stimulation produced a decrease of emotional outbursts. Might this be an effective way to at least partially reverse a history of HTCR deprivation in young children? Could this treatment derail some of the neurological damage observed in psychopathy? Does enough funding exist to continue the studies?

Erik Erikson's Psychosocial Developmental Theory—Implications for Psychopathy & Antisociality

Early Childhood Stage (Ages Six to Seven Years)

We now present a template consisting of relevant aspects of Erik Erikson's developmental stage theory of psychosocial "crises" that are useful in observing antisocial parenting elements.

Erikson's *psychosocial developmental theory* underlies curricula in many nursing programs in North America and in much of the world. It recognizes a series of *emotional crises* vital to address at sequential stages. When successfully resolved, individuals leave the stage *emotionally enhanced.* They move to the next stage *with a positive emotional experience to build on.* When unsuccessfully resolved, individuals *suffer emotional deficits* (as well as neurological glitches) entering subsequent stages; they move on with less emotional stamina, resilience, and neurological functioning.

Growth and emotion slowly build upon each other with success. Effective, nurturing experiences help resolve the emotional "crises." Children enter each succeeding stage emotionally robust. Fragile self-esteem receives nurturing along the way. However, deficits experienced in earlier stages tend to sabotage succeeding stages. When emotional needs are not being actualized, children often display behavioral red flags associated with oppositional defiant behavior.

It is instructive to locate the acting out behavior observed in **ODD (Oppositional Defiant Disorder)** at one end of a deviancy continuum while observing the withdrawn, brooding, and reclusive behavior characterized by the **schizotypal personality disorder** at the other pole. Both abnormal behavioral types—*oppositional defiant and schizotypal*—characterize potential criminal behavior. As personality studies demonstrate, the oppositional defiant category of deviance gravitates toward violence and psychopathy.

Trust versus Mistrust—Implications in Psychopathy

Erikson believes the *infancy stage* (birth to first year of life) produces the first emotional crisis that needs resolution. Its form is *trust versus mistrust.* This initial stage acts as the foundation for the construction of all other stages. Care, nurturing, parental attentiveness, and competence in meeting infants' needs prepare the child to trust caregivers. The emotional benchmark of trust acts as buoyancy to budding self-esteem, as infants move into Stage Two (second year of life). In this stage, effective *toilet training strategies* produce a sense of *autonomy.* The opposite is an overly demanding or impatient training regime that translates into internalized feelings of shame and/or doubt. Freud theorized that "anal retentive" personality characteristics erupted at this stage, producing *obsessive-compulsive* behavior leading to OCD, or obsessive-compulsive disorder, later in life.

The third year through the fifth year of life produces the *play stage* and a crisis of *initiative versus guilt.* The very essence of *motivation* may arise in this critical stage. The *sequencing of behavior* arises in this stage, where children learn "what comes next" after termination of such events as play, study, eating, or sleeping. Smart, involved parents teach children to pick up toys, get homework before

TV, and keep their rooms organized and clean—all strategies considered normal. If they do otherwise, parents promote **learned helplessness** in their children, who often live like "slobs" in unkempt bedrooms or apartments.

It is easy to see how unfulfilled emotional crises start to implode around the increasing dysfunctional life of the developing psychopath. Through antisocial parenting, parents fail their children by age five by systematically replacing trust, autonomy, and initiative with *mistrust, shame, doubt, and guilt.* It is not surprising to discover the start of oppositional defiant behavior in preteens as emotionally abusive and hateful parenting increases. The child has not felt loved or valued through any of the first three stages of development.

Stage Four (ages six through puberty) sets up a crisis of *industry versus inferiority.* Loving and competent parents who display nurturing, attention, and care for children produce a sense of achievement and productivity (industry) with the successful resolution of this stage. Normal children gain a *sense of purpose.* On the other hand, antisocial parenting produces the dominant feelings of frustration, anger, and emotional deviance. Feelings of inferiority consume the child.

Incarcerated criminals often mention feeling "inferior" or "powerless" as the motivation to commit serial crimes. The same motivation extends into "Rambo" adolescent killers.

The Central Importance of Mothering

Many developmental psychologists believe that the most important adult figure in early childhood development (up to age three) is the mother. A strong *emotional sentiment* that develops during this time is a sense of belonging and love, or its lack. Today, modern neuroscience believes *moms are primarily responsible for wiring their babies' brains.*

Evidence from incarceration interviews with serial offenders shows the relationship with antisocial mothering was uniformly *cool, distant, unloving, and neglectful,* and characterized by a lack of *emotional warmth.* Infants who eventually grew up to become serial killers *never internalized maternal love* from an early age.

Internalizing abuse—sexual, verbal, mental, and/or physical—continually showed up in interviews with serial killers.

It is no wonder children grow up angry and oppositional defiant when they have an emotionally absent mother and often a literally absent father. They have received practically no behavioral limits, or by contrast, they have received nearly intolerable boundaries. This condition has produced a nihilistic "monster" incapable of caring for and nurturing others.

Comprehending the world in *egocentric* ways characterizes the early developmental stages of childhood development in both normal and psychopathic children. However, as normal children develop, they become less egocentric and more empathetic. They can see the world from another's perspective. By contrast, the psychopath becomes more egocentric. He often hides his fragile ego behind arrogance and inflated confidence. Listen to the way he talks about himself; it always gives him away.

The renowned Swiss developmental psychologist, Jean Piaget (1896–1980) demonstrated egocentricity with his famous "Three Mountains" experiment. The researcher seated children across a small table from a doll and asked them to judge a papier-mâché model of three mountains, as it would appear from the doll's perspective. Piaget selected the responses from a number of cards depicting different angles of the mountains in relation to the different perspectives around the table. Pre-operational children (age two to seven) chose the picture of the mountain *from their own perspective,* not the doll's, an example of egocentric thinking.

In contrast, the older, concrete operational children (age seven to twelve) most often chose the correct picture, the way the mountains looked from the doll's side of the table (Piaget and Inhelder, 1956). Metaphorically, adolescent psychopaths have the same egocentricity as preoperational children. Life is all about them.

The lethal combination of *superficial normalcy* observed in many psychopaths, paired with *emotional immaturity and egocentricity,* with increasing focus on *sexual perversity,* explains why sexual psychopaths are so dangerous to unsuspecting victims.

Strong, influential adult role models never emerge in the first seven years of life in the slowly "simmering" development of psychopathic personality. Of interest is Freud's theory that much of personality is "set" by age six or seven and becomes highly resistant to change, a view that reinforces the *irreversibility* of psychopathy.

Childhood Stage (Ages Eight to Twelve Years)—Implications in Psychopathy

A boy entering puberty needs a strong, stabilizing figure in his life; boys need a loving father. More than half of all serial killers studied (over eighteen of the original thirty-six respondents) saw their fathers leave the home during this stage. Absent fathers produce anger, embarrassment, and, worst of all, *loneliness* during this stage. *Isolated* and *lonely* feelings from lifelong *emotional scarcity* characterize a salient feature of psychopathy.

Also damaging is the fact that preadolescent sexuality and fantasies do not connect to another person. Rather, at approximately twelve to fourteen years of age, they emerge as *autoerotic, rape fantasies.* A pronounced fascination with pornography, fetishism, voyeurism, and compulsive masturbation in mid-to-late adolescence may exacerbate perverse fantasies. Both garden variety and sexual psychopaths are immature and socially incompetent when entering adolescence; that is, they lack the social skills required to foster normal interpersonal relationships. They have many short-term relationships that end, according to their egocentric view, because of "the other person."

Some "becoming" psychopaths who launch from this developmental stage feeling inferior remain painfully introverted and shy (Edmund Kemper types), but some appear extraverted with a gift of gab (Ted Bundy types), which masks their inner loneliness, deviance, and emotional desperation.

Deviant Sexual Fantasies

Deviant sexual fantasies spill over into the minds of adolescents on the brink of entering adult sexuality who are far from feeling like proactive, independent adults. Deviant sexualized fantasies further erode any hope of developing normal social and sexual skills with consenting adults, further exacerbating resentment and oppositional defiance for not having nurturing experiences from competent parenting.

According to Ressler, loneliness, isolation, and sexual "daydreaming" provide an emotional platform for cruelty to animals and other children, truancy, setting fires, fighting, and assaulting teachers. Later, as young adults, they are job "hoppers" and unstable in the workplace and underachievers in school. This is further evidence of their greatest fear that others will discover their incompetence and feelings of inferiority. Later adolescence (fourteen to eighteen years of age) produces compulsive masturbation, lying, promiscuous sex, and nightmares. Adolescent psychopaths sleep very poorly and wake up chronically tired.

The possibility of acting out deviant fantasies becomes an obsession as they enter young adulthood. They perceive the world as a cruel place. With little restraint on oppositional deviant behavior in middle childhood, sexual themes of dominance, molestation, manipulation, and revenge fuel aggression and the need to "act out." As young adulthood transcends the teen years, the time draws near to unleash rapacious behavior on "deserving" victims.

The nihilistic and egocentric mindset of rapists and murderers allows them to *depersonalize* victims as "objects" in order to fulfill their sexual fantasies. Deviant cognitive maps set up neural pathways in the brain with tainted, sadistic, and rapacious sexuality.

Stressors (or environmental "triggers") provide the final push into *rapacious* behavior, i.e., preying on others. The loss of a job, a relationship breakup, or financial problems can trigger the actualization of deviant fantasies. The "straw that broke the camel's back" may be something minor in relation to what normal people adjust to in everyday life.

Magical thinking enters the mindset of human predators diagnosed with paranoid schizophrenia, a psychotic thought disorder. For example, serial killer Richard Chase, the "The Vampire of Sacramento," believed his own blood would "magically" turn to powder if he did not drink his victims' blood. Yet, as we have shown, the vast majority of serial predators are not mentally ill; they know exactly what they're doing. What's more, they're compelled to keep doing it.

Adolescent Stage (Ages Twelve to Eighteen)—Implications in Psychopathy

The crisis of *identity versus role confusion* characterizes this stage of development. Personality becomes more or less integrated under the strong influence of peer pressure and parental expectations.

What identity resides inside the mind of the young psychopath? Speculative logic dictates *role confusion,* isolation, frustration, anger, and egocentricity. Ted Bundy may have envisioned a legal career after high school, but his rapacious fantasies compelled him to do otherwise.

PREDATOR PROFILE 6-2

Richard David Berkowitz
The "Son of Sam"

Ashleigh Portales

Birth Name
Richard Falco
Time Span of Crimes
July 29, 1976 to July 31, 1977
Offenses Prior to Serial Killing
1. Arson
2. Attempted Murder

Quoting Berkowitz
"My parents did the best they could . . . I don't know why I didn't respond . . . I became so self-destructive."

Victimology & Modus Operandi

Berkowitz stalked and preyed upon young couples parked in cars. He focused primarily on the female due to the deep hatred and resentment he harbored toward all women. His birth mother had not wanted him; his adopted mother abandoned him (through death); and his stepmother shunned and despised him. As an adolescent and young adult, he had been unable to have a normal, reciprocated relationship with a woman, and was forced to rely upon fantasy. The only real sexual encounter he had during this time resulted in the contraction of a venereal disease from a hooker. Upon capture he said, "I must slay women for revenge purposes to get back at them for all the suffering they caused me."

Birth of "The Son"

Richard David Berkowitz was born on June 1, 1953; the *illegitimate* child of an affair between Betty Broder and Joseph Kleinman, a married real estate agent with whom Broder had begun a relationship after her husband left her for another woman. Broder already had a daughter, Roslyn, from her previous marriage. When she informed Kleinman of her pregnancy, he told her to abort the child if she wanted the relationship to continue. She decided that she couldn't afford to keep the child, so she immediately gave him up for adoption. He became the son of a Jewish couple, Nathan, or "Nat," and Pearl Berkowitz, who couldn't have children of their own. Though grateful, the couple was not particularly the social type, and they imparted their *isolationistic* quality to their son.

Not Able to Socially Bond

Berkowitz grew up a *loner, a self-conscious introvert,* who always felt different from other people and thus uncomfortable around them. He was nervous around the opposite sex and became tongue-tied whenever he attempted conversation. Physically larger than other children his age, he was the neighborhood *bully,* often assaulting kids with little or no provocation. At home, he was hyperactive and largely undisciplined.

In 1965 and again in 1967, his adoptive mother re-contracted cancer. Due to their lack of empathy and poor communication skills, neither Nat nor Pearl prepared Berkowitz for the severity of Pearl's disease. This mistake resulted in serious emotional damage to him, as she wasted away in a cancer ward suffering through chemotherapy treatments. Her death broke his heart. David was only fourteen when his entire world turned upside down. Before Pearl became sick, the family had planned to move to the high-rise development of Co-Op City in the Bronx to escape the growing danger of gang activity in their community. However, by the time their new apartment was complete, Pearl had passed away, and David and his father moved into it alone. Things went steadily downhill from there for young David. His grades greatly suffered, and the changes destroyed his faith in God. He began to believe that his mother's death was part of a larger plot to leave him utterly alone in the world.

Home Alone

In 1971, Nat remarried. The woman was not engaging to his son, and a negative relationship developed with her that compounded his already *strong aversion to women.* The "straw that broke the camel's back" occurred when, without notice, his father and stepmother moved to a retirement home in Florida, leaving him alone to fend for himself (a similar incident later emotionally wrecked the life of Jeffery Dahmer when his mother and younger brother moved away from the family home and left him "home alone").

David began to drift aimlessly about, spending less and less time in the real world in favor of his own increasingly *rich fantasy life.* Part of his fantasy included a romantic relationship with Iris Gerhardt, who considered Berkowitz nothing more than a friend. He made an uninspired attempt at a few college courses at Bronx Community College, mostly to appease his father, but soon lost interest and decided to pursue a new past-time: the United States Army. During his three years of service, he was shipped to Korea, where he became a marksman. Although fellow soldiers converted Berkowitz from Judaism to the Baptist faith, he soon lost interest in religious pursuits. David also experienced *his first and only consummated sexual relationship with a woman,* a Korean prostitute who left him with nothing more than a venereal disease.

When Berkowitz returned home from the Army in 1974, his father was extremely upset that his son had abandoned the Jewish faith. He made life very uncomfortable for David, spurring in him a desire to find his birth family. He contacted the Bureau of Records, learned that his real name was Richard Falco of Brooklyn, and used an old telephone directory to track down his mother and older sister, Roslyn. He left a letter in his mother's mailbox pleading to see her. To his shock, he received a phone call a few days later. The family had an emotional reunion, and David soon found himself staying in his sister's home along with her husband and their children. By all outward appearances, things appeared well for David and his new "family," but like so many times in his emotional roller coaster of a life, they soon changed. Roslyn noticed her brother frequently complained of severe headaches, yet he resisted her efforts to get him to seek help. Also, visits with his mother became increasingly brief. By all appearances, he fast grew bored with "family life." After a few months, David left his sister's house and moved into a small room in a tiny apartment in the Bronx and took a job as a security guard at IBI Security. David always had lived life on the outskirts of society, and he didn't change. In November of 1974, he wrote a letter to his father in Florida, which reflected his *antisocial view of the popular culture from which he always had felt isolated:*

> *"It's cold and gloomy here in New York, but that's okay because the weather fits my mood—gloomy. Dad, the world is getting dark now. I can feel it more and more. The people, they are developing a hatred for me. You wouldn't believe how much some people hate me. Many of them want to kill me. I don't even know these people, but they still hate me. Most of them are young. I walk down the street and they spit and kick at me. The girls call me ugly . . . they bother me the most. The guys just laugh. Anyhow, things will soon change for the better. "*

After writing this letter, he cocooned himself in his small apartment, leaving only to get food. His behavior showed developing *schizotypal features,* perhaps with *schizophrenic episodes,* evidenced by bizarre messages he scrawled around a small hole in one of the walls of his apartment:

> *"Hi. My name is Mr. Williams and I live in this hole. I have several children who I'm turning into killers. Wait til they grow up. My neighbors I have no respect for and I treat them like shit. Sincerely, Williams."*

David later told his psychiatrists that during this time *voices of demons* said they wanted him to "provide blood for them." On Christmas Eve, 1975, Berkowitz decided to listen. He drove to Co-Op City armed with a large hunting knife, on the prowl for a victim. The demons chose a young Hispanic woman whom he spotted leaving a supermarket. Berkowitz later claimed he heard them say, "She has to be sacrificed." After parking his car, he began to follow her. When he caught up to her, he stabbed her in the back but was disappointed when she did not fall dead to the pavement. Fearing discovery, Berkowitz ran back to his car and headed for home. On the way, he crossed paths with fifteen year old Michelle Forman. He parked his car and immediately began to follow her. He overtook the girl and stabbed her in the head and the back until she screamed and fell to the concrete, writhing in pain. Again, Berkowitz retreated to his car.

Forman, his second victim of the night, had a punctured lung, but she managed to stagger into the apartment building where she lived with her parents. She recovered after seven days of hospitalization. Police never learned of the first attack, and the victim outlined in Berkowitz's later confession remained unknown. His demons satisfied for the time being, Berkowitz treated himself to a hamburger and fries.

Howling "Demonic" Dogs

After the attacks, Berkowitz kept his job as a security guard but moved from his Bronx apartment to a two-family home in Yonkers owned by Jack and Nann Cassara. Soon, however, he began to be tormented by the Cassaras' German Shepherd dog. Other neighborhood dogs promptly answered its frequent howls. In Berkowitz's increasingly delusional state of mind, demons lived within the dogs "ordering him to go hunting for the blood of pretty young women." After only three months at the Cassaras' home, David moved out without asking for the return of his security deposit and found an apartment in Yonkers. Regarding the Cassaras, he later told investigators:

> "When I moved in, the Cassaras seemed very nice and quiet. But they tricked me. They lied. I thought they were members of the human race. They weren't! Suddenly the Cassaras began to show up with the demons. They began to howl and cry out, 'Blood and Death!' They called out the names of the masters! The Blood Monster, John Wheaties, General Jack Cosmo."

Later he told prison psychiatrists he did not find relief at the Yonkers location. His neighbor, Sam Carr, had a black Labrador, Harvey, whom Berkowitz tried to kill by hurling a Molotov cocktail at the dog. Berkowitz contended: "Carr was the host of a powerful demon named Sam who worked for General Jack Cosmo." Berkowitz claimed he was this "Sam's" son. According to him, "Sam was the Devil, and only God could destroy him at Armageddon" (paranoid delusions with religious themes often are part of a clinical diagnosis of paranoid schizophrenia). According to later testimony, Berkowitz had not pleased his demons by failing to kill with a knife, so they convinced him to use a gun—a big gun—a .44 Bulldog.

That June, he contacted his friend Billy Dan Parker who lived in Houston, Texas and convinced Parker to buy him a gun. The result was a .44 Charter Arms Bulldog five-round revolver, purchased for $130.

Arsonist

For a period of time, Berkowitz set fires all around New York and kept track of his handiwork in a journal. Authorities later discovered that Berkowitz always had dreamed of being a fireman but never could pass the qualification test. Perhaps arson was the closest he thought he ever could come to this career. In addition to the sexual gratification an arsonist gets from setting fires, David also liked the fact that he could control the destruction of both property and people. He felt emotionally stimulated by the image of a body carried from a burning building, and he fantasized about fiery plane crashes. According to his journal, Berkowitz set a staggering 1,488 fires throughout New York. Before long, however, his "demonic hunger" craved more than fire.

Time for Bloodshed: Body Count Rises

Bloodshed resulting in death satisfied Berkowitz and his demons. He began to feed his appetite on the evening of July 29, 1976, the day before he quit his job as a security guard in favor of becoming a taxi driver—the perfect job for seeking prey. The .44 caliber was about to become the talk of New York City.

Eighteen year old Donna Lauria and her friend, nineteen year old Jody Valente, were sitting in Jody's car outside Donna's home in the Bronx, chatting about boyfriends. When the conversation ended, Donna opened the passenger door to get out. As she did, she noticed a young man standing on the curb a few feet away holding a brown paper bag. As she watched, he reached into the bag, pulled out a gun, and dropped into a crouching position. Puzzled, Donna asked Jody, "What does this guy want?" A bullet struck her in the right side of the neck, quickly followed by another into her elbow. She fell to the sidewalk as another bullet hit Jody in the thigh. Jody lurched forward onto the horn, which alerted Donna's father who was already on his way downstairs to walk the family poodle. Hearing the shots and the horn, he rushed outside to find the girls bleeding profusely. Donna was dead before the ambulance reached the hospital, and doctors treated Jody for her wound and hospitalized her for hysteria. She provided police with a description of the killer as a "clean-shaven white male with curly dark hair, about thirty years old."

Jody was sure that she had not known the attacker and that he was not a disgruntled ex-boyfriend of Donna's. Several neighbors stated that they had seen a yellow car parked a few spaces behind Jody's but that it had disappeared by the time police arrived. During confession, David claimed he had not wanted to kill Donna and her friend, but the "demons made him do it." Yet he admitted that once it was over, *he felt both pleasure and exhaustion*. He stated, "Sam had been pleased," and had promised, "Donna will some day be your bride . . . when she arose from the dead to join him."

Twelve weeks later, the next shooting occurred some distance away from the first, so no connection was made. The burrow of Queens (where the middle class thrived) sat across the East River from the Bronx. It was the complete socioeconomic opposite of the scene of the first murder, which had taken place in the Bronx. Eighteen year old Queens College student Rosemary Keenan and twenty year old record salesman Carl Denaro had just left a bar and driven to a quiet spot where they could be alone. Carl, who had shoulder-length brown hair, was riding in the passenger seat, which probably saved Rosemary's life, as Berkowitz assumed he was a female. David drew the .44 caliber Bulldog from his belt, approached the car, and emptied the barrel into the passenger window. One bullet hit Denaro in the back of the head as he ducked forward to avoid the flying glass. The bullet chipped off a piece of his skull that surgeons later replaced by a metal plate. It required two months of treatment but he survived.

Berkowitz found his next victims on November 27, 1976. Eighteen year old Joanne Lomino and her sixteen year old friend Donna DeMasi were sitting on the front steps of Joanne's home around midnight when a man walking down the opposite side of the road caught their attention. As soon as he spotted them, he abruptly turned, crossed the street, and headed straight for the pair. He pretended to ask for directions but quickly pulled out his gun and began firing. The first shot hit Joanne in the spine and paralyzed her from the waist down, and the second shot entered the base of Donna's neck, an injury that required a three-week recovery.

As the two stumbled into the bushes, Berkowitz followed, emptying his gun, though the remaining three shots missed both victims. Right after the last shot, the gunman fled, but not before being spotted by a neighbor, who described a man with long blonde hair to the police, a description the girls backed up. Police suspected a connection between the shootings, but the discrepancies in the shooter's appearance puzzled them.

The trail went cold until the shooting of thirty year old John Diel and his twenty-six year old fiancée Christine Freund on January 29, 1977. They were returning to their home in Queens from a late dinner when the "Son of Sam" struck again. A bullet to the head caused Christine's death at the hospital a few hours later. Ballistics experts linked the four crimes together, but they could not explain why the descriptions of the suspects did not match, and they began to speculate that more than one shooter might be using the same gun.

The next to die was nineteen year old Virginia Voskerichian, a student at Columbia University who was returning to her home in Queens on March 8, 1977. She encountered Berkowitz on the street and stepped aside to allow him to walk past her. Rather than complying, he pulled out his gun and shot her directly in the face. The bullet entered her upper lip and knocked out several teeth before lodging in her skull. She died instantly. Witnesses reported that a man in a black ski mask ran away from the scene. Ballistics matched the bullet recovered from Voskerichian to the other four shootings, and newspapers began running stories of "The .44 Killer."

Task Force Assembled

A task force formed to apprehend the killer, but it faced the difficulty of sifting through an average of *three hundred tips* each day. News of the task force reached Berkowitz, who, true to the model of a psychopath, desired media attention and recognition for his crimes. He made sure to get the publicity he felt he so deserved when he shot and killed eighteen year old Valentina Suriani and her twenty year old boyfriend Alexander Esau on April 16, 1977. They were "making out" in their borrowed Mercury Montego. Berkowitz's first two shots hit Valentina in the skull, resulting in immediate death. The next two entered the top of Esau's head and resulted in his death two hours later.

At the scene, Berkowitz left a white envelope, void of fingerprints, a few feet away from the car in the middle of the road where it could not be missed. Inside was a letter addressed to Captain Joe Borrelli, the deputy of Timothy Dowd, who was leading the Omega Task Force. Written in bold block capital letters, it read:

> *DEAR CAPTAIN JOSEPH BORELLI,*
> *I AM DEEPLY HURT BY YOUR CALLING ME A WEMON-HATER. I AM NOT. BUT I AM A MONSTER. I AM THE "SON OF SAM." I AM A LITTLE BRAT.*
>
> *WHEN FATHER SAM GETS DRUNK HE GETS MEAN. HE BEATS HIS FAMILY. SOMETIMES HE TIES ME UP TO THE BACK OF THE HOUSE. OTHER TIMES HE LOCKS ME IN THE GARAGE. SAM LOVES TO DRINK BLOOD.*
>
> *"GO OUT AND KILL," COMMANDS FATHER SAM.*
>
> *BEHIND OUR HOUSE SOME REST. MOSTLY YOUNG—RAPED AND SLAUGHTERED—THEIR BLOOD DRAINED—JUST BONES NOW.*
>
> *PAPA SAM KEEPS ME LOCKED IN THE ATTIC TOO. I CAN'T GET OUT BUT I LOOK OUT THE ATTIC WINDOW AND WATCH THE WORLD GO BY.*
>
> *I FEEL LIKE AN OUTSIDER. I AM ON A DIFFERENT WAVELENGTH THEN EVERYBODY ELSE—PROGRAMMED TO KILL.*
>
> *HOWEVER, TO STOP ME YOU MUST KILL ME. ATTENTION ALL POLICE: SHOOT TO KILL OR ELSE KEEP OUT OF MY WAY OR YOU WILL DIE!*
>
> *PAPA SAM IS OLD NOW. HE NEEDS SOME BLOOD TO PRESERVE HIS YOUTH. HE HAS HAD TOO MANY*

HEART ATTACKS. "UGH, ME HOOT, IT HURTS, SONNY BOY"

I MISS MY PRETTY PRINCESS MOST OF ALL. SHE'S RESTING IN OUR LADIES HOUSE. BUT I'LL SEE HER SOON.

I AM THE "MONSTER"—"BEELZEBUB"—THE CHUBBY BEHEMOTH.

I LOVE TO HUNT. PROWLING THE STREETS LOOKING FOR FAIR GAME—TASTY MEAT. THE WEMON OF QUEENS ARE THE PRETTYIST OF ALL. IT MUST BE THE WATER THEY DRINK. I LIVE FOR THE HUNT—MY LIFE. BLOOD FOR PAPA.

MR. BORRELLI, SIR, I DON'T WANT TO KILL ANYMORE. NO SUR, NO MORE, BUT I MUST, "HONOUR THY FATHER."

I WANT TO MAKE LOVE TO THE WORLD. I LOVE PEOPLE. I DON'T BELONG ON EARTH. RETURN ME TO YAHOOS.

TO THE PEOPLE OF QUEENS, I LOVE YOU. AND I WANT TO WISH ALL OF YOU A HAPPY EASTER. MAY GOD BLESS YOU IN THIS LIFE AND IN THE NEXT.

I SAY GOODBYE AND GOODNIGHT.

POLICE: LET ME HAUNT YOU WITH THESE WORDS:

I'LL BE BACK!

I'LL BE BACK!

TO BE INTERPRETED AS—BANG, BANG, BANG, BANG, BANG—UGH!!

YOURS IN MURDER
MR. MONSTER

Authorities kept the letter secret, but they allowed one journalist, Jimmy Breslin, to see it. He leaked clues about it in his column in the New York *Daily News,* which prompted editors to publish it in its entirety. It is possible that, despite several misspellings, Berkowitz intentionally spelled women "wemon" to rhyme with "demon." Though much remained a mystery, the letter made one thing clear. The killer either was insane or wanted to be thought insane and sent the letter to set up an insanity defense.

Though police found fingerprints, they were only partials from the very tips of the fingers and not suitable to match, which suggested the writer had held the paper by its very edges to prevent prints. What investigators did not know was that one week earlier, Berkowitz had sent an anonymous letter to his neighbor, Sam Carr, concerning his black Labrador. Two days after the murder, on April 19, another letter came in the same all uppercase handwriting as the first, stating:

I HAVE ASKED YOU KINDLY TO STOP THAT DOG FROM HOWLING ALL DAY LONG, YET HE CONTINUES TO DO SO. I PLEADED WITH YOU. I TOLD YOU HOW THIS IS DESTROYING MY FAMILY. WE HAVE NO PEACE, NO REST. NOW I KNOW WHAT KIND OF A PERSON YOU ARE AND WHAT KIND OF A FAMILY YOU ARE. YOU ARE CRUEL AND INCONSIDERATE. YOU HAVE NO LOVE FOR ANY OTHER HUMAN BEINGS. YOUR SELFISH, MR. CARR. MY LIFE IS DESTROYED NOW. I HAVE NOTHING TO LOSE ANYMORE. I CAN SEE THAT THERE SHALL BE NO PEACE IN MY LIFE, OR MY FAMILIES LIFE UNTIL I END YOURS.

Carr called police, who remained reluctant to act until ten days later when Carr phoned again to report gunshots coming from his backyard. There he discovered Harvey, the black Labrador, bleeding, and a man wearing jeans and a yellow shirt running away. Carr rushed the dog to the vet, who saved him. An investigation ensued, but no one thought to connect the incidents because authorities had not released the Borrelli letter yet.

Soon after the Borrelli letter, journalist Jimmy Breslin received a letter, written again in the same block lettering, all capitals:

"HELLO FROM THE CRACKS IN THE SIDEWALKS OF NYC AND FROM THE ANTS THAT DWELL IN THESE CRACKS AND FEED IN THE DRIED BLOOD OF THE DEAD THAT HAS SETTLED INTO THE CRACKS.

HELLO FROM THE GUTTERS OF NYC, WHICH IS FILLED WITH DOG MANURE, VOMIT, STALE WINE, URINE, AND BLOOD. HELLO FROM THE SEWERS OF NYC WHICH SWALLOW UP THESE DELICACIES WHEN THEY ARE WASHED AWAY BY THE SWEEPER TRUCKS.

DON'T THINK BECAUSE YOU HAVEN'T HEARD FROM ME FOR A WHILE THAT I WENT TO SLEEP NO, RATHER, I AM STILL HERE. LIKE A SPIRIT ROAMING THE NIGHT. THIRSTY, HUNGRY, SELDOM STOPPING TO REST; ANXIOUS TO PLEASE SAM.

SAM'S A THIRSTY LAD. HE WON'T LET ME STOP KILLING UNTIL HE GETS HIS FILL OF BLOOD. TELL ME, JIM, WHAT WILL YOU HAVE FOR JULY 29? YOU CAN FORGETABOUT ME IF YOU LIKE BECAUSE I DON'T CARE FOR PUBLICITY HOWEVER, YOU MUST NOT FORGET DONNA LAURIA AND YOU CANNOT LET THE PEOPLE FORGET HER EITHER. SHE WAS A VERY SWEET GIRL.

NOT KNOWING WHAT THE FUTURE HOLDS, I SHALL SAY FAREWELL AND I WILL SEE YOU AT THE NEXT JOB? OR SHOULD I SAY YOU WILL SEE MY HANDIWORK AT THE NEXT JOB? REMEMBER MS. LAURIA. THANK YOU.

IN THEIR BLOOD AND FROM THE GUTTER,

'SAM'S CREATION.'44

HERE ARE SOME NAMES TO HELP YOU ALONG. FORWARD THEM TO THE INSPECTOR FOR USE BY N.C.I.C.:

'THE DUKE OF DEATH'

'THE WICKED KING WICKER'

'THE TWENTY TWO DISCIPLES OF HELL'

'JOHN WHEATIES'—RAPIST AND SUFFOCATOR OF YOUNG GIRLS

P.S.: J. B. PLEASE INFORM ALL THE DETECTIVES WORKING THE SLAYINGS TO REMAIN."

The *Daily News* published the letter again, save for the list of names reserved for the N.C.I. C. Police had requested their omission, as they felt they could contain some vital clues to the killer's identity. This time, however, Berkowitz had not taken enough care. Partial fingerprints appeared large enough to make a match. Though they did not aid in the initial investigation, they confirmed Berkowitz's identity as "The Son of Sam" upon his arrest.

By request of New York City Mayor Abraham Beame, a team of forty-five psychiatrists, headed by Dr. Martin Lubin, former head of forensic psychiatry at Bellevue, viewed the letters and produced a *psychological profile* of the offender. They told the police the man they sought was "a *paranoid schizophrenic,* who may have considered himself possessed of a demonic power." They were also certain the killer was "a loner who had difficulty with relationships, particularly relationships with women" (one of the characteristics of schizotypal personality disorder).

Before police got a chance to use the profile, another murder occurred. On June 25, 1977, three weeks after the publication of the Breslin letter, seventeen year old Judy Placiso, a classmate and funeral attendee of Valentina Suriani, was celebrating her graduation at a discotheque in Queens. There she met Salvatore Lupo and decided to leave with him. As they talked in the car, Berkowitz approached and began firing. The first bullet passed through Salvatore's wrist and the skin of Judy's neck. The next one hit her skull but did not penetrate; a shot to her right shoulder followed. The bleeding victims exited the car seeking help. Both survived the attack, but neither had gotten a clear view of the perpetrator. The only witness, three blocks away, reported seeing a "stocky white male" running away.

A Hippie Type Man

The anniversary of Donna Lauria's murder came and went without incident, despite the killer's insistence about it in his letter, but police suspected this was not the end of the killer's reign of terror. Their suspicions became confirmed on August 1, 1976, when twenty year old Stacy Moskowitz and her date, Bobby Violante, decided to take a walk along Shore Parkway after seeing a late movie. As they sat and talked, they noticed a "hippie type" leaning against a wall, but he soon walked away. They walked to their car, got inside, and started "making out." A shattering window and gunfire interrupted them. Two bullets hit Violante in the face, blinding him and shattering his eardrums. Stacy suffered fatal wounds and died thirty-eight hours later in the hospital.

Tommy Zaino had witnessed the entire incident in his rearview mirror sitting in a car parked a few yards in front of Bobby and Stacy. He reported that a "stocky man with stringy hair" approached the car from behind, pulled a gun, crouched, and began to fire. As the shots rang out, Tommy's girlfriend, Debbie Crescendo, asked, "What was that?" He said, "Get down. I think it's Son of Sam." Zaino looked again to see the man running away and noticed on his dashboard clock that it was exactly 2:35 A.M. In addition, several people in the area had noticed a yellow VW parked near the entrance to the Shore Parkway playground.

Earlier in the day, the car had followed a girl on a bicycle. At the time of the killings, another couple seated on a park bench heard the shots and then observed a man in a denim jacket, carrying what they thought might be a cheap nylon wig, jump into the car and drive away. The woman had commented to her boyfriend, "He looks like he just robbed a bank." Farther away, a nurse also had heard the shots and looked out her window to locate the disturbance just in time to see the VW speed away. In the process, it recklessly cut off another motorist and almost caused an accident. The other driver involved became so angry he followed the VW for several blocks before losing track of it. He later described the VW's driver as having stringy brown hair.

The police believed the mystery of the mismatched identities solved. There was only one shooter who decided to wear a wig on some occasions.

Or, was there another explanation?

A Critical Break

The most crucial witness would not come forward for fear of retaliation from the shooter. Her information would end "The Son of Sam's" reign of terror. Forty-nine year old widow Cacilia Davis had returned from a date at around 2:00 A.M., and she and her date sat in the car, triple parked on a one way street. Because of this, Davis was watching for police cars approaching. Sure enough, she soon noticed a police officer place a ticket on the windshield of a yellow Ford Galaxie parked next to a fire hydrant. Davis hurried from the car to her apartment, noticing along the way that the driver was a young man with dark hair. A few minutes later, Davis decided to take her dog for a walk in the park. On her way home a few minutes later, she noticed a dark-haired young man in a denim jacket crossing the street. Davis made a mental note that he resembled the man she had seen driving the yellow Ford Galaxie. Though Davis feared the "Son of Sam" might try to come after her in order to eliminate a witness, friends convinced her to come forward. When she did, she met police resistance. No record existed of a traffic ticket that night and, after all, their suspect drove a VW. But Davis persisted, and ten days later police found the *misfiled parking ticket.* It had NY license plate number 561-XLB and was a yellow Ford Galaxie registered to a person who lived at 35 Pine Street, Yonkers—a man named "David Berkowitz."

What Lies Beneath "The Son of Sam"

Four days later, detectives decided to check out the Berkowitz residence, and they started with his Ford Galaxie. Peering into the windows, they noticed a duffel bag in the back seat with a rifle butt sticking out of it. While not illegal, it aroused the suspicions of Detective Ed Zigo, who opened the car door and examined the gun. The semi-automatic Commando Mark III was not usually seen in the possession of the average, law-abiding citizen. Detective Zigo next opened the glove box, a move that sealed Berkowitz's fate. He found a letter inside addressed to Deputy

Inspector Timothy Dowd, the detective in charge of the Omega Task Force, threatening an attack on Long Island.

Zigo rushed to the nearest pay phone, called headquarters, and told Sergeant James Shea, "I think we've got him." Six hours later, Berkowitz walked out of his Yonkers apartment and got into his car. As soon as he shut the door, Detective William Gardella knocked on the window and pointed his gun at Berkowitz's head. "Freeze!" he ordered. "We're the police!" Berkowitz just looked at him and flashed an eerily calm smile. Another officer, Dectective John Falotico, entered the car from the passenger side, put his revolver to Berkowitz's head, and forced him out of the car. As Berkowitz placed both hands on the roof of the car, Falotico asked him, "Who are you?" Turning to the detective with a childish smile, he replied, "I'm Sam."

Back at headquarters, Berkowitz was more than happy to tell his story. In fact, he was so cooperative that the interrogation took only half an hour. He related his story of the demon voices he heard coming from his neighbor's dog ordering him to kill. Leaving the interrogation room, Sergeant Joseph Coffey summed up his feelings. "I feel sorry for him. The man is a f***ing vegetable."

Defense psychiatrists classified Berkowitz as a *paranoid schizophrenic.* The prosecution's equivalent contended that, while he showed paranoid traits, they did not affect his ability to know right from wrong. He declared him "fit to stand trial."

"The defendant is as normal as anyone else," stated Dr. David Abrahamsen, "maybe a little neurotic." Ultimately, the clinical analysis would not matter, for in the end Berkowitz pled guilty to all the "Son of Sam" shootings. They sentenced him to three hundred and sixty-five years in prison.

The Demon Voices Silenced

The case of David Berkowitz did not end with his incarceration. In 1979, FBI profiler Robert Ressler met with Berkowitz on three different occasions in Attica Prison. He learned that they had allowed Berkowitz to keep a scrapbook of newspaper clippings relating to the crimes, and that he often used these to revisit the scenes to indulge his fantasies. From the very beginning of his interviews, Ressler made it clear that he did not buy the story of the demon voices, and soon he got the truth from Berkowitz. He confessed that he carefully *concocted the entire story in order to prepare for an insanity defense.* He admitted to Ressler that his real reason for shooting women was "out of *resentment toward his own mother, and because of his inability to establish relationships with women.* "

He further revealed that stalking and slaying women aroused him sexually and that, if he did not find a victim on any given night, he would revisit the scene of an earlier murder. Ressler quotes from his interviews that "it was an erotic experience for him to see the remains of bloodstains on the ground and police chalk marks." He would contemplate the grisly mementos and masturbate in his car.

Berkowitz also confessed that he had desired to attend the funerals of his victims but had feared police would place the services under surveillance. He admitted to sending the letters in an attempt to attract media attention, thereby feeding his *pathological desire for recognition.* He admitted that he had liked the moniker "Son of Sam" and even had fashioned a logo after the name.

On July 10th of the same year, an unidentified inmate in his Attica cellblock attacked Berkowitz with a razor, slashing him from the left side of his throat to the back of his neck. The injury required fifty-six stitches. Had the cut gone slightly deeper, he would have died. Berkowitz would not identify his attacker, but later he speculated whether the occult had attempted to insure his silence.

The One and Only Son?

There is an interesting twist. New evidence suggests that "Sam" may have had more than one "son." Investigative journalist Maury Terry got confused by all of the contradictions between witness statements and the actual timeline of events, as well as the conflicting accounts of the shooter's appearance. For instance, some had seen a short, stocky man with short dark hair (obviously Berkowitz), while others reported a taller, "hippie" type with stringy blonde hair. Terry re-interviewed all the witnesses, who reiterated their original stories, but stated that they believed Berkowitz was really the shooter; they just had mistaken their recollections.

Not satisfied, Terry dug deeper. What he found was enough to write a book, *The Ultimate Evil.* The evidence he unveiled made a surprisingly strong case that Berkowitz may not have acted alone. According to Berkowitz, he never actually had met his neighbor, Sam Carr, a fact Carr confirmed. Yet the letter writer had revealed an obsession with Sam Carr. Terry then learned that Carr had two sons, John and Michael, both of whom hated him. Furthermore, John's nickname was "Wheaties," (the name mentioned in the Breslin letter). Wheaties was a hippie-type with stringy blonde hair.

In addition, it was common knowledge that dogs obsessed Berkowitz. Terry learned that, in the year that the "Son of Sam" killings occurred, eighty-five German Shepherds and Dobermans, "dead and skinned," had appeared in Yonkers and in Walden, New York, only an hour's drive away. A local teenager informed Terry that, as far as he knew, a satanic cult that held sacrificial ceremonies in the woods nearby had committed the Yonkers dog slayings.

Terry recalled that the "Son of Sam" letters had made obvious satanic references. He set out to find John "Wheaties" Carr, but he learned with a shock that John had died in Minot, North Dakota, the victim of what authorities ruled a suicide. To Terry, however, the evidence suggested otherwise. Supposedly, Carr had shot himself in the mouth with a rifle in his girlfriend's bedroom, yet on the boards skirting the bed beside his body appeared the scrawled letters S.S.N.Y.C. in blood. If a man shot in the head died immediately, who had written the letters? It appeared to Terry and to several police officers in the area that someone had killed Carr; they had probably beaten him to the ground where he wrote the letters in his blood before his assailants finished him off. What did the letters mean? A possible interpretation is: "Son of Sam, New York City." In addition, the numbers 666 appeared written in blood on Carr's hand, which convinced Terry of a cult connection.

Police files revealed that Carr in fact had become involved with an occult group and knew David Berkowitz. His girlfriend, Linda O'Connor, reported that when news of Berkowitz's arrest as the "Son of Sam" broke on the television news, Carr remarked, "Oh, shit."

In fact, John had visited New York on at least four and possibly five of the dates of the "Son of Sam" murders, and he bore a striking resemblance to police composites of the shooter. In January 1978, Carr had driven the fifteen hundred miles from Minot to New York, intending on an extended stay. However, only two weeks later, he phoned his girlfriend to tell her the police wanted him, which prompted a quick trip back to Minot. There he opened a bank account, rented a post office box, and visited the local Air Force base about continuing to receive his service disability checks. His actions do not seem commensurate with a man intending to take his own life, yet two days later they found his body in O'Connors' bedroom.

Terry's discoveries drew so much attention that John Santucci, the District Attorney of Queens, reopened the case for further investigation. Terry began to track down people who had known Berkowitz and became surprised by how many friends he found of the supposed "loner."

Berkowitz had met Michael Carr, John Carr's drug-addicted brother, the year before the killings began (1975) at his Bronx apartment, where he attended a party held by a witchcraft coven John belonged to. The name of the coven was *The Twenty Two Disciples of Hell,* and the Breslin letter referred to it. The coven was the reason Berkowitz eventually moved to the Yonkers address only two hundred yards away from Michael. Terry decided he needed to speak with Michael Carr and tracked him to the same literally "dead" end he had met with his brother, John. Michael had died on October 4, 1979 when his car collided with a streetlamp at 75 mph. However, the road showed no skid marks, and his sister suspected murder. It did appear that someone had forced him off the road or shot a tire out.

In February that same year, Berkowitz personally had called a press conference to renounce his claims of hearing demonic voices during the murders, admitting it all a carefully constructed insanity hoax. Berkowitz began to correspond with Terry by letter and to give him verifiable facts that supported the new investigation. One such fact was that, shortly after the murders began, Berkowitz had gone to an interview for a job at a dog pound. Though the pay was no incentive, he claimed, "There was another way in which I was getting paid, somebody needed dogs! I guess you understand what I'm trying to say."

In another of his letters, Berkowitz instructed Terry to "Call the Santa Clara (California) Sheriff's Office . . . Please ask the sheriff . . . what happened to ARLISS, PERRY." What Terry discovered was unlike anything he could have imagined.

Perry Arliss's murder took place on October 13, 1974 in the church at Stanford University. According to the *Crimes and Punishment: Murder Casebook:* "The body was naked from the waist down, and spread eagled in a ritualistic position, with an altar candle inserted into the vagina. Her arms were folded across her chest, and between her breasts was another candle. Her jeans were so arranged across her legs that the result was a diamond pattern. She had been beaten, choked, and stabbed with an ice pick behind the ear." Berkowitz revealed to Terry things never publicized about the murder, suggesting that he knew who killed her. In his book, Terry attributes the murder to a West Coast satanic cult with whom Charles Manson once associated.

Terry's Conclusions

The conclusion drawn by Terry is that Berkowitz committed three of the "Son of Sam" killings—those of Donna Lauria, Alexander Esau, and Valentina Suriani (two separate shootings). Evidence, including witness descriptions, suggests that the others were John Carr's victims. According to one informant, the shooter in the black mask was a female member of the coven. Reasons for the various murders also began to surface. Christine Freund had angered a coven member. Stacy Moskowitz, whom Terry theorized John Carr had killed, was the subject of a secretly recorded snuff film. This is why they had chosen the car parked under the streetlamp over one parked in the dark: to allow for better lighting. The supposed leader of the satanic cult in New York was Roy Alexander Radin, a "showbiz tycoon" who relocated to California in 1982. However, when Terry attempted to contact him, he learned of Radin's murder during the following year. Someone had dumped his body in Death Valley with a defaced Bible lying nearby.

Terry makes it clear at the end of his book that, while John and Michael Carr and Roy Radin are dead, many of the members of *The Twenty Two Disciples of Hell* still live. He provides descriptions of more recent murders, which suggest the coven still may be active. Did David Berkowitz act alone? Those who may know for sure are either dead or not talking. Yet, true to the clinical dictates of psychopathy, Berkowitz revels in the glory he received in the press for the crimes, regardless of whether he was truly the perpetrator or not.

Aftermath

David Berkowitz's first parole hearing took place on July 9, 2002, at Sullivan Correctional Facility in Fallsburg, NY. Forty-nine year old Berkowitz attended, although he had failed to appear at the hearing scheduled the month before. When Commissioner Irene Platt inquired why he had not shown up the previous month, he replied, "I had a lot of anxiety, and I thought it would be best for the families that I not come at all and I, after a lot of soul searching and a lot of praying, I just decided it would be best to just come and face you and apologize. I'm not seeking parole. I don't feel that I deserve parole."

Commissioner Platt then asked why he felt this way. Berkowitz answered,

> "Well, for the crimes that were committed and the people that are suffering today because of my actions. I know they have a lot of pain and hurt that will probably never go away. I wish that I can go back and change the past. I can't, so I have to come to terms with this and realize that I'm here in prison."

The Commissioner then focused on Berkowitz's crimes and inquired about the process of *victim selection* and the motivations behind the crimes. Berkowitz ultimately offered no explanation. "Ma'am, I'm sorry. I don't know. I don't understand what happened. *It was a nightmare.* I was tormented in my mind and in my spirit. My life was out of control at that time, and I have nothing but regret for what happened." When asked to elaborate on that "torment," he stated, "It was just my mind was not focused right. *I thought I was a soldier for the devil and all kinds of crazy things.* I had things like the satanic bible that I was reading. I just got stupid ideas out of it. I'm not pushing the blame on anything. I take full responsibility, but I just at the time things got twisted."

When the Commissioner suggested that Berkowitz really had not dealt with the severity of his actions, he replied, "Ma'am, in all honesty I really haven't. I still struggle with coming to grips with the things of the past. There are still issues that I have to deal with. I'm not there yet."

Thankfully, they formally denied parole at that time, as well as at the June, 2004 hearing that followed.

CHAPTER SEVEN

Psychology of Perversion

Child, Adolescent, & Female Killers

"Child abuse or neglect, unstable or erratic parenting, or inconsistent parental discipline may increase the likelihood that Conduct Disorder will evolve into Antisocial Personality Disorder."

—DSM-IV-TR (2002)

Psychology of perversion is aptly na[illegible] as the behavior of women and children who become perpetrators of [illegible]ce is examined in Chapter Seven. Kids—many too young to drive—[illegible]er kids and adults in homeplace and schoolplace violence. The rem[illegible] of the chapter provides an overview of Females Who Kill, such as so[illegible]lack Widows" and "Angels of Death."

As we have s[illegible]ow metamorphosis from normalcy to perversion is a byproduct o[illegible] toxic, *predatory parenting* that produces deep emotional scars. *B[illegible]motional scars lies a brain rewired toward violence and impulsivity,* ofte[illegible]g individuals with considerable neurological damage.

[illegible]hough small children who are abused and mistreated scream, cry, [illegible]ing to incompetent "caregivers" when authorities arrest them in their [illegible]esence and take them from the home. They cry: "Please don't put my daddy in jail!" "Who's going to take *care of me?*" "Care" may be the only word the child knows, but CPS (child protective service) and police investigators define this kind of care as *incompetence, neglectful, and abusive.* Authorities know the direction of these kids' lives. They know every single social agency will fail them. Young girls may end up in depressing *co-dependent* relationships or as drug-addicted prostitutes. She may turn to petty crime or become a killer as she is filled with self-hatred. Often, the boys become violent criminals and/or sexual psychopaths—the serial killers of society.

Violent psychopathy starts in the home, but the predictable outcome of violence ends in many places and has many faces.

Normal brain development becomes perverted and abnormal due to hateful toxic parenting. High resolution brain scanning technology documents the extent of the trauma by showing damage in "cool-coded" blue that denotes minimal blood flow and activity. Violent psychopathy "hides" in cerebral tissues such as the:

- temporal lobes
- cerebellum
- amygdala
- prefrontal lobes—the center for restraint and social appropriateness

High resolution brain scanning continues to evolve. It shows clear demarcation between normalcy and abnormality in neural centers. The brains of kids in primary school who have a propensity to kill classmates can appear neurologically different. We predict these sophisticated brain scans will show a decreased networking of restraint centers that broadcast to prefrontal lobes and associated cortices via the thalamus and other neural pathways. What do we expect with restraint muted?

We can envision a time in the near future when children will have to show evidence of brain scans as well as a vaccine record before registering for public schools, unless all schools are equipped with metal detectors. Authorities using "cool-coded" scans that show damaged prefrontal lobes will send those children off to clinical facilities with bars on the windows.

In the meantime, the *preponderance of the evidence* is far too consequential to ignore. It includes the weight of all that neuroscience and experimental psychology know about human development, cognition, and affective (emotional) centers in the brain, as well as how this knowledge stacks up against sexualized perversion, violence, and deviant cognitive mapping. Classic, long-standing animal studies both in North America and the UK, with decades of replication in labs around the world, document the *neuropsychological ramifications* of deficits in touch, bonding, cognitive mapping, and cerebellar inactivity that points to neurological damage and psychopathy. And now we have brain scans.

Cool-Coded Brain Scans

Forensic neuroscience by way of *high resolution brain scanning technology* in the courtroom shows the telltale cool-coded blue of *prefrontal lobe damage* in human predators. At least we have a reason why their behavior is unstable and violent. We understand why they will never stop.

This chapter is an **apocalyptic** wake-up call that speaks to core issues of women and children in North American society. Social historians remind us that civilizations start declining and freefall into **decadence** when mothers and children do not receive protection from emotional and physical abuses, or when mothers and children become so violent themselves that the very essence of the family unit is in danger of becoming a breeding ground for violence and psychopathy. Until families learn to be competent, skillful, loving, and respectful of human dignity, and schools become empowered with ***in loco parentis*** to do their jobs with authority, it will only get worse.

The only reason kids and adolescent killers do not progress into the ranks of serial killers is that they get caught red-handed by authorities in schools or in homes by shocked parents. The most shocking statistic is how young some of them are—eleven to eighteen years old!

Young Killers

Young, prepubescent and adolescent killers reside along the psychopathy continuum in line with serial killers, rapists, and other *malum in se* criminals (literally, a crime "evil in itself"). They ambush and kill their parents in their own homes. They bring weapons to school to kill classmates, teachers, staff, janitors, and principals. They kill with guns, knives, baseball bats, brickbats, and bare hands—and seldom show remorse. Some, like the teenagers at Columbine, go out in a blaze of suicidal glory in the context of guerilla warfare. This occurs not in some faraway country defending democracy, but in the halls of suburban high schools amid middle-to-upper-class values in decent communities.

Born and Reared in Pop Culture USA—Valueless, Impulsive, and Violent

The only thing that prevents young killers from killing again is capture during or shortly after they kill. **Pop Culture USA** births them, and nurtures them.

It is important to understand that popular culture (*pop culture* for short) is never about *competence, morality, decency, or ethical behavior.* Pop culture is about *appearances, superficiality, hedonism, recreational drug use, and* **pseudo-intellectualism.** At best, pop culture is *pretentious, flashy,* and *seeks immediate BandAid® solutions.* At its worst, it is the harbinger of thugs, criminals, and psychopaths, masquerading as normal, everyday citizens.

Christianity and Catholicism, once moral compasses to righteous behavior, have become reduced to ritualistic **dogma.** The major religions of North America have apparently lost out in a tug of war with *pop culture hedonism.* As statistics of criminal behavior pile up against wholesome morality, the evidence for failure of morality is right in our faces every time we purchase a newspaper or watch the evening news. Many citizens choose to bury their heads in the sand of denial.

Behaving Badly

Some priests, preachers, youth ministers, and televangelists who *behave badly* have all but driven the stake into the heart of religion as an ideology to be taken seriously. Recently documented examples of clergy behaving badly include child porn sites found on church computers, priests molesting altar boys, sixteen year old girls or boys having sex with youth ministers, and pastors having sex with the secretaries and the flock. The message from pulpits is clear: Preach one thing and practice another. Put on your Sunday best. If "you look good, and you're in a good place, you must be good." Yes, we know the argument—they're just people like everyone else.

Regardless, the damage done by **hypocrisy** is especially cruel to young minds that lack perspective, maturity, and experience. The attitude "we sin like everybody else, but we're saved" reeks of insincerity and impropriety. This mindset is especially dangerous to sexual psychopaths who often operate out of religious obsession. They "know" they have God's forgiveness no matter how much they "rip up" victims. They have forgiveness by a higher power.

Neurotheology

A truly spiritual lifestyle requires the use of the higher centers of the brain in providing restraint, morals, ethics, and a certain amount of compulsivity in weighing consequences (see William James' *Varieties of Religious Experience,* Collier, 1966, for an excellent discussion about *religiosity* versus *spirituality*). The argument some **neurotheologists** make is compelling. Perhaps society needs less **ostentatious** religion and more profound spirituality to combat pop culture's powerful influence.

In contrast, the brain's lower centers (brainstem and midbrain limbic regions) have neural centers dedicated to pleasure. These places jazz and stimulate revelers who seek **pan-hedonism:** sexuality, drugs, and dangerous innuendo in relationships. The message is particularly seductive for the young and beautiful—the mannequins of pop culture, full of hopes and dreams. Hedonistic **aphorism** abounds: "If you look good, if you have any measure of fame, or if you make a lot of money, *you must be good.* " When an interview takes place with a pop culture "icon" or a celebrity, America listens as if expecting to hear something profound.

What do we expect to hear from self-centered narcissists who have embraced the very essence of pop culture as their ticket to fame? What we usually get is variations on *pseudo-intellectualism.* If you use a big word, you must be smart. For example, how many times have you heard celebrities use the word **"epiphany?"** It's one of their favorite words. Spending their entire lives in a career devoted to playing other people may explain why some have so much trouble "finding themselves."

Pop culture looks really good as surface shine, but just don't "pop the hood" because chances are you're "buying" a lemon.

YAAVIST Society—Components of Pop Culture USA

According to Jacobs (1992), the following components of pop culture provide product manufacturers with an influential launch pad for *psychological manipulation* through all forms of **consumerism.** The list is by no means exhaustive. However, our real concern is the less recognized *psychological impact* on the minds and behavior of young adolescents and adults. YAAVIST society pertains to:

1. **Youth.** A familiar slogan by a soft drink manufacturer states: "Think Young. Drink Pepsi." The over-forty crowd, especially middle-aged and senior-aged individuals, loses out to the media blitz of fresh-faced young bodies. *Forget*

mind. Forget personality. Think sexualized bodies. The message inflates self-importance and body type as indicators of desirability and even success. Many young adolescents and young adults don't feel they "match" the almost unattainable visual images in magazines and in electronic media. The "Y" in YAAVIST is all about *appearances, superficiality, and physicality* as the banner headlines of pop culture. You're in the "in crowd" with them, enjoying the superficial allure, the full banquet table afforded adolescent youth.

2. **Attractive.** Closely related to youth is attractiveness—looking beautiful or handsome. This equates with sexuality and desirability. Reality TV cashes in on the pop culture's obsession with youth and attractiveness. The message from media advertisers is: "If you don't have it, then spend the money necessary to get it." The alternative is exclusion from YAAVIST Society. Liposuction. Breast Enhancement. Rhinoplasty. Facelift. Tummy Tuck. The standard for beauty in North America is practically unattainable with the airbrushed perfection of pop culture mannequin-like models. YAAVIST showcases individuals with garden variety psychopathy. For aging psychopaths and everyone else, next stop: Botox® injections and red Corvettes.
3. **Visual.** Since visual imagery is the "language" of pop culture, we expect to be blitzed with youthful and attractive bodies and young faces. The visual message is disturbing: "If you look good, you must be good." *A glance establishes trust.* Generally, females considered serial killer Ted Bundy somewhat attractive. He approached females on college campuses and asked help to carry books to his car armed with a sling around his arm (hiding a crowbar), decked out in a business suit, and blessed with an engaging personality and a "gift of gab." Those who complied . . . died. Searchers never found some of the bodies. Pop culture dazzles with visual appearances and surface shine. YAAVIST is like a low-calorie health drink, filling for the moment, but lacking substantive nutrition.
4. **Isolation.** Not wanting to be left out, young males and females clamor to bars and nightclubs, hoping to find conversation, companionship, maybe sex, and even love. Where else? Everyone wants to be an **extravert**—a people person—whether or not they naturally fit the typology. Trying to be someone they're not, they shun the **introvert** label with its connotation of being a pathological loner. This couldn't be farther from the truth. The brain's *RAS* (reticular activation system) determines the "amplitude" of incoming sensation that determines the *attitudes of introversion versus extraversion. Serial* killers interviewed on death row expressed feelings of *isolation from mainstream interaction. They feel rejected by those they fantasize about and eventually kill.*
5. **Success.** Dress. Cars. Jewelry. Pop culture caters to those with money to buy whatever consumable products they perceived keep them young, attractive, sexy, thin, and desirable. We need go no further than the billion-dollar-a-year profiteers of diet schemes who promise perfect bodies as the quickest way to "have our dreams come true." As everyone knows, if a person drives an expensive sports car and wears nice clothes, he or she *must be successful, therefore, desirable.*
6. **Thin Body.** The most recognizable image of pop culture is a *thin, sexualized body,* the perfectly proportioned "mannequin" that youth seeks to emulate.

Skipping meals, smoking cigarettes, and taking harmful "weight loss" drugs rob important vitamins and minerals from blood and tissue and cause catastrophic health problems in later life. The focus of YAAVIST is here and now. Forget tomorrow.

In conclusion, the anthem of this ultimate pop culture of YAAVIST Society is *hedonism*. The automotive industry, cosmetic and clothing manufacturers, perfume and cologne manufacturers, vacation programs, diet plans, health clubs, and "healthy" prepackaged "nutritional" foods all line up in TV-land, magazine ads, and billboard advertisement with one goal: *to convince us to spend our hard-earned money on products unconditionally guaranteed to keep pace with the young, the beautiful, and the desirable.* Results underlie YAAVIST thinking: Put *successful* looking, *visually young and attractive* people together with *thin bodies* and you have *sex* and plenty of it.

PREDATOR PROFILE 7-1

Aileen "Lee" Carol Wuornos

Ashleigh Portales

Birth Name
Aileen Carol Pittman
Time Span of Crimes
December 1989 to October 1990
Date of Death
October 9, 1992; Florida
Offenses Prior to Serial Killing
Forgery
Prostitution
Assault
Trespassing
Carrying a Concealed Weapon
Armed Robbery
Quoting Wournos
"I'm one who seriously hates human life and would kill again."

Preferred Prey

Wuornos hitchhiked up and down the Florida highways, prostituting along the way to support herself and her girlfriend. Each of her victims was a "john" carrying a substantial amount of cash that she shot at some time during their sexual encounter. Wuornos' victim typology probably arose from the fact that *her own father abandoned her, her grandfather and his friend raped and beat her from a very young age,* and she began *prostituting* by her mid-teens. At every stage of her life, at best men had discarded her, and at worst they had abused her severely. This perversely nurtured within her a *hatred of males in general and in society as a whole.*

A Monster Is Made

Aileen Carol Pittman entered a world that did not want her on February 29, 1956. Her mother, Diane Wuornos, had married Leo Dale Pittman when she was just fifteen; she could not have made a worse choice. *Pittman was the perfect psychopathic personality to father a serial killer.* His favorite pastime as a child involved tying two cats together by the tail and throwing them over a clothesline to watch the ensuing fight. He repaid his grandmother's kindness and benevolence by beating and abusing her as soon as his grandfather died. After marrying Wuornos in Rochester, Michigan, the couple had two children: Keith, born in 1955, and Aileen, whom Diane was carrying when she divorced Leo because she was afraid of him. Soon after, police incarcerated Leo for child molestation. He committed suicide in 1969 by hanging himself in his prison cell.

Meanwhile, seventeen year old Diane found the rigors of single motherhood unbearable and *abandoned* her children in 1960. That same year their maternal grandparents, Lauri and Britta Wuornos, adopted them, and the children made the proverbial leap out of the frying pan and into the fire. The two raised the children as their own; *Aileen did not know until she was twelve that they were her grandparents.* But they were not loving and attentive grandparents. Both *drank heavily* and were *oblivious* to the fact that the children set fires with lighter fluid until, at the age of six Aileen burned her face, leaving permanent scar damage.

Lauri was a *very strict disciplinarian,* as the information in Sue Russell's book, *Lethal Intent,* evidences:

> "When she was made to pull down her shorts and bend over the wooden table

> in the middle of the kitchen, when the doubled-over belt flew down onto her bare buttocks, little Aileen railed against her grandfather. Sometimes she lay face down, spread-eagled naked on the bed, for her whippings."

At the age of eight, Aileen's grandfather's friend raped her. When she told her grandfather, he beat her for what she said. Though her acquaintances doubt the story, Aileen claims that she had sex with her brother at an early age. It is, however, certain that very early on the damaged girl became *hypersexual* and *sexually promiscuous.* In 1971, at age fourteen, she became pregnant, and her *family banished her* to a home for unwed mothers where she had to give her son up for adoption. In hindsight, this was probably for the best, but it nevertheless must have traumatized Aileen. Soon after, Aileen quit school, ran away, and began hitchhiking and prostituting.

That same year, Britta Wuornos died of liver failure, but many, including Aileen's biological mother, suspected that Lauri had killed her. While he may not have induced the trauma, he did refrain from calling an ambulance when she went into convulsions because he did not have the money to afford it. Britta soon was followed in death by Keith, who succumbed to throat cancer at the young age of twenty-one, and by husband Lauri, who committed suicide.

With no family left to return to, twenty year old Aileen headed for Florida. Sixty-nine year old Lewis Fell, the wealthy president of a yacht club, picked her up there while she was hitchhiking. It was love at first sight for Fell, and the couple's 1976 marriage made the society pages in local newspapers. Raised in an environment of hatred and torment, Aileen could not recognize a good thing when she had it, as *no one ever had taught her how to love in the first place.* Acting the only way she knew how, she mistreated her new husband and embarrassed him by getting into bar fights and getting arrested for assault. Aware that he had made a mistake, Fell had the marriage annulled after little more than a month.

Aileen spent the next ten years *drifting from one destructive relationship to another* and engaging in *various criminal offenses, abusing both alcohol and drugs* and *attempting suicide* at least once along the way in 1981 when a particularly special boyfriend decided to end their relationship. Planning to kill herself, she got drunk and then bought a gun, but she changed her mind in the process and held up a convenient store clad only in a bikini instead. She received a prison term of three years for this but was released after only eighteen months. She then moved in with one of her many prison pen pals.

Aileen's lifestyle had begun to take its toll on both her mind and body when she entered a gay and lesbian bar in Daytona in 1986. Here she met twenty-four year old Tyria Moore, who bore an uncanny resemblance to the father Aileen had never met. Tyria had a stocky build, reddish-brown hair, and freckles. Ty was looking for love, and Aileen simply was *seeking an emotional bond with anyone who would accept her as she was,* so they made a perfect match. Aileen was not exclusively a lesbian but rather a bi-sexual, as evidenced by her many consensual relationships with men. She later said that "her greater love [for Tyria] wasn't sexual."

The two began a relationship that satisfied the *abandonment fears manifested by Aileen's borderline personality disorder.* Tyria quit her job as a hotel maid and followed Aileen from one cheap motel to another, letting Aileen support her by turning tricks. As Aileen aged, the amount she could get from each "john" gradually declined until her wages could not support the couple. She needed a better source of income, which she felt would keep Tyria by her side.

Monster's Rampage

Aileen's first victim was Richard Mallory, the middle-aged owner of an electronics repair shop in Clearwater, Florida. During confession, Wuornos claimed that each of her seven victims had threatened, assaulted, or raped her. This showed *evidence of the psychopathic tendency to blame everyone else for one's own problems.* While this claim is blatantly false, in Mallory's case it was most likely true. His was probably the "straw that broke the camel's back" in relation to

Aileen's lifetime of sexual, physical, mental, and emotional abuse. Mallory had a reputation for embarking on frequent "binges" of alcohol and sex, often abruptly closing his business and disappearing for days at a time and then returning in a state of paranoia. At one time, Mallory served a ten year prison term for a sexual violence conviction. He was close to no one, and thus no one missed him when he disappeared on the first of December 1989. In fact, no one even knew he actually disappeared until they found his 1977 Cadillac outside Daytona.

Twelve days later, on December 13, his badly decomposed body appeared wrapped in a rubber-backed carpet runner along a dirt road near I-95 in Volusia County, Florida. Mallory had suffered three shots from a .22 caliber weapon. Autopsy revealed that Mallory had lived an agonizing ten to twenty minutes before dying. Police first suspected a local stripper who, ironically, worked under the pseudonym Chastity. She initially confessed to her boyfriend that she had partied for several days with Mallory and then killed him. However, police soon debunked her statements, and she admitted she only had confessed because she was angry with her boyfriend and wanted to get back at him.

After that, the trail went cold until, on the first of June, they found the naked body of an unidentified male in the woods of Citrus County, Florida, approximately forty miles north of Tampa. It took six days to identify the deceased as forty-three year old Sarasota resident David Spears. The heavy equipment operator last was seen on May 19 when he informed his boss he was traveling to Orlando. He never made it. They found his body inside his truck along I-75, shot several times with a .22, with doors locked and license plate missing. Near his body lay a used condom.

The next victim appeared even before the identification of the previous one. On June 6, thirty miles south of Spears' body in Pasco County, someone found a body a few miles off of I-75. The advanced decomposition made fingerprinting impossible, and authorities could not estimate the time of death. There were nine bullets in the remains, and although decomposition had damaged them, too, the police determined that they came from a .22 caliber weapon. The body later became identified as Charles Carskaddon, and authorities in Pasco County, aware of the two murders in Citrus County and the similarities between the three crimes, contacted law enforcement there and shared information. They agreed to stay in touch.

Before another body appeared, Aileen and Tyria destroyed their anonymity and became linked to the murders in a way they never had planned. On the Fourth of July, Rhonda Bailey was sitting on her front porch when a 1988 gray Pontiac Sunbird carrying two women came careening off State Road 315 and plowed into a large clump of brush. Two women emerged from the car screaming profanities at each other and pitching empty beer cans into the woods.

The blonde woman, whose arm was bleeding from a cut suffered in the crash, did most of the talking and begged Bailey not to call the police because her father lived just up the road. Though the accident had smashed a window in the car and various other damages, the blonde woman and her brunette companion said very little, got back into the car, and managed to drive it out of the brush, only to abandon the barely functioning vehicle not far down the road. Suspicious and concerned, Bailey had called police despite the women's pleas. Responding fireman Herbert Hewett passed the two women, who were walking, and stopped to ask if they were in the accident. The blonde cursed at the man, saying no, and they didn't want any help. Hewett drove on, and the women disappeared into the distance.

Police found the car where the women had left it, but they discovered bloodstains throughout the interior and noticed that the license plate was missing. They ran the VIN number through the computer and learned that the Pontiac belonged to sixty-five year old retired merchant seaman Peter Siems. Since retiring, the man had devoted most of his time to a Christian outreach ministry, but he had disappeared on July 7 after leaving his home in Jupiter, Florida to visit relatives in Arkansas. Though investigators held little hope of finding Siems alive, they sent out a nationwide bulletin containing sketches of the two women provided by Bailey and Hewett.

The next body found belonged to Troy Burress, a deliveryman for Gilchrist Sausage who had not returned from or completed his route on July 30. Burress' wife had reported him missing at two o'clock in the morning, and Marion County sheriff's deputies found his truck along State Road 19, unlocked and with the keys missing. There was no sign of Burress. A family on a picnic in the Ocala National Forest found him five days later, approximately eight miles from his truck. The Florida heat and humidity had hastened decomposition severely, and Burress' wife had to identify her husband's remains by his wedding ring. Autopsy revealed that two .22 bullets had killed him.

Fifty-six year old Dick Humphreys was the next to disappear on September 11, one day after his thirty-fifth wedding anniversary. The former Alabama police chief disappeared on his way home from his last day of work at the Sumterville office of Florida's Department of Health and Rehabilitative Services, where he served as a protective investigator specializing in abused and injured children. He was to transfer to the Department's Ocala office the next day. They found his body the next evening in Marion County, killed by seven shots from a .22, one of which they never recovered as the bullet had passed through his wrist and disappeared. They found his car at the end of the month in neighboring Suwanee County.

The seventh and last victim was sixty year old Walter Gino Antonio, a trucker, security guard, and member of the Reserve Police. His nude body appeared on November 19 on a Dixie County logging road less than twenty-four hours after he was killed by four .22 shots. Five days later, his car showed up on the other side of the state in Brevard County.

Hunting the Monster

Captain Steve Binegar, commander of the Marion County Sheriff's Criminal Investigation Division, was aware of the murders committed in both Citrus and Pasco Counties and had begun to formulate a theory with the prime suspects being the two women from the gray Pontiac. He contacted the press in an attempt to locate them, and soon papers across Florida were running the story along with the previously created police sketches. Leads began to pour in from all over the state, and by mid-December police had identified the women as lesbian lovers Tyria Moore and Susan "Lee" Blahovec, the dominant one of the pair and a truck stop prostitute. The tip that finally paid off came from the Port Orange, Florida, Police Department, which had tracked the couple from September to the present. They informed Binegar that the two were staying on and off at the Fairview Hotel in Harbor Oaks, and that Blahovec registered under the alias Cammie Marsh Greene. She had returned alone from their last departure and stayed until December 10. Driver's license checks uncovered criminal records for Moore and Blahovec, but Greene came back clean. Police also discovered that the photograph on the Blahovec license did not match the one for Greene. However, the Greene ID proved the most useful.

A check of area pawnshops revealed that Greene had pawned a camera and radar detector, both belonging to Richard Mallory, along with a set of tools in Ormand Beach that disappeared from David Spears' truck. On both occasions, she had left a requisite thumbprint that later matched an outstanding warrant and weapons charge for a Lori Grody through the National Crime Information Center (NCIC). The print also matched a bloody handprint recovered from Mallory's car. After combining responses from Michigan to Colorado to Florida, the police determined that Lori Grody, Susan Blahovec, and Cammie Marsh Greene were aliases for Aileen Carol Wuornos.

On January 5, 1991, investigators began an intensive undercover search for Wuornos. Officers Mike Joyner and Dick Martin, posing as Georgia drug dealers "Bucket" and "Drums," noticed Wournos at the Port Orange Pub three days later. They planned to build their case against her gradually before her arrest so she had no chance of getting out on bail. Yet their plans almost failed because local police unexpectedly entered the pub and took Wuornos outside. Joyner quickly called back to the command post that operated out of the Pirate's Cove Motel and consisted of officers from six jurisdictions. They determined that no information leak had occurred; Port Orange police just were doing their job. The local police got the message and did not arrest her. Wuornos returned to the bar where Joyner and Martin began a conversation with her. They bought her a few beers and even offered to give her a ride home, but she declined, leaving the bar around 10:00 P.M. Two Florida Department of Law Enforcement officers who began following Wuornos in their cruiser with the headlights off once again placed the task force's efforts in jeopardy.

Another quick call from the command post preserved the operation; Wuornos made it to her next destination, a biker bar called *The Last Resort.* There, "Bucket" and "Drums" met up with Aileen, and they drank together until just after midnight when the two left her in the bar. Wournos, however, did not leave; instead, she slept on an old car seat at the bar on what would be her last night of freedom.

The next afternoon, Aileen's new friends returned to *The Last Resort,* both wearing wires to transmit their conversations back to the command post. They had planned to make the arrest later that afternoon but decided to proceed early when they learned that *The Last Resort* was preparing for a barbeque and would soon be crowded with bikers. "Bucket" and "Drums" offered to take Aileen back to their hotel room to clean up, which she agreed to. As she opened the door to leave, Larry Horzepa of the Volusia County Sheriff's Office approached and informed her of her arrest for the outstanding warrant on Lori Grody. No one mentioned murder, and the press received no information, as police still had no murder weapon and no compounding testimony from Tyria Moore.

Authorities located Moore on January 10 in Pittston, Pennsylvania, where she was living with her sister. Authorities flew to Pennsylvania to interview her. They charged her with nothing, but informed her of her rights and made sure she understood the definition of perjury. Then they just sat back and let her talk. She confided that she had known of the murders since "Lee" came home with Richard Mallory's Cadillac. Lee had told her that she had killed a man that day, but she had not seemed overly upset, nervous, or drunk. Moore told police she had told Lee not to say anything else. "I told her I didn't want to hear about it. And then any time she would come home after that and say certain things, telling me about where she got something, I'd say I don't want to hear it." She said she didn't want to know any more than she had to because she was afraid she might feel compelled to inform authorities of Lee's actions. She didn't want to do that. "I was just scared. She always said she'd never hurt me, but then you can't believe her, so I don't know what she would have done."

The next day, Moore returned to Florida with the investigators, after agreeing to aid authorities in extracting a confession from Wuornos. To accomplish this, they put Moore in a room at a Daytona motel where she would make recorded calls to Lee in jail with the story that she had gotten some money from her mother and had returned to Florida to get the rest of her things. Additionally, she was to tell Wuornos that authorities had questioned her family and that she was afraid the murders would be pinned on her.

Moore made her first call on January 14, with Wuornos still under the impression that she was held solely on the charge against Lori Grody. Wuornos tried to calm Moore's suspicions about the murders by saying, "I'm only here for that concealed weapons charge in '86 and a traffic ticket, and I tell you what, man, I read the newspaper, and I wasn't one of those little suspects." Aware that the jailhouse monitored phone calls, Wuornos attempted to speak in code. "I think somebody at work—where you worked at—said something that it looked like us. And it isn't us, see? It's a case of mistaken identity." Moore continued to call for three days, becoming increasingly insistent that the police were after her.

Gradually it became clear that Wournos was aware of what was going on and knew they expected a confession. She told Moore to "just go ahead and let them know what you need to, what they want to know, or anything, and I will cover for you, because you're innocent. I'm not going to let you go to jail. Listen, if I have to confess, I will." And, on the morning of January 16, that is exactly what she did.

The Monster Speaks

Throughout the duration of her confession, Wuornos emphatically maintained two points. First of all, she insisted that Moore had no part in any of the crimes. The second point, which she adamantly insisted, was that each of her killings was pure self-defense. She contended that every victim had assaulted or attempted to assault her in some way. It seemed to the detectives that Aileen made up her story as she went along, often backtracking when she realized she had said something incriminating. Several times during her confession, public defender Michael O'Neil advised Wuornos to stop talking. He finally asked in complete exasperation, "Do you realize these guys are cops!" Aileen replied, "I know. And they want to hang me. And that's cool, because maybe, man, I deserve it. I just want to get this over with."

Even when the confession ended, Wuornos did not stop talking. Book and movie offers poured in and, not knowing there was a Florida law against criminals profiting in such a manner, Aileen believed she would make millions from her story. With her face plastered across every local and national news broadcast, she felt like the famous celebrity she always had longed to be. She told her story to anyone who would listen, including the jailers, refining the tale each time to cast herself in a more flattering light. It soon became clear to Lee that she had to keep murdering to get money to keep up her relationship with Tyria, the only person she felt ever had cared about her.

After killing a man, she often would return to the couple's current motel room, spoils in hand, promising to pay the rent, buy beer, take them on a trip to Seaworld, and do whatever else Tyria desired. The fact also came to light that, during the period when she had murdered three men in as many weeks, she had feared losing her lover to the sister visiting from Ohio. The two were close and the belief that *MO is a constantly evolving element of serial crime,* she confessed that she had learned from each murder and had begun to carry what she called a "kill bag." This contained, among other things, her .22 pistol and Windex to wipe away fingerprints and cover her tracks.

One thing was evident to all who spoke to her: *Aileen hated people.* Not just men, but all people in general. *She felt the world had done her wrong, and she had a whole lifetime of hurt and pain to avenge.* She had *no regard for the sanctity of human life,* a fact that she could not hide no matter how hard she tried.

The Mother She Never Had?

A response to Aileen's publicity that shocked the law enforcement community came from forty-four year old Arlene Pralle, a "born again" Christian and proprietor of a horse breeding and boarding facility in Ocala. After seeing Wuornos' picture in the newspaper, she wrote a startling letter to the serial killer. "My name is Arlene Pralle. I'm born again. You're going to think I'm crazy, but Jesus told me to write you." Contained within was her telephone number, and Aileen made her first collect call to her newfound friend on January 30. Pralle cautioned Wuornos to be wary of her public defender as well as the media and everyone else involved, as they were only trying to profit from her story. As a result, Wuornos requested new lawyers, which Pralle promptly provided.

Pralle made many subsequent television appearances and gave interviews to magazines like *Vanity Fair,* to whom she described her relationship with Aileen as "a soul binding. We're like Jonathan and David in the Bible. It's as though part of me is trapped in jail with her. We always know what the other is feeling and thinking." Another publication quoted her as saying, "If the world could know the real Aileen Wuornos, there's not a jury that would convict her." Pralle spoke to anyone who would listen about Aileen's "true, good nature" and even arranged interviews for Aileen herself with reporters she thought sympathetic.

During each interview, the women stressed the trials and tribulations of Wuornos' troubled childhood, and each time the cold-blooded killer further embellished her stories, placing blame on anyone other than herself who seemed convenient. Pralle completely embraced Wuornos as her own child, obviously unaware of how deep into deception an *extreme-severe psychopath (with antisocial personality disorder and borderline personality disorder)* can go. She and her husband officially adopted her on November 22, 1991, because "God told me to."

Justice for the Monster

Wuornos' new attorneys, funded by Pralle, acquired a plea bargain in which their client would plead guilty to six counts of murder in exchange for six consecutive life sentences. Only one state attorney held out for the death penalty. Wuornos went to trial for the murder of Richard Mallory on January 14, 1992. A parade of witnesses, both personal and professional, testified against her. Florida law also allowed presentation of evidence of the six other murders to the jury in order to establish a *pattern of behavior.* Yet her claim of self-defense probably received the most damage by the excerpts played from her videotaped confession. "I took a life," spoke Wuornos' image on the screen. "I am willing to give up my life because I killed people. I deserve to die."

Much to her attorneys' dismay, Aileen insisted on testifying on her own behalf, and the prosecution's cross-examination subsequently destroyed every shred of credibility she had. As they exposed her lies, she became agitated and angry, invoking her Fifth Amendment right against self-incrimination twenty-five times. The jury left for deliberations on January 27 and returned less than two hours later with a verdict of guilty of first-degree murder. As the jury filed out of the courtroom after the verdict was read, Aileen screamed from the defense table, "I'm innocent. I was raped! I hope you get raped! Scumbags of America!"

These comments still lay fresh in their minds when the court asked the jury to recommend a sentence. Their unanimous verdict chose the electric chair and was read to the court on January 31. This occurred in spite of the defense's claim that Wuornos was nothing more than a "damaged, primitive child." At this phase of the trial and for the rest of the proceedings, Pralle noticeably was absent. Her relationship with Wuornos had soured, and in the end Pralle did not even know her "daughter's" execution date.

Two months later, Wuornos pled no contest to the murders of Dick Humphreys, Troy Burress, and David Spears on the grounds that she wanted to "get right with God." In a rambling statement to the court, she stated, "I wanted to confess to you that Richard Mallory did violently rape me as I've told you. But these others did not. [They] only began to start to." Then, in a complete change of mood, she turned to the Assistant State Attorney and hissed, "I hope your wife and children get raped in the ass!" She then received three more death sentences to which she responded with an obscene gesture accompanied by the utterance, "Motherf***er!"

In June of the same year, she pled guilty to the murder of Charles Carskaddon and received her fifth death sentence in November. Three months later, in February 1993, she received her sixth and final death sentence for her plea of guilty to the murder of Walter Gino Antonio. No charges ever occurred in the case of victim number seven, Peter Siems, as they never recovered his body.

In a letter to the Florida Supreme Court, Wuornos wrote, "I'm one who seriously hates human life and would kill again." After reading the letter, the Court ruled in April 1993 that Wuornos could fire her attorneys and stop all appeals processes. They also allowed her to choose her manner of death, and she opted for lethal injection in favor of the electric chair.

In 2002, Florida Governor Jeb Bush issued a stay of execution pending mental examination. After being interviewed by three psychiatrists, all of whom declared that Wuornos was sane and fully understood her sentence and why she would be executed, the stay was lifted and her sentence was carried out on Wednesday, October 9, 1992 at 9:47 A.M. at the Florida State Prison near Starke, Florida. Wuornos was forty-six. Here last words were: "I'd just like to say I'm sailing with the Rock and I'll be back like *Independence Day* with Jesus, June 6, like the movie, big mother ship and all. I'll be back."

Aftermath

Aileen Carol Wuornos, though not the first woman ever to kill, was the first to fit fully the FBI's definition of a *modern sexual predator.* Though she was not male, a substantial argument exists that she may have had more masculine tendencies since she functioned as the dominant half of a lesbian relationship. Research has found that the brains of gays and lesbians differ from their heterosexual counterparts.

This raises the question: "Was Aileen really that different from a Ted Bundy or an Arthur Shawcross?" The fact that her crimes had a deeply sexual undercurrent cannot be ignored; she picked up all of her victims while prostituting, and most bodies were naked. Additionally, a gun often is seen as a phallic symbol in such cases, with the firing of the shot likened to ejaculation. Without a doubt, Aileen Wuornos was unlike anything America has seen before or since.

News media have immortalized and even idolized her, exaggerating her prostituting severely by claiming that she serviced two hundred and fifty thousand clients. For this to be true, she would have had to have sex with thirty-five men every day for twenty years! Yet, she appears on her own edition of the series of "Serial Killer Trading Cards" and has appeared as the subject of several books and movies, and even an opera, *Wuornos,* written by Carla Lucero, which debuted in 2004.

The most well known cinematic blockbuster, appropriately titled *Monster,* was released in 2003. It starred Charlize Theron as Wuornos and Christina Ricci as her lesbian lover, whose name they changed to "Selbie." Due to superb makeup and practically flawless acting, Theron bore an eerie resemblance to Wuornos. She delivered an outstanding performance that earned her an Oscar for Best Actress. It is ironic that, in her death, Aileen became, by proxy, the glamorous movie star she always wanted to be.

Like Pulling Wisdom Teeth

The real challenge adoptive parents face is making children feel safe when suspected abuse issues exist from birth parents. It would be wonderful for caring adults to intercept a child headed for psychopathy, but frankly, the odds are against it. As you recall, loving parents adopted Jeremy Skocz, but later he sought out his birth father, a convicted felon, on his own.

Can dysfunctional children ever escape the orbit of deviance?

Ironically, in our blatantly permissive society, teachers and counselors fear parents suing them; they may have strong suspicions, but they must wait to offer assistance until a child is hurt or killed. Child Protective Services (CPS) knows the drill very well. Investigators have observed the craftiness and dishonesty of psychopathic parents. They roll the dice every time they leave a child with parents who are under CPS investigation.

Loving and nurturing parents talk to their children about everything. By this example, they are encouraging reciprocity from their children. In contrast, antisocial parents spend no time in productive talk; rather they bully and humiliate their children.

The people who should be intervening in children's lives to help with the myriad of adjustments required with everyday problems are the very people that are preying upon them hatefully. By age two, most victims of antisocial parenting are already "lost children."

Morally "Fuzzy"

Then, there's morality. A permissive society like Pop Culture USA is subject to a variety of interpretations, including the latest expression for propaganda, **pop semantics:** "Whatever I say or appear to be in public is true, regardless" (whether or not it is actually true). Interestingly, the guru of popular semantics is, of course, former President Bill Clinton, known for "I didn't inhale" (marijuana)—or—"I did not have sex with that intern." After eight years of hearing "spin" from his morally ambivalent staff—"after all Bill and Monica are both adults"—it is no wonder American citizens are a little **"morally fuzzy."**

Incompetent Parenting—Far From Toxic, Predatory Parenting

Antisocial parenting aside, what should society expect from whiny, pampered kids raised by *incompetent* (yet caring) parents who lead them to believe society owes them something because they are so special, even though they haven't accomplished anything? Incompetent parents, who show love and concern but who have little or no parenting skill, still are far from the kind of toxic parents who wreak emotional and developmental devastation. Violent psychopathic kids hate everybody and everything in authority. The really sick ones worship at the "altar" of violent video games—not just playing them but becoming obsessed by them to the point of seeking to create personalized versions. They collect bizarre Web sites of deviant sexuality and get jazzed by violent, decadent, and depressing songs, just like the Columbine "Trenchcoat" killers Eric Harris (eighteen years old) and Dylan Klebold (seventeen years old).

In a movie, Rambo stalked his enemies with a pop culture persona and pumping testosterone. "Rambo" killers—like Harris and Klebold—seem to step out of video games with black high-top boots, trench coats, and guns blazing, seething with **ethnocentrism.** Some prefer less of a grand entrance as they kill ex-girlfriends, peers, gays, or lesbians—anyone different—in vacant fields and leave the bodies behind for vermin and vultures. No person, especially one so young, has this coming.

Young, adolescent killers reared in Pop Culture USA become compulsive liars and compulsive cheaters. The culture nurtures them with its hedonistic "values," and sexual immorality, impulsivity, and violence. It teaches them to live by the creed, "It's ok to do it, as long as you don't get caught." Right now, neuroscientists are showing that they suffer from *irreversible neurological brain abnormalities.* Normal kids may *play* violent video games occasionally, but the games *obsess* kids prone to impulsivity and violence. They get jazzed by violence in all its forms in a dizzying cocktail of surging chemistry from their own brains' pleasure pathways.

Some pre-teens are so neurologically dysfunctional that they cannot separate lies from truth, fact from fiction. Two questions arise regarding cognition or the thinking and thoughts that occur in **internal dialogue.** How does the *quality of thinking* become conditioned? And, does the quality of thinking influence behavior?

Tolman: Making Sense Out of Mazes

Edward Chance Tolman spent most of his career at the University of California, Berkeley. Tolman coined the term *cognitive mapping* as a consequence of studying *behavior* in rats. Tolman's perspective—**purposive behaviorism**—brings cognitive theory into the midst of behaviorism. In his famous "cognitive mapping" study, he placed one group of rats in *various* "start" locations within the maze with food *in the same location* for each trial. The rats quickly learned the maze.

Another group of rats had food placed in *different locations* (with the same exact left and right turns as the other rats.) The rats with food in the same location performed much better than the group with food placed randomly in different locations. Tolman believed the rats had learned the *exact location* of the food due to experience and cognitive anticipation, rather than just a series of left and right turns. In other words, the group developed an internal, *goal-directed cognitive map* for the location of the food. The other group performed less effectively because it lacked an internal cognitive "locator."

Just like rats, humans are *goal-directed,* and they seek the *shortest path to reach goals.* Tolman contends this set of conditions is due to the development of powerful *cognitive maps of learning.* Due to cognitive "maps," we find our cars on a crowded parking lot by noticing landmarks; we can visualize a double-decker strawberry shortcake sundae with whipped topping in stark visualization to the point where we salivate; and we can race over to the downtown courthouse in our hometown without looking for road sign locators. Thus, the development of cognitive mapping is no longer speculative, even in humans. Empirical demonstrations show its reality. Deviant cognitive maps from antisocial parenting, coupled with neurologically damaged brains, can have a devastating effect on individuals targeted as victims.

Rapacious Neurocognitive Maps

The *quality of thinking* influences behavior in powerful ways (and vice versa as behavior influences thinking). It is instrumental in the development of happy, well-adjusted children as well as in the rapacious mindset of adolescent killers, female killers, and male serial rapists and murderers. Imagine the impoverished and twisted cognitive maps of adolescents *obsessed* with guns, pipe bombs, explosives, sexually deviant Web sites, violent video games, perverse song lyrics of doom, death, and destruction. What do we expect will happen?

Expecting psychopathic kids to be honest with anyone amounts to delusional thinking. They embrace *hidden emotional agendas*. Such kids are sure to respond with complete deception to any reference about peers as co-conspirators, or to experiences such as making pipe bombs, or to explosives, deviant Web sites, or music. They have intense *emotional investment* in these scurrilous activities.

Early in development, psychopaths master the art of manipulation and lying. They lie right to your face. They look straight into your eyes and never blink. They offer a demeanor to society that is extremely effective at deception. As a society, we had better start looking behind the smiling faces or blunt affect (lack of emotional expression) that typifies severe psychopathy.

Young killers often start with petty crimes—vandalism and theft—and finish with major violence in home and school, just like Harris and Klebold did. It is not conceivable that modern neuroscience can undo their neurological brain irregularities resulting from *systematic abuse and neglect* in antisocial families. Perverted cognitive maps have taken over as cognitive "focusers." Society is not safe until young killers weed themselves out by committing violence, having shootouts with police, and landing in juvenile detention or prison.

On the other hand, a child surrounded by emotionally nourishing activities with attentive, supportive, and loving family members and friends (even with an occasional touch of incompetent parenting) develops a far different set of cognitive maps. The behaviorist John B. Watson thought so as well, as his famous view of rearing children shows:

> "Give me a dozen healthy infants, well-formed, and my own specific world to bring them up in and I'll guarantee to take any one at random and train him to become any type of specialist I might select—doctor, lawyer, artist, merchant, and yes, beggar-man and thief, regardless of his talents, penchants, tendencies, abilities, vocations, and race of ancestors." (Watson, 1966)

The picture is a grim one. Homeplace violence. Schoolplace violence. The face of violence lives in many locations. For example, in Austin, Texas, a popular, bright, and vivacious sixteen year old girl decides to break up with her jealous and possessive boyfriend. Despondent, the boyfriend pulls a knife and stabs her to death in a hall of Reagan High School.

Several years ago, two high school seniors—a boy and a girl—living near Fort Worth, Texas conspired to kill a female classmate because the boy's girlfriend was jealous. Now known as "The Cadet Murder" the female killer thought her boyfriend and the female victim had sex. They lured the victim away from the safety of her home, shot her, and bashed her head in. Adrianne Jones died in a vacant field. Before the secret became known, our country's most prestigious service academies accepted both killers—Diane Zamora and David Graham. Now, careers and lives ruined, they sit in federal prison. Recently, Zamora asked permission to marry a fellow inmate.

In 1979, long before the decade of accelerated schoolplace violence arrived in the 1990s, a seventeen year old female entered an elementary school and shot eight children and a police officer with a rifle she received as a Christmas present. Two men lost their lives trying to shield the children. After the six-hour siege ended, she told authorities, "I don't like Mondays." In 1987, a twelve year old boy brought a pistol to school and killed the boy who tormented him. Then he killed himself. Then, the 1990s arrived.

Violent Vignettes

In 1993 in Kentucky, a seventeen year old male used his father's pistol to shoot his teacher in the head and later a janitor. He held his classmates hostage. He disliked his teacher's comments about his book report. It reviewed Stephen King's novel *Rage,* which had a similar scenario.

In 1995, three separate incidences of schoolplace violence occurred. The first took place in California. A thirteen year old male, distraught over having to wear a school uniform, stole a shotgun from a friend, returned to school, and shot the principal in the face. Then he committed suicide. In South Carolina, a sixteen year old male shot a teacher in the face and then killed himself because the school had suspended him for making "a vulgar hand gesture." In Tennessee, a seventeen year old male fatally shot a teacher in the face then killed another student and teacher. The reason? Academic problems and a traffic accident the day before.

In 1996, a thirteen year old honor student shot and killed two students and a teacher with a deer rifle he brought from home. One of the students he killed had called him "a nerd." His parents were going through a divorce, and he suffered from depression. He thought Oliver Stone's *Natural Born Killers* and Stephen King's *Rage* were "cool." He watched the Pearl Jam video "Jeremy"—about an outcast who commits suicide.

Four incidences of schoolplace violence occurred in 1997. In Alaska, a sixteen year old student killed one student and a principal and wounded two other students with a shotgun. He had grown tired of being called pejorative names. The perpetrator's father was known as the "Rambo of Alaska" due to his own violent behavior. In Mississippi, a sixteen year old male student stabbed his mother to death and then shot nine students at school, two of whom died, including his ex-girlfriend. He referred to himself as a "satanic assassin." The boy's idol was Adolf Hitler. Before the shooting, he tortured and killed his dog. In Kentucky, a fourteen year old boy shot and killed three students, wounding five others. He enjoyed the violent video games *Doom* and *Quake*. When wrestled to the ground following the shootings he cried, "Kill me now!"

In 1998, three violent schoolplace events occurred. A thirteen year old male and an eleven year old male fatally shot four students and a teacher and wounded ten others. All victims were female. They stole seven handguns and three rifles from relatives. In the stolen van, police found sleeping bags, a radio, and a stuffed animal. In Pennsylvania, a fourteen year old suicidal boy shot and killed a teacher and wounded two students and a teacher. In Oregon, a fifteen year old murdered both his parents, who were teachers. He drove to school and fatally shot two students and wounded twenty-three others. He had grown tired of being teased and felt angry with his parents and with the school. Indicative of some school administrators and school boards' denial of student violence, his middle school yearbook named him "Most Likely to Start World War III." His topic for a speech class was *How to Build a Pipe Bomb.*

In 1999, two violent schoolplace events occurred. In Georgia, a high school student injured six students with a rifle. He was on anti-depressant medication following a breakup with his girlfriend and failing grades. They found directions on making bombs and explosives in his room. Then, in Colorado there was Columbine, the incident that raised the ante for schoolplace violence.

The Columbine Massacre

A cursory sketch follows in this section regarding what is known about the perpetrators of Columbine. The massacre at Columbine High School stands as a reminder of the impact Pop Culture USA has on neurologically damaged brains. It exemplifies arrogance, strategizing, bold maneuvering, and the sheer number of the assailants' victims (thirteen dead, twenty-three injured, and countless victims emotionally scarred for life).

The teenage perpetrators were loners whom their peers called names: "inbreeds," "dirtbags," and "faggots." They associated with a group of peer malcontents known by the cool pop culture moniker "trench coat mafia." Harris and Klebold obsessed over sexually deviant Web sites. They did not merely play video games; rather they *obsessed* to the extent that one of the killers customized a segment for his own enjoyment.

A few days before the killings, the Marines rejected Harris due to his psychiatric medication prescription. Harris created his own Web site for his tortured psyche, including directions on how to make a pipe bomb and the cryptic message, "goodbye to all on April 20th"—the date of the Columbine massacre. Looking back in 20/20 hindsight, prescient "red flags" appeared everywhere (as they always do) from home to school to Web sites in virtually reality. Who's paying attention to our children?

Both boys had prior felony convictions for breaking into a van. A parent complained about the Web site and harassment against his son to the police, who forwarded a memo to the school. *The school took no action.*

The *Chicago Sun* was the only newspaper in the country that didn't carry the story on its front page. It had an established policy that contended media hype contributes to increased crimes. Due to the onslaught of other media coverage, however, recently researchers listed the story as the third largest in this century (Zinna, 1999).

A *killing team* of two perpetrators to reinforce each other, using both bombs and guns to extract the most damage, was a tactical departure from most instances of schoolplace violence. Both boys had studied carefully the school's layout and knew exactly where the greatest concentration of students would be at the 11:00 A.M. strike hour—the cafeteria and library. They calmly talked and joked with each other in a blatant disregard for life as they stalked victims, killing those hiding under desks and fleeing for their lives. And, just as quickly as it had begun, the rampage ended.

Schoolplace Offender Profile

In 1998, the FBI developed the *schoolplace offender profile,* intended as a general measure of *violent risk potential.* Profilers caution that schoolplace violence is evolving and perhaps escalating. The following list of fourteen characteristics represents the type of behavior characteristic of young killers:

1. White male, seventeen years of age or younger, lacks discipline
2. Fascination with firearms and/or explosives
3. Cruelty to pets
4. Believes mother (or other family member) disrespects them
5. Seeks to defend narcissistic view of self
6. A depressed suicidal ideation turned homicidal by a triggering event—failed romantic relationship, lack of support from family, rejection, or revenge
7. History of expressed anger or acts of physical aggression at school
8. May have been influenced by satanic or cult type belief system
9. May listen to music lyrics that promote violence
10. May appear sloppy or unkempt in class
11. May feel isolated from others
12. May have a history of mental health treatment
13. May feel powerless; may commit acts of violence to assert power over others
14. May have openly expressed a desire to kill others
15. Presents low-self esteem

Compiled from *After Columbine: A Schoolplace Violence Prevention Manual* (1999, Zinna).

Today, *home schooling* looks better all the time when compared to suburban schools, their doorways lined with metal detectors in an attempt to keep our children from being massacred. But who's going to stay home with the children when both parents work? The behavioral dynamics of so-called latchkey children—children left alone and unsupervised before and/or after school—are a glaring part of the problem. Some do just fine. But, there are always some who don't.

Once Upon a Time

It is not necessary to run a sociological or historical survey comparing the *zeitgeist* of the 1950s to the 1960s to the 1990s and 2000s. The current generation's parents or grandparents know the time frame. Once upon a time, students respected teachers and coaches. Teachers controlled the classroom. In many American families, moms stayed home and raised the kids; dads worked—great arrangement. *Values, morals, and structure were central to schoolplace milieu* where students stood to the Pledge of Allegiance and then followed with a morning prayer in the classroom. Parents did not flock to the school to demand this tradition stop. Once upon a time, atheists filed no lawsuits to stop any procedures in the classroom.

Few students were disruptive, certainly not by today's standards. If they were, the boys (and some girls) got "licks" by a wooden paddle for inappropriate behavior. In fact, if boys didn't get a few paddles from the true "board" of education, classmates considered him too much of a Momma's boy, but his behavior was not hateful or rapacious. Now parents are ready to sue if anyone paddles their little darlings. With little if any meaningful discipline at school or home, why do we become shocked when kids take matters into their own hands?

Harking back to the past, family budgets and economics made sense in the 1960s. A new Mustang automobile cost approximately $2,500 or less. A new four-bedroom home with approximately 1,800 square feet cost under $20,000. Once upon a time, the numbers worked.

Families had dinner time to socialize and share the activities of the day (a few risk-takers ventured away from the table into the living room to watch TV with their food on solitary TV trays). At first, at least, most family members chose the comforting mix of food and conversation. Who has time to sit down at a meal today with family members? Tabulating how much our children spend a month on gas to and from fast food restaurants and the cost of the food might bring on a migraine.

Today, **cynicism** is rampant in schools. Students openly laugh at and ridicule teachers. "Sue me," they say. Parents seem in a quandary about what to do. Parents of normal kids are in denial about the severity of pseudo-sapiens—kids with wrecked neurological systems clandestinely stalking their children. Incompetent parents act shocked in parent-teacher conferences that their kids act out in class even though they know they have emotional issues. Constant fighting, insufficient money, and/or impending divorce hit the kids right between the eyes every morning. Who will they live with? What's going to happen? Will they ever see their friends again? Antisocial parents, who have severely scarred their children emotionally by abuse, emotional neglect, and hateful parenting, never show up for conferencing. The kids may or may not attend school regularly.

In every way, teachers represent the "Rodney Dangerfield" of education—*they get no respect.* Students bully teachers and fellow students they consider "nerds." Students control the classroom. Decency, morals and ethics went away a long time ago. Some political historians contend the meltdown occurred as a "trickle down effect" from the top leadership of our country beginning in the late 1970s when Nixon told his country, "I am not a thief!" Trickle down continued years later with Clinton's finger wagging to a national audience—"I did not have sex with that intern." Yes, Nixon was a liar and so was Clinton. Then, some observers say we must expect "minor" dishonesties from politicians. Who's left to blame for moral ineptitude? Are litigious atheists really responsible for the moral decay in our homes and schools?

Recently, an elementary school kid brought a gun to school. He showed it off; miraculously, the gun didn't discharge. The story goes that the kid's father is long on money, and the school is short on memory. School officials did nothing over the protests of his classmates' parents. As a society, we already know you can get away with murder if you have money and influence.

Tongue-in-cheek, some nutritionists say the emotional problems students have in school is due to all the fat they eat in school cafeterias, especially fried foods and pizza. Many students become obese before the sixth grade. Projections indicate that one in three kids will become diabetic by age ten; one in two will become diabetic among black and Hispanic children. True, the high content of fat in fried foods clogs arteries and sensibilities. What about the brain?

Then there's the argument that all the toxins students breathe in from carpet fibers and poor air circulation found in even the most modern school buildings cause them to "act out." Is there a connection between the alarming record number of allergies in children and airtight classrooms and carpeting? Voluminous studies show that fluorescent lighting causes hyperactivity in some children and inattentiveness in others. Still, classroom lighting is fluorescent. It's cheaper.

Sadly, as statistics show, perpetrators of schoolplace violence litter the thirteen to eighteen year old age group of middle school and high school aged kids. Critics of public education contend that, academically, *high school is a mind-numbing experience.* They maintain that if you take time to ask kids whether or not high school prepared them for college, the first thing they do is laugh. The second thing they do is ask, "Are you kidding?" No one seems to mind.

Perhaps years ago, some high schools believed they had a mission to be a prep school and prepare students for college. This is a logical mission statement. Few argue today that high school is the last grasp for extra-curricular activities, sports, and socializing before the seriousness of college.

Critics further maintain that *high school is mostly about academic boredom.* Everybody knows only nerds study. "No pass, no play" has become a national joke no one takes seriously. Social historians are at a loss to explain the tectonic shift away from the "prep" school mission to idyllic Hormone High.

In today's home, both parents usually work. Budgets are always tight, as most families live paycheck to paycheck, hand to mouth. High interest rates and irresponsible spending bloat credit cards. Today, a new Mustang costs $40,000 dollars or more, and a new four-bedroom eighteen hundred square-foot home costs approximately $180,000 (or $100 a square foot to build.) While salaries have gone up, many citizens' "standard of living" reflects the 1970s' standard. Who has time for parenting? Stress is over the top with parents working, bills to pay, and kids to raise. *Emotional burnout* is a silent epidemic in North America.

We use examples from Eriksonian stage theory to present compelling arguments about the development of adolescent killers primarily across three stages:

- Stage One: *Trust versus Mistrust.*
- Stage Four: *Industry versus Inferiority.*

- Stage Five: *Identity versus Role Confusion.* In Stage Five, we highlight damaging effects of horrific parenting.
- Serial killers emerge in Stage Six, where the *emotional crisis* of intimacy versus isolation tips the scales toward sexualized violence.

Stage One: Trust versus Mistrust

Trust and Competent Parenting

No one expects to get an answer back from a one year old infant when asked, "Do you trust me?" But, the answer is there nonetheless in body language. The result is predictable when loving parents attend to and nurture infants, put them on a predictable schedule of sleep and feeding, hold, rock, love, touch, talk to, and play with them. When parents devote quality time away from the busy adult world, children feel safe, secure, and loved, and they trust their caregivers. They feel valued. Even mothers who retain their jobs (as observed in various *Mommy Tracks* available with many employers), they choose to breast feed their infants for at least three or four months to insure her infant's immune system gets nature's boost. A supportive husband, competent sitters, and attentive grandparents give her needed breaks.

Effective parenting is exhausting as well as exhilarating. The couple and infant *feel like family.*

Infancy Mistrust

Contrast the above example with the neurologically tweaked development of the infant during his or her first year of life. Antisocial, hateful parenting conveys through body language alone a far different message to the child. First, mom may have used drugs during pregnancy—legal as well as illegal. If she smoked cigarettes on a regular basis, the infant is born addicted to nicotine; if she drank alcohol on a consistent basis, the infant is an alcoholic addicted to ethanol. We have read newspaper accounts and TV documentaries of the horror of "crack babies." *Addicted mothers give birth to addicted infants.* Addicted infants cry and cry for weeks as they battle addiction and the awful withdrawal. This further alienates insecure mothers. Addicted infants don't thrive well, and they don't appear to feel secure and happy. They don't feel and look like they trust their "caregivers."

Antisocial father figures may have walked (or more descriptively run away) from responsibility, as psychopaths are prone to do, leaving the mother alone, tense, and afraid. She communicates negative feelings to her unborn infant. Perhaps her family disowned her or made it clear that "you made your bed, now sleep in it." If the father remained in the picture, he might be immature, ambivalent, and/or battling some kind of addiction himself.

In contrast to the nurturing couple, incompetent couples do *not feel like family.* They may feel trapped, cheated, or unlucky. Why did this have to happen to us? Immature and incompetent male "fathers" shun their pregnant girlfriends so they can keep partying with their equally immature and incompetent "friends." Some have poisoned their pregnant girlfriends to "get rid of the problem."

New mothers may feel uncomfortable with breastfeeding, so bottle feeding will have to do. Finding a formula that agrees with her child is often a "hit and miss" proposition. She may not be sure what to do when her baby cries. Often, supportive grand parenting is absent, and she cannot afford competent sitters. Within months, very young mothers who have little or no support systems are tense, nervous, and exhausted. Mothers who raise their children alone often have limited funds and may have limited understanding of effective parenting skills. It is an exercise in futility to routinely prop a bottle under a folded diaper to feed the infant lying flat on its back so the exhausted mom can take a short nap. The slightest movement of the infant's head in the wrong direction will displace the nipple from mouth to ear, if the nipple has not collapsed already. Crying acts like a demonic "alarm clock" to the stressed-out mother.

Stage Four: Formal Schooling

A traditional benchmark accomplishment of normalcy in our society is the successful transition from the home milieu to the school milieu. Society considers it dysfunctional when students make a poor transition, become truant, and/or oppositional defiant. Something is missing. In stage Four, *Industry versus Inferiority* (six years to twelve years of age), normal children come to expect *predictability* from teachers and *formal schooling* as well as from parents. Social and academic success comes in stages or rather grade levels, as normal kids manage tasks related to the level of learning expectations. Dysfunctional children *never like school;* mostly, they are weak academically and aloof socially and seldom achieve more than "average" accomplishments.

Elementary and middle school teachers see the "red flags" of dysfunctional parenting as troubled children act differently than normal ones. Evidence shows that home schooling is an effective alternative to public or private education. Recently, the author spoke at a commencement exercise of home-schooled students. Talented children performed vocally and on musical instruments while devoted parents gleefully watched the evening's highlights. Families gathered together on stage to celebrate promotion to the next grade level. *There is no question that home schooling provides an effective alternative to the "cookie-cutter" factory of public school education.*

Stage Five: Twelfth Year through Eighteenth Year

In this stage, even with limited perspective and maturity, normal children come to see themselves as *unique individuals* with "upsides and downsides." They understand the "pecking order" of high school with its predictable cliques and stereotypes. The children of some involved parents have received good advice and find it amusing: *High school is not the real world.* Other kids find the high school pecking order unfair, irritating, or humorous, but only the kid with *features* of psychopathic personality is at the extreme: He either loves it (too much) it or he hates it (too much).

Features of Psychopathy

In fact, some of the most popular high school students have *features* of psychopathy. This does not mean they will become serial killers. It does mean they might be very successful in business, especially sales. *Psychopathic personality stretches across a wide continuum that includes mild, moderate, and severe features.* Sexual psychopaths (serial killers) are at the most extreme end of the psychopathy continuum. Chapter Eight covers the "mild" and "moderate"—or garden variety psychopaths—in more detail.

Besides being physically attractive, individuals with mild psychopathic features display egocentricity (self-centeredness), narcissism, extraversion with a "gift of gab," superficial charm, emotional shallowness (that ironically guards against becoming discouraged by other's comments), and a self-styled fabrication of reality to suit themselves. Other less attractive and more introverted students, who are less popular in high school, often "blossom" as *late bloomers* in college. Immaturity and lack of perspective aside, many of the "red flags" of psychopathy clearly are visible by high school. Some students with mild features of psychopathy simply grow out of them as they mature and embrace the higher expectations of the college experience. Those not willing to adapt, adjust, or modify their "high school" mindset to the rigors of higher education exemplify personality *rigidity* characterized *by faulty coping skills,* which are psychopathic traits.

Dropping out of high school has many causes, but severe psychopathy cannot be ignored.

Along with the process of ego integration, normal kids at this age start to trust their own "gut" feelings. They see a completely different world than the budding psychopath does. With parental encouragement, they become connected or involved in some aspect of school or community that enriches and expands their personalities. They genuinely want to help others.

The loner psychopath may feel completely depersonalized (Erikson refers to it as *role confusion*). Young killers feel detached, isolated, frustrated, and angry. He already may have displayed antisocial behavior and has to report to a probation officer, or he may beaten up a string of dysfunctional girlfriends, committed random acts of vandalism, gotten expelled from school, and/or set fires. He's one "triggering" event away from committing his first act of violence.

Females Who Kill

Female Serial Killers

Boys and young adult males exposed to hateful and abusive antisocial parenting display violent behavior with pronounced oppositional defiance and behavioral deviance. In contrast, adolescent and young adult females typically "turn on themselves" and become sexually promiscuous, drug-addicted, passive, or dependent with **borderline** or **histrionic** personality features. As noted earlier, *Personality Disorder NOS (PD NOS)* is usually the most accurate clinical definition for the male variety. This is due to the presence of multiple personality disorders, exacerbated by poly-addiction. While female serial killers experience similar dysfunctional childhoods and abusive backgrounds, they show a pattern of not acting out aggressively like the male variety. Instead, they direct rage inward as *self-hatred.* Female killers tend to be spree killers, "heat of passion" killers, or retaliation killers. They often become chronically depressed and fall into addiction.

An example of heat of passion rage took place in 2003. A thirty-something female dentist repeatedly ran over her husband in her Mercedes-Benz in front of the hotel where he and his young girlfriend met for weekly sexual liaisons. She testified when convicted of murder and given a twenty-year prison sentence, "I didn't mean to run over him three times, I was aiming at his car" (maybe she should have studied to be an ophthalmologist).

In the 1990s, Long Island "Lolita" Amy Fisher attempted to murder her seedy boyfriend's wife by shooting her in the face when she answered the door. In the 1980s, Lorena Bobbitt severed her husband's penis while he slept. She said he was abusive. Over a year later, after her divorce, a boyfriend filed an assault charge against Lorena Bobbitt. Given new life with an enhanced penis, John Wayne Bobbitt turned to *Pop Culture USA* and became a porn movie star.

Male serial killers kill strangers, while females often kill people they know—husbands, nursing home residents, hospital patients, coworkers, kids in the neighborhood, or their own children.

Motives, Methods, Typology, and Profiles

Researchers who reviewed the literature agree that female serial killers are rare. According to one source, only eight percent of all serial killers in America are female; however, the same eight percent accounts for seventy-six percent of all female serial killers worldwide. The number of male serial killers increased fifty-seven percent since 1970, while in the same time period, the number of female serial killers increased by one hundred thirty-eight percent! Yet, in ***Mindhunter*** (1996), former FBI profiler John Douglas states, "The fact remains women do not kill *in the same way* or in anywhere remotely near the numbers men do."

What motivates female serial killers? What happens in the minds of women—known for "maternal instincts" and nurturing—to turn them into murdering predators? How does the MO of a female killer differ from that of male serial killers? Does she display signature? What characterizes her *modus vivendi?*

Interviews with female serial killers (Alarid, Marquart, Burton, and Cullen, 1996) uncovered the disturbing statistic that eighty-six percent of all female serial killers work in tandem with either a male or female *accomplice* (known as the **secondary follower role**). This fact accounts for a much different dynamic from the killer who acts alone. The average amount of time before apprehending a female serial killer is *eight years,* while it takes approximately half that time to catch men.

Kelleher & Kelleher (1998) describe the female serial killer as *successful, careful, precise, methodical, and quiet.* Hickey (2002) first consistently identified the female serial killer's favored *methods* and *motivation* (some killers used more than one method and displayed more than a single motive, raising the percentages to over one hundred percent).

Method Used:	
Poison	80%
Shooting	20%
Bludgeoning	16%
Suffocation	16%
Stabbing	16%
Drowning	5%
Motive:	
Money	74%
Control	13%
Enjoyment	11%
Sex	10%
Drugs	24%

Typology

As previously noted, scant research exists into the psychological or neurochemical aspects of female serial killers. According to Kelleher & Kelleher, the male serial killer's typology of *organized/disorganized* is noticeably absent from female serial killer typology. Consequently, researchers developed a deductive nine-point categorization based on the assessment of the two typologies of female serial killers:

- females who act alone
- females who act in partnership with another (or others)

1. Females *acting alone* include:
   a. Black Widow
   b. Angel of Death
   c. Sexual Predator
   d. Revenger
   e. Profit/Crime
2. Females *acting in partnership* include:
   a. Team killers
   b. Insane killer (question of sanity)
3. Unexplained Killers
4. Unresolved Killers

A compelling argument exists for serious changes to the "unsolved" and "acting alone" categories. For example, the "unsolved" category simply does not supply enough information to attribute the crimes examined to a female killer at all, and for all practical purposes, it is best to omit it. Furthermore, the *acting alone* category (as defined by Kelleher & Kelleher) clearly *represents the organized* type of killer (as observed in the male variety). Therefore, it seems logical to modify the typologies as follows:

1. Acting alone (organized)
2. Acting alone (disorganized)
3. Acting in partnership

Acting Alone—Female Organized Serial Killer

This typology includes killers who are more mature, careful, deliberate, socially adept, and highly organized. These women usually attack victims in their homes or places of work. Favored specific weapons include poison, lethal injection, or suffocation.

FEMALE PREDATORS

11 Sketches of Females Who Kill

"Black Widow"

"Black Widows" usually begin their career after age twenty-five. They progress to a cycle of systematically killing spouses, partners, family members, or anyone with whom they have developed a *personal relationship*. The typical cycle is six to eight victims over a period of ten to fifteen years. *Poison is the preferred weapon,* or they may use other lethal substances intended to mimic medication. The salient motive is greed, where the widow benefits from an inheritance or life insurance policy.

Margie Barfield, Dianna Lumbrera

Margie Velma Barfield, a fifty-three year old grandmother, committed her crimes from 1969 to 1978. She *killed seven husbands, fiancés, and her own mother* in North Carolina. She burned some victims to death while they slept, making it appear as though they were smoking in bed. She arranged prescription drug overdose for some of the victims and used arsenic to mimic gastroenteritis in others. She was executed by lethal injection in 1984, the first woman executed in the United States since 1976.

Dianna Lumbrera systematically *suffocated six of her own children* in Fort Worth, Texas from 1977 to 1990. The MO consisted of rushing each child (who already was dead) to the hospital and blaming the medical staff for not resuscitating the child. Initially, officials suspected *Sudden Infant Death Syndrome,* or "crib-death." Eventually, hospital officials became suspicious. Authorities tried, convicted, and sentenced her to three life terms.

"Angel of Death"

The "Angel of Death" begins her career around age twenty-one. She works in a localized medical setting such as a hospital, a nursing home, or an extended-care facility, where death occurs regularly and where administering life-sustaining or life-terminating medications is part of the routine. She easily disguises murder in her camouflaged role of caregiver. Such an offender *enjoys the power of "playing God"* by selecting "who will live and who will die." The typical cycle is eight victims over a one to two year period. Angels of Death tend to brag about their crimes, which often brings about their capture. The offender who moves from one facility to another may kill many more victims before discovery.

Genene Jones

Genene Jones was a twenty-seven year old licensed vocational nurse (LVN) who specialized in the terminal care of pediatric patients. Jones was a highly mobile killer. She began her career in large metropolitan hospitals in San Antonio before migrating to other hospitals. She may have killed as many as forty-six patients from 1978 to 1982. Jones' MO consisted of injecting either the heart medication Digoxin® or the respiratory paralytic Pavulon® into infants. Then she "discovered" the patient's distress. She enjoyed the attention she received from grateful co-workers and parents for her skill at resuscitating infants. After the CDC investigated her, she relocated to a physician's office in Kerrville. She went to trial in 1984 after the last patient she injected died en route to the hospital. The child had come to the clinic for routine immunizations. Jones' method included having the parent leave the room during the immunization by stating that the child would cry less. She then injected the child with a lethal injection. Amazingly, persons at Texas institutions illegally shredded her employment and medical records to prevent further investigation and potential lawsuits.

Sexual Predators

Female sexual predators are so rare as to be an anomaly. Researchers are seeking worldwide for other offenders that fit this typography. Sexual predator Aileen Wournos may represent the prototype of all future female sexual predators.

Aileen Wournos

Aileen Wournos, a thirty-three year old prostitute living in Florida, killed seven men who solicited sex (authorities accused her of two more murders, but the cases remained unproven). At some point in each sexual encounter, she perceived the sexual encounter as abusive. As a remedy, she shot each victim repeatedly with a .22 caliber handgun, robbed him, and hid the body in the woods along Florida's I-95 corridor. Her victims included men from all walks of life: a known rapist, two retired or reserve police officers, one missionary, two truck drivers, and deliverymen. Her apprehension came from a careless paper trail she left from pawning items she collected during the robberies. She claimed self-defense due to a childhood of physical and sexual abuse at her trial in 1992. Interestingly, her lesbian lover provided testimony against her, disproving her claim of self defense. Authorities sentenced her to the electric chair. "Monster," a movie starring actresses Charlize Theron as Wournos and Christina Ricci as her lesbian lover, came out in summer 2003.

Revenge Killer

The common revenge killing is a *crime of passion.* However, the female revenge serial killer is very rare. The deep, pathological anger that drives the killer makes the revenge serial killer unique. There is little or no cooling-off period, implying an obsessive attachment to single-minded revenge. The revenge killer usually begins at about age twenty-two. Victims may include family members and/or symbols of an organization the killer deems offensive, such as an abortion clinic. The usual pattern is three to four victims over a period of less than two years, although some killers may operate for up to five years. Although the revenge killer usually can exert enough control to conceal the crime, she also is careless and demonstrates a lack of planning. Interestingly, such killers show great remorse when captured.

Martha Wise

Martha Wise, known as "the Borgia of America," was a thirty-nine year old widow who fell in love with a younger man her family opposed. She poisoned three of her family members one by one before other relatives became suspicious and reported her to authorities. She confessed to the murders, to several attempted murders, and to the arson of the church, which had resisted her marriage ceremony. Her defense at trial was "the devil made me do it." She received life imprisonment.

For Profit/Crime Killer

Female serial killers motivated by profit or crime are also extremely rare. Some appear to be organized contract killers for hire. Some have set up localized scam operations to rob victims. *Authorities consider them the most intelligent, resourceful, and careful of all female serial killers.* Their careers begin at about age twenty-five or thirty and may last for ten years. They kill from ten to twenty-five victims before their apprehension. This type of killer uses a variety of methods and acts dispassionate about her murders. *Sig-*

nificantly more female killers of this genre exist worldwide than in the United States.

Madame Popova, Anna Marie Hahn

Madame Popova perpetrated a murder-for-hire service in Russia specializing in killing cruel husbands for a fee. Authorities credit her with killing over three hundred victims by means of poison, weapons, her own hands, or a hired assassin. Russian police apprehended her when a remorseful widow turned her in. A firing squad executed Madame Popova for murders committed over an extended period from 1879–1909.

Anna Marie Hahn, twenty-six year old immigrant from Germany, worked in the Cincinnati area as a live-in caretaker for elderly men. Over a period of about five years, she bilked five victims out of their assets. She was also an expert in the utilization of poisons. She used different poisons to kill each victim. Her downfall came at the hands of suspicious bank examiners. Even though she utilized a mercy killing defense, in 1938 she became the first woman in the history of Ohio to die in the electric chair.

Acting Alone—Disorganized Female Serial Killer

Severe *psychotic disorders* such as schizophrenia and *substance abuse* disorders characterize this type of female killer. Investigators should observe prudent skepticism in labeling these extremely rare murderers "disorganized." "Women come from the same backgrounds as males in the disorganized typology. Girls are even more subject to abuse and molestation than boys."

Question of Sanity—Insane Killers

Finding a female killer insane is rare. While "Angel of Death" predators who offer the psychological defense of **Munchausen Syndrome by Proxy** may most successfully implement the insanity defense, most killers with a question of sanity still become convicted of first-degree murder.

Bobbie Sue Terrell, Andrea Yates

Bobbie Sue Terrell, a twenty-nine year old diagnosed schizophrenic, worked the nightshift as an employee of a nursing home. She gave insulin injections to each of twelve victims in St. Petersburg, Florida. After the murders, she mutilated herself, called police, and alleged an attack by a serial killer. Police arrested her when they discovered her psychiatric background. The court judged her insane and sentenced her to sixty-five years in a clinical forensic prison.

On June 20, 2001, Andrea Yates, a thirty-seven year old "post-op" nurse, murdered her five children by drowning them one by one. She had a long history of psychotic episodes, a diagnosis of schizophrenia, and recently had received the diagnosis of *postpartum depression* after the births of her fourth and fifth children.

Yates' mother remarked that her daughter always seemed isolated as a teenager. Her father was a teacher and a perfectionist. Yates consistently overachieved academically and made great efforts to please her father. After the birth of her first child, she began having homicidal visions that prominently featured knives and stabbings. She attempted suicide on several occasions and frequently had auditory hallucinations. Her family history includes two siblings with depression and another sibling diagnosed as bipolar. Despite a strong insanity defense, on March 13, 2002, a Houston jury took just forty minutes to find her guilty of the deliberate murder of three of her five children. She received a life sentence in prison.

Unexplained Motive—Neighborhood Killers

Neither the female serial killer nor the authorities seem able to discern a motive for these murders. In many instances, drug abuse may provide the "trigger."

Christine Fallin, Audrey Hilley

Christine Falling, a seventeen year old babysitter from Perry, Florida, murdered at least five neighborhood children by suffocation from 1980 to 1983. She stated she heard voices telling her to commit the murders and to prevent others from hearing the victims' scream. She received life imprisonment for her crimes but will become eligible for parole in 2007.

Audrey Hilley, a forty-two year old housewife from Alabama, poisoned several neighborhood children and family members with arsenic. Three victims died, including some of her family, and several more became seriously ill. She claimed to suffer from bouts of "alternative consciousness" where she became her twin sister. The court judged her insane. She became a model inmate in prison, but after her release, she repeated her macabre behavior, which earned her a life sentence.

Acting in Partnership—Female Serial Killers

Characteristics of this typology of female killers acting in partnership include killers who are younger, aggressive, and vicious attackers. They sometimes appear disorganized and display evidence of a lack of careful planning. They attack victims in diverse locations and often use torture before using guns and/or knives to kill their victims. In most instances, the female killer seems more of a follower than a leader, a dependent personality type rather than the instigator.

Team Killers

The team killer type represents *one-third of all female serial killers,* and subdivides into three team subtypes. Each category has an average murder count of nine to fifteen victims. They use a wide variety of weapons, including guns and knives.

The three team categories are: (1) male-female, (2) female-female, and (3) family. Male-female teams are the most common and the most sexualized. The female member is usually about twenty years of age. She will have a short career of one to two years as a team killer.

Bonnie Parker and Clyde Barrow, Caril Ann Fugate and Charles Starkweather, Debra Brown and Alton Coleman

Bonnie Parker (sixteen years old) and Clyde Barrow (twenty-one years old) began a crime spree in Dallas, Texas that extended from 1930 to 1934 into six states—Oklahoma, Kansas, Missouri, Iowa, Arkansas, and Louisiana. Eventually, they formed a gang, stealing cars, robbing banks and grocery stores. They shot and killed sixteen people, which included thirteen police officers. Reportedly, Bonnie enjoyed putting a few extra bullets into the corpses of police officers. Actor/director Warren Beatty made a movie of the team's exploits.

In 1958, Caril Ann Fugate (age fourteen) and Charles Starkweather (age nineteen) embarked on a one-month crime spree in Nebraska and Wyoming. Originally they came from Lincoln, Nebraska. They shot and killed eleven people, including Caril's family then ate and had sex for three days in the same house with her family's dead bodies. Charles raped at least one female victim before shooting her, and Caril reportedly mutilated this same victim's genitals in a jealous rage. Police captured them in a high-speed car chase. Charles died by electrocution in 1959. Caril spent twenty years in prison. Two Hollywood movies based on the team killers' exploits appeared, *Badlands,* and *Natural Born Killers.*

In 1984, Debra Brown (age twenty-one) and Alton Coleman (a twenty-eight year old ex-con), an African-American, common law couple from Chicago, targeted African-American victims in a seven-week crime spree. The killings took place in Illinois, Indiana, Michigan, and Ohio. They brutally raped and murdered eight of the victims, who ranged in age from seven to forty-one. Alton had a history of being an aggressive "pansexual." The couple made the FBI's most wanted list. Each received multiple death sentences in three states.

Female Killer Teams

Female team killers are typically twenty-five years old. They kill for two to four years until apprehended.

Gwendolyn Graham and Catherine May Wood

In 1987, Gwendolyn Graham (age twenty-four) and Catherine May Wood (age twenty-five), a lesbian couple, achieved sexual thrills in their murders of five elderly female patients at a nursing home where they worked. Suffocation of the victims was their MO. Their signature consisted of making love immediately afterwards to enhance the intensity and thrill of sex. Wood was the submissive one, while Graham was the dominant, sexually exploitative partner. When Graham left her lover to take a job out of state, Wood confessed to the crimes. Graham received life imprisonment while Wood received incarceration for twenty years.

Unsolved Killers—UNSUB (Unidentified Subject)

UNSUB "The Butcher of Kingsbury Run," UNSUB Angel of Death "Muscle Relaxant" Murder

From 1935 to 1938, the murders of twelve men by castration and decapitation occurred along a stretch of railroad between Cleveland and Pittsburgh. The UNSUB became known as "The Butcher of Kingsbury Run." Was the murderer a male or a female serial killer or team killers? At the time, popular sentiment speculated that just one woman did the decapitation murders.

In 1975, five murders occurred by injection of a muscle relaxant. The murderer somehow administered the drug by intravenous tube, even when hospital officials closely observed the patients. The FBI and CDC also became involved in the investigation. They believed that a technologically savvy "Angel of Death" operated in the hospital, but they could never identify any suspects.

Family Team of Killers—Female Serial Killers

Female killers are about twenty years old when a family team of killers begins. The family teams last about a year before their apprehension.

McCrary Family Killers

The McCrary Family raped and shot twenty-two victims, mostly young female employees of stores they robbed from Florida to California from 1971 to 1972. Family members consisted of Carolyn and Sherman McCrary, their son Daniel, and their daughter and son-in-law, Ginger and Ray Taylor. They repeatedly raped the victims in the car, shot them, and threw the bodies out the window of the speeding car. Police caught them after a shootout in Santa Barbara, California. The FBI linked them to twelve additional unsolved homicides, although they received conviction for only ten murders.

PREDATOR PROFILE 7-2

From Mother to Murderer The Saga of Susan Smith

Ashleigh Portales

In recent years, society has witnessed an alarming increase in homicides against children perpetrated by the victims' own mothers. Such an abomination of the natural human order raises serious questions and concerns within the public. Why does this happen? Are these murderous mothers sane or mentally ill? Who is ultimately to blame? The answers vary from case to case; further complicating efforts to understand such crimes because there exists no single catalyst to murder that would help to calm the ensuing fears of the dark side of the human psyche.

America faced one such horror on Thursday, November 3, 1994. After nine days of searching for her "kidnapped" children, South Carolina mother Susan Smith confessed to sending her own car into John D. Long Lake with her two sons, Michael, age three, and Alexander, age fourteen months, strapped in car seats. The subsequent investigation and murder trial attempted to answer the most disturbing question of all: Why would a mother kill her own children? The answers, heartless and devastatingly cold, shocked the nation.

Little Girl Lost

It was 1960 in Union, South Carolina, when twenty year old Harry Ray Vaughan married seventeen year old Linda, who already was pregnant from a previous relationship. Together the couple raised Linda's son, Michael, and added two of their own children, a son, Scotty, and a daughter, Susan Leigh. Susan was born in their industrial hometown on September 26, 1971. The firefighter turned textile mill worker and the homemaker made a turbulent match, creating a dysfunctional home life for the three children from the start. Harry was an alcoholic obsessed with the notion that his wife was unfaithful. Arguments often escalated into physical assaults right in front of the children, who felt tense and frightened by the behavior.

Before Susan was old enough to attend preschool, her half-brother Michael attempted suicide by hanging himself. This resulted in treatment at Duke University Medical Center and many subsequent stays at other residential mental health treatment facilities. Susan also displayed side effects of her tumultuous environment. A friend's mother recalled the girl's demeanor as "unusual and sad . . . Susan would stare in space, like she wasn't there." In 1977, when Susan was six, her mother filed for divorce, sending Harry into a downward spiral of depression and drinking. Five weeks after the divorce finalized, police received a call to Linda's home on a domestic violence report. They arrived in time to see Harry strike Linda. Soon after his arrest, Susan's father committed suicide, mortally wounding himself with a gunshot to the abdomen. He called 9-1-1 but died soon after emergency surgery. He was thirty-seven. Even though he had disrupted Susan's life, his death created a huge void in her. Susan had enjoyed a special closeness to her father and spent the rest of her life (apparently) in futile attempts to compensate for the lack of paternal love.

Susan's mother, however, wasted no time mourning. In fact, she already had remarried just two weeks after her divorce from Harry, suggesting that he was correct in his suspicion of infidelity. Her new husband was Beverly "Bev" Russell, a prominent local businessman with several daughters from a previous marriage. Russell was a South Carolina State Republican executive committee member who served on the advisory board of the Christian Coalition. He moved the family to his exclusive home in Union's Mount Vernon Estates with all the trappings of success.

However, his credentials as a stepfather proved much less distinguished. Desperate for masculine attention, Susan soon began to compete with her mother for Bev's affections, and he took full advantage of the situation. He began molesting Susan as she neared her sixteenth birthday. At first Susan pretended to be asleep during the molestations. Later she told her mother she "wanted to see how far he would go." This was the first of many inappropriate reactions displayed by Susan throughout her life.

Though Susan filed complaints against her stepfather, Linda quickly discontinued family therapy sessions and stifled any further reports. She chastised Susan for tarnishing the Russell name in the community and involving Social Services, more concerned with her family's reputation than her daughter's mental and physical health. With no support from her domineering mother, she sought the advice of a school guidance counselor, who had to report abuse claims to Social Services the following year as the abuse continued. An investigation revealed Bev often had fondled Susan, French-kissed her, and placed her hands on his genitals. Though Linda told investigators Bev had not denied the accusations, she pressured her daughter not to press charges and, in an agreement between Bev's attorney and the prosecution sealed by the judge, charges never did get filed. From there the abuse continued, and Susan attempted to hide it behind academic success. She was a member of the Beta, Math, Spanish, and Red Cross Clubs. Throughout high school, she also volunteered annually for the Special Olympics and served as a candy striper at the local hospital. Hiding her troubles behind an outgoing persona won her the title of "Friendliest Female" in her senior year of high school.

That summer Susan began working at the local Winn-Dixie, where she continued her distorted desire for male companionship by beginning an affair with an older married co-worker. Soon after the affair began, Susan became pregnant and had an abortion. The older co-worker then discovered Susan was carrying on a relationship with another employee as well and subsequently broke things off. Depressed, she attempted suicide by overdosing on aspirin and Tylenol. The doctors discovered during her hospital stay that she had made a similar attempt at age thirteen. After a month of recovery, Susan returned to work. She became involved with David Smith, a stock clerk and former high school classmate with whom she had kindled a friendship before her suicide attempt.

A Cinderella Story?

David Smith came from an equally unhappy home. His father, a Vietnam veteran and Wal-Mart manager, had grown to detest his mother and her devotion to the deprivation and isolation favored by the religion of Jehovah's Witnesses. Similarly, David rejected the religion and moved in with his great-grandmother next door.

During the summer of 1990, David and Susan began what David viewed as "casual dating," despite the fact that he was engaged to long time girlfriend Christy Jennings. In January 1991, Susan told David she was pregnant. When Christy found out, she quickly broke off her relationship with Smith. The couple decided to marry despite several obstacles because neither desired an abortion. Though Susan saw marriage as a safe and stable existence, she would have to give up her dreams of a college education. In addition, Susan had grown up in the city, a fact that meant she was "better" than a country kid like David within the small town of Union. Therefore, it disgusted Susan's parents that their daughter now had to marry "beneath" her upbringing. However, the wedding between twenty year old David Smith and nineteen year old Susan pushed forward on March 15, 1991, despite the death of David's brother eleven days prior from Crohn's disease. Linda had insisted the wedding take place for fear Susan would begin "to show" (her pregnancy) if they waited any longer.

It was customary in the small southern town for young people to marry right out of high school and to begin having children. Yet, this practice often gave couples overwhelming financial situations and many marital stressors. It was no different for the Smiths. Their first conflict arose over housing. David had renovated a small home on his great-grandmother's property for several years in preparation for the day he would live on his own. However, after Bev and Linda saw the house, Susan refused to live in it, calling it no more than a "tinroofed country shack." David compromised, and the couple moved in with his great-grandmother. They also squabbled over money. Susan spent their salaries on material things in an attempt to grasp a lifestyle beyond her reach.

As a result, she often borrowed money from her mother to pay the bills, a fact that put more stress on the already strained relationship between Linda and David.

Michael Daniel Smith was born on October 10, 1991, and Susan returned to work at Winn-Dixie. This was another problem area for the couple, as Smith was Susan's boss at the grocery store. Trouble reared its ugly head soon as both engaged in extramarital affairs. The couple had separated several times by their third anniversary. During one of several attempts at reconciliation, Susan again became pregnant and told David their marriage would succeed only if they had their own home. So, Linda and Bev provided the down payment, and they bought a small home down the street from David's great-grandmother.

During her pregnancy Susan constantly complained that she was "fat and ugly" and shut David out completely. Lonely, he began an affair with co-worker Tiffany Moss. Jealous and angry, Susan often appeared at Winn-Dixie where coworkers recalled her screaming at David every time he talked to another woman.

When Susan gave birth to Alexander Tyler Smith on August 5, 1993 by emergency Cesarean section, the couple again tried reconciliation, but they abandoned the effort three weeks later. Susan felt she could not work under David and found a new job at Conso Products, working her way up to assistant to the executive secretary for J. Carey Findlay, the company's president and CEO. Susan enjoyed the job and the glimpses she got of the expensive lifestyle she always wanted.

She began a relationship with twenty-seven year old Tom Findlay, the wealthy son of her boss and the town's most eligible bachelor. Because she desired to live the upscale life Findlay could afford her, Susan filed for divorce in September of 1994. Unbeknownst to both Smith and Findley, Susan also was carrying on simultaneous sexual relationships with her stepfather, Bev Russell, and Tom's father, J. Carey Findlay. Susan's dreams shattered when Tom, believing she was too needy and possessive, ended their relationship in a "Dear John" letter dated October 17, 1994. "You will, without a doubt, make some lucky man a great wife. But unfortunately, it won't be me . . . You have some endearing qualities about you, and I think you are a terrific person."

"But, like I have told you before, there are *some things about you that aren't suited for me, and yes, I am talking about your children.* " He further cited their contrasting socioeconomic backgrounds as an insurmountable hurdle. He complimented Susan on her recent enrollment in night classes at the local community college, but he chastised her behavior at a hot tub party he had recently thrown where Susah had kissed and fondled a friend's husband while both were naked in the hot tub. "If you want to catch a nice guy like me one day, you have to act like a nice girl. And you know nice girls don't sleep with married men." Findlay added insult to injury by choosing to type the letter on a word processor rather than to write it out by hand. This gave it the feel and appearance of a business document rather than a personal letter whose words would trigger great emotion. Despite her efforts, Findlay refused to resume the relationship. Susan took days off from work to stay home and drink, filled with anxiety and struggling with depression.

Slaughter of the Innocents

All the conflict and confusion Susan had experienced throughout her life came to a head on

October 25, 1994. Tuesday began just like any other for Susan Smith and her two boys. She fed and dressed the children before she dropped them off at daycare and drove herself to work. She went to lunch with a group of coworkers, including Tom Findlay, but she seemed unusually quiet and withdrawn. Around 1:30 P.M., Susan requested to go home early. When her supervisor, Sandy Williams, asked if something was wrong, Susan expressed she was "in love with someone who doesn't love me." When questioned about the identity of that person she replied, "Tom Findlay, but it can never be because of my children." Upset, Susan returned to her desk and called Findlay in his office at 2:30 P.M. She requested he meet her outside to talk and confided that David Smith was about to reveal embarrassing information in divorce court, including accusations of tax fraud and "having an affair with your father."

Rather than evoking sympathy from Findlay, it induced shock, prompting him to tell Susan that while their friendship could remain intact, "our intimate relationship will have to stop forever." Undeterred, Susan sought Findley out in his office and attempted to return the Auburn University sweatshirt she had borrowed, but Findlay insisted she keep it. After picking up her sons from daycare, she returned to work again, determined to rekindle her relationship with Findlay. Co-worker Sue Brown watched Michael and Alex while their mother again confronted her former lover, but he quickly banished her from his office. Susan was upset and told Sue she "may just end it," and then drove home with her children around 6:00 P.M. Later that evening, Sue Brown was having dinner at a local bar with a group of friends that included Tom Findlay when the waiter brought her a cordless phone. Susan Smith was on the line desiring to know if Tom had asked about her, to which Sue replied "No." At 8:00 P.M., Susan dressed the boys, ages three years and fourteen months, put them in the car seats, and began driving around town. She later said, "I had never felt so lonely and sad in my life."

Her children never were seen alive again.

Around 9:00 P.M., Shirley McCloud heard a wailing from the direction of her front porch one-fourth of a mile from John D. Long Lake. Upon investigation, she discovered Susan Smith hysterically screaming, "Please help me! He's got my kids and my car! A black man has got my kids and my car!" McCloud's son dialed 9-1-1, and the call went to the Union County Sheriff's Office at 9:12 P.M. When Sheriff Howard Wells arrived, she told him, "I was stopped at the red light at Monarch Mills and a black man jumped in and told me to drive. I asked him why he was doing this and he said 'Shut up and drive or I'll kill you.'" She stated that the intruder ordered her to drive northeast of Union for approximately four miles until "he made me stop right past the sign," a sign that Shirley McCloud confirmed was the sign for John D. Long Lake near her home. Susan continued, "He told me to get out. He made me stop in the middle of the road. Nobody was coming, not a single car. I asked him, 'Why can't I take my kids?'" She contended that her assailant replied, "I don't have time!" and pushed her out of the car at gunpoint. "When he finally got me out, he said, 'Don't worry, I'm not going to hurt your kids.'" She then stated that she laid on the ground as the man drove away, tortured by the sounds of her sons calling out to their mommy for help. She told police that, after an unknown interval of time, she got up and began to run until she reached the McCloud residence. She then used the restroom and called her mother, whom she was unable to reach, before calling David at work.

Sheriff Wells, who knew Susan through his friendship with her brother, Scotty, immediately began a full scale search for the missing boys and called the South Carolina Law Enforcement Division, or SLED, for additional assistance. The fact that Smith wore a gray and orange Auburn University sweatshirt was among the notes he took from this initial interview.

Concerned family members and friends flooded the McCloud home, prompting Sheriff Wells to suggest they meet elsewhere. Susan volunteered her mother's home and, while riding there with David, she informed him that Tom Findlay might come to see her and requested that David keep his cool in such an event. Such an untimely comment struck David as odd, but he put it out of his mind in the search for his children.

Suspicious Minds

Divers began scouring a section of John D. Long Lake while Susan met with police sketch artist Roy Paschal. Paschal used the physical description she gave to produce a sketch of an African-American male, approximately forty years old, wearing a dark knit cap, dark shirt, jeans, and a plaid jacket. Divers surfaced empty handed, but Paschal felt he had more luck. He found it odd that Susan's description was very vague on important points and quite detailed about many smaller insignificant aspects of the carjacker's appearance.

An additional analysis of the sketch came from Jeanne Boyton, a cognitive graphic artist who had provided the sketch of Richard Allen Davis. Davis later received conviction of the 1993 kidnapping and murder of twelve year old California girl Polly Klass. Boyton believed that if Smith actually was carjacked as she claimed, she would have relayed far more details about the perpetrator. In addition, she felt the body positioning in the drawing was incorrect and found the suspect's passive and emotionless expression very suspicious. When she attempted to speak with Smith, Smith refused. Boyton had worked over seven thousand criminal cases and found such behavior atypical. In her professional opinion, Susan Smith was involved in the disappearance of her children.

As the investigation continued, more holes began to appear in the story. Susan claimed that on the evening of their disappearance, she had fixed dinner, but that her sons were "fussy" and refused to eat. Smith confirmed this fact, as he had called during this time and heard the boys in the background. However, she then stated that she had put the boys in the car and gone to Wal-Mart, per Michael's request. Yet no one working or shopping in the store could place Smith and her "fussy" children there on the evening in question. Furthermore, she claimed she was carjacked at the red light at the Monarch Mills intersection while on the way to a friend's house. That friend, however, was not expecting Susan and her children and was not even home at the time. Most troubling was the fact that the light at Monarch Mills intersection is permanently green unless triggered by a car on the cross street. Therefore, if the streets were vacant as Smith had claimed, a red light would not have stopped her.

Susan altered her story when confronted with the information and said she really had driven around town aimlessly with her sons in the backseat, but she didn't want to say so for fear of sounding suspicious. Another matter of speculation was how a black man could drive around for days with two white babies and go unnoticed, especially with the increased presence of law enforcement searching for the boys.

Throughout the investigation Susan never spent a moment alone, constantly comforted by David, her family, and friends, a welcome change from the isolated depression she had experienced prior to her sons' disappearance. Though Tom Findlay never visited, he did call with condolences, yet Susan quickly steered the conversation toward their shattered relationship. Findlay never called again. When co-worker Sue Brown came to comfort Smith, Susan asked only when Tom was planning to visit.

A deluge of national news media descended on the town of Union, requesting a public address from the Smiths. A camera shy and blatantly devastated David, with Susan at his side, made this statement:

> To whoever has our boys, we ask that you please don't hurt them and bring them back. We love them very much . . . I plead to the guy please return our children to us safe and unharmed. Everywhere I look, I see their play toys and pictures. They are both wonderful children. I don't know how else to put it. And I can't imagine life without them.

Both David and Susan submitted to lie detector tests on October 37, 1994. While David's results revealed he knew nothing about his sons' fate, Susan's test was inconclusive. She showed the greatest level of deception when the tester asked her, "Do you know where your children are?" A frustrated sheriff called in Agent David Caldwell, a criminal profiler of the SLED, to interview Susan. He confronted her about her relationship with Tom Findlay and the reason for their breakup. "Did this fact play any role or have any bearing on the disappearance of your children?" Caldwell asked, to which Smith replied, "No man would make me hurt my children. They were my life." Smith's word choice indicated to Agent Caldwell that she knew her children already were dead. Later in the day in another interview he questioned Smith about her story's many inconsistencies. He wondered why she had lied about the trip to Wal-Mart and, addressing the boys' reported fussiness, asked, "Is that why you killed them?"

Caldwell was shocked by her response as the previously quiet and timid woman who often repeated, "God look after my babies," slammed her fists on the table and shouted, "You son of a bitch! How can you think that!" before storming out of the interrogation room yelling, "I can't believe you think I did it!" The FBI Agent who had administered Susan's polygraph had noted she made "fake sounds of crying with no tears in her eyes." Caldwell had noticed much of the same throughout his interviews.

The Union County Sheriff's Office sought confirmation with a profile of a homicidal mother, convinced they had stared a killer in the face all along. What they received could have had Susan Smith's name on it. The profile described:

> a woman in her twenties, who grew up or lived in poverty, was undereducated, had a history of either physical or sexual abuse or both, remained isolated from social supports, had depressive and suicidal tendencies, and was usually experiencing rejection by a male lover at the time she murdered.

Additionally, the mother might find herself enmeshed with her children and show an inability to define her boundaries as separate from her children. The depression in the mother was often correlated with a blurring of boundaries. A mother's biological ties, her strong role expectations to be a mother, her significantly greater care giving responsibilities, her isolation in carrying out those responsibilities, and her greater tendency toward depression and self-destruction were likely to result in her becoming trapped in enmeshment with her children. During a homicidal act, a mother may view a child as a mere extension of herself rather than as a separate being. A mother's suicidal inclination may often be transformed into filial homicide.

Spurred by this new information, divers again searched areas of John D. Long Lake, combing the murky water in vain. They also searched the north and south sides of Highway 49 but uncovered nothing. What investigators did not know was that they had received misinformation.

According to Court TV's "Crime Library,"

> Experts had made a tremendous error when they told the divers to assume that anyone trying to hide a car would drive it into the water at a high rate of speed. None of the experts considered that a driver might simply let a car roll from the edge of the banks into the water. It is easy to envision that a car driven into a body of water at a high speed would go further than a car driven slowly, in reality, the opposite occurs. The faster that a car hits the water, the more waves it creates which stops the forward momentum of the car. A car driven at a high rate of speed into water simply drops and sinks at the edge of the body of water. Because the Mazda had been rolled into the lake at a slow speed, it had drifted out much further from the edge of the water, nearly one hundred feet. Divers searched the edge of the water, while the Mazda remained submerged.

Their inability to locate the vehicle plagued investigators who continued to question Susan daily, using meticulously scripted and choreographed routines designed to extract a confession. They wanted to develop trust from Susan without pushing her too hard, which they felt would cause her to either shut down or to commit suicide. In multiple polygraphs, Susan routinely failed the question, "Do you know where your children are?" As a result of his many interviews with her, Agent Caldwell produced a psychological profile of Susan Smith that described a "cool, cunning woman with a strong drive to succeed." Provided with a copy of Tom Findlay's letter to Susan by Findlay himself, Caldwell developed a possible motive: "that greed and ambition had pushed Susan to rid herself of her children by murdering them." Caldwell subsequently suggested several interrogation strategies that might result in confession. He planned to increase media pressure on Susan by airing a segment on *America's Most Wanted* about her boys' disappearance. He arranged for a group of Union's most influential ministers to hold a press conference appealing to the carjacker to turn himself in. He even proposed creating a fake newspaper containing a story about a mother who had killed her children, served a short term in prison, and married a wealthy doctor upon release, suggesting that the hope of a new and different life might prompt a confession.

However, these strategies never were put into effect for, on November 3, 1994, nine days after her sons' disappearance, Susan Smith, tired and worn from multiple interrogations, confessed to their murder.

Fall from Grace

Before her confession, David and Susan sat holding hands in the living room of Bev and Linda Russell for an interview on CBS This Morning. When asked if she had played any part in her sons' disappearance Susan replied, "I did not have anything to do with the abduction of my children. Whoever did this is a sick and emotionally unstable person" (Smith could not have described herself more accurately). At 12:30 that afternoon, Sheriff Wells sent for Susan to hold an interrogation at a safe house away from the press, in the Family Center of the First Baptist Church. This time Susan changed the location of the carjacking from "Monarch Mills" to "Carlisle." Wells informed her that this could not be true because he had placed undercover drug officers at that intersection on the night of the abduction, and they had seen and heard nothing. He also told Susan she would have to retract her statement to the media concerning the black carjacker because it was causing racial tension in the community. To his surprise, Susan requested the Sheriff pray with her. In closing, Wells said, "Lord, we know that all things will be revealed to us in time." Looking at her, he said, "Susan, it is time." Smith hung her head repeating, "I am so ashamed, I am so ashamed." She then asked Wells for his gun so she could kill herself. When he asked why she would want to do that, she replied, "You don't understand, my children are not alright." She continued to relate how depressed and isolated she had felt driving her Mazda the night of October 25 and how she had desired to end her life with suicide. She contended that she had planned to leave her sons with her mother but came to the conclusion that even her mother could not help her "escape her loneliness."

Susan appeared to sob while other investigators entered and aided in taking her written confession. In stark contrast to the emotions she was trying to convey stood her carefully constructed script with rounded letters and little hearts drawn in place of the actual word "heart." Smith wrote that she originally had driven to John D. Long Lake with the intention to kill herself but had decided her sons would be better off with her and God than without a mother and altered her plan so they could all die together. She confessed she had stopped the car with the emergency brake several times during its descent into the lake before finally exiting the vehicle and, "overcome with grief, loneliness, and pain," reached inside and released the parking brake. This caused the car to roll forward, down the boat ramp, and into the lake. She told investigators she was sorry, that she loved her sons, and had not intended to harm them. She claimed she "wanted to undo it" after the car sank into the water but realized it was too late and began to plan her alibi as she ran to the McCloud's house. She also said it was "very difficult" to keep her secret while a search ensued and her family and David endured nine days of agony. However, she pinpointed the exact location where the car had gone under, which meant she had waited around to see her sons die.

Divers soon found the overturned vehicle in eighteen feet of water where visibility did not exceed twelve inches. Diver Steve Morrow testified at trial that all the windows in the car were rolled up. He said that he saw a "small hand against the window glass" and that he could see the boys "in car seats hanging upside down." It took approximately forty-five minutes to pull the car from the lake and flip it right side up. When they did so, the dome light came on, suggesting Susan intentionally had turned on the light to afford a better view of her children as they died. The next day autopsy revealed that Michael and Alex had not drowned until the car was submerged completely, a process that forensic investigations revealed would have taken a full six minutes. Subsequent reenactments revealed that the rear of the car was rising while the front filled with water. Thus, Michael and Alex had faced the lake as it slowly rose to swallow their little lives.

In an interesting aside, a study of child murders in the U. S. by the National Center for Missing and Exploited Children found that "mothers who murdered their children disposed of their bodies in a distinctively womb-like manner." According to the data collected, "some victims were submerged in water and others were found carefully wrapped in plastic." Additionally, "all the victims' bodies were found within ten miles of their family home."

Susan wrote a letter to her estranged husband after completing her written confession. Several times she repeated, "I'm sorry," but she interspersed such apologies with complaints that her own feelings were "getting lost in the midst of everyone else's sorrow."

The Trials of Motherhood

Three days after their discovery, the bodies of Michael and Alex received burial together in a single casket at a private graveside service. The boys lay next to their uncle, Danny, who had died eleven days before their parents' wedding. Meanwhile, police were holding their mother at the York County jail, having waived her right to bail and charging her with their murders. Her parents mortgaged their home to hire attorney David Bruck who specialized in death penalty cases. Judy Clarke, a federal public defender from Washington State, assisted in the defense. Clarke later defended Unabomber Ted Kacyznski in 1997. Union County Solicitor Thomas Pope was the prosecutor, at thirty-two the youngest prosecutor in the state of South Carolina. He previously was an undercover drug agent for SLED and had tried only one murder case in which a father confessed to smothering his son. In that case, he had accepted a plea for a prison sentence of eight years. Pope requested that Smith be subjected to an impartial psychological evaluation, but Judge John Hayes denied the motion due to Bruck's concerns. Since Bruck had not yet declared Smith would offer an insanity plea, he feared that Pope could use information obtained in the evaluation against his client should Pope choose to seek the death penalty.

Susan awaited trial at the Women's Correctional Facility in Colombia, South Carolina where she was placed on twenty-four hour suicide watch for the next eight months. Smith was allowed only a Bible, a blanket, and her glasses. She wore a paper gown and was monitored via a closed circuit television camera all day, aided by the bright cell lights that never went off. During this time, she requested a visit from David who complied but left when Susan would not give him an answer as to why she had killed their children. She also received a letter from stepfather Bev Russell dated June 18, 1995, Father's Day. In it Russell wrote, "My heart breaks for what I have done to you . . . I want you to know that you do not have all the guilt for this tragedy." To the sane observer, it is easy to see from whom Susan learned the gift for exhibiting *faux* emotion.

On January 16, 1995, Solicitor Pope filed a notice of intention to seek the death penalty, citing the aggravating circumstances that she had killed two people during one act and that her victims were under the age of eleven. In a not uncommon turn of events, Susan's pastor, Mark Long, with the permission of David Bruck, held a press conference a few days prior to trial revealing that Susan had "undergone a jailhouse Christian conversion and baptism." While the intent was clearly to deter a death sentence, the ploy backfired. As a whole, the public doubted the validity of her "redemption" due to its startlingly convenient timing. Additionally, it seems quite ironic that, through an immersion baptism, Susan felt she could be saved and begin a new life when water was exactly the thing that had doomed her children and ended their lives.

Authorities issued a gag order, and trial began on July 10, 1995 in Susan's hometown of Union. David Bruck had not requested a change of venue, believing that if he could gain the sympathy of Susan's neighbors he could spare her life. The jury selection was complete by the sixteenth. The jury consisted of nine men (five white and four black) and three women (two white and one black), along with two alternates. Opening statements were to begin on the eighteenth of July, but a bomb threat received at the Union County Courthouse caused the complete evacuation of the building and delayed proceedings until the next day, when police apprehended the man who phoned in the threat.

On the morning of Wednesday, July 19, 1995, Special Prosecutor Keith Giese, Pope's assistant, began his opening statement by saying, "For nine days in the fall of 1994, Susan Smith looked this country in the eye and lied. She begged God to return her children to safety, and the whole time she knew her children were lying dead at the bottom of John D. Long Lake." He contended Smith had killed her children in order to remove the "obstacle" they were to the relationship she desired with Tom Findlay. "This is a case of selfishness," he told the jury, "of I, I, I, and me, me, me." In closing, he asked jurors to "hold onto their common sense in the weeks ahead, because they would come to see Susan Smith as a selfish, manipulative killer who sacrificed her children for love of the son of a rich industrialist."

Defense assistant Judy Clarke countered by charging jurors to look "into their hearts, and through that softer focus, find a disturbed, childlike figure who, after a lifetime of sadness, just snapped . . . When we talk about Susan Smith's life, we are not trying to gain your sympathy, we're trying to gain your understanding." She talked of Susan's ever-present sense of failure, even failure at her own suicide attempts. Regarding the murders, she proposed that, "at the last second, her body willed itself out of the car, and she lived and her toddlers died." She addressed Susan's nine days of deception by saying that her "lie was wrong. It's a shame, but it is a child-like lie from a damaged person." In stark contrast to the defense's picture of that child-like individual sat Susan Smith, looking well beyond her twenty-three years. Dependant on Prozac, Smith sat quietly behind the defense table reading mail or fiddling with small objects in her hands. Smith's eight months of inactivity during incarceration had added a significant amount of weight to her frame. Beneath her conservative dress and wire-rimmed glasses was a face virtually void of emotion, save when discussion of her sons produced brief, discreet "tears."

The prosecution's witnesses were the first to testify. Among them was Sheriff Wells who conveyed how he had tricked Susan into confession with a story about undercover agents at the intersection where she claimed she was hijacked. Wells' testimony ended the first day, after which Judge Howard removed juror Gayle Beam, the only black woman, and placed her in contempt of court for failing to disclose her recent plea of guilty to credit card fraud in her initial questionnaire. Beam later admitted her daughter had filled out the forms for her.

With an alternative juror in place, day two of the trial began with the testimony of SLED agent Pete Logan, sketch artist Roy Paschal, and FBI agent David Espie. They all said that they had doubted Susan from the beginning for various reasons, among which were her lack of real tears and her vague description of the assailant. Tom Findlay then took the stand, and the "Dear John" letter he had written to Susan, found in the car along with the boys' bodies, was entered into evidence. However, Findlay also assisted the defense's argument upon cross-examination when he stated, "The pleasure she got from sex was not physical pleasure. It was just in being close, being loved." Findlay was followed by three of Susan's co-workers from Conso who testified that, on several separate occasions, she had expressed to them that "she wondered how her life would be different if she had not gotten married and had children at a young age." The state ended its case with the testimony of Dr. Sandra Conradi, the pathologist who had autopsied Michael and Alex Smith. Her statements were cut short because David Bruck had stipulated her comments to the identity of the victims and drowning as the cause of death. Judge Howard did not allow the jury to see pictures of the decayed bodies, fearing they would be too prejudicial.

Next the defense took the floor. Doctors had found Susan not insane prior to the trial, so her attorneys' only defense was to plead that "Susan was suffering from severe mental depression and that the murders were a failed suicide in which Susan planned to drown herself as well as her sons." In the months before trial, a team of psychiatrists had evaluated Smith. Dr. Seymour Halleck, hired for the defense, led the team, joined by Dr. Donald Morgan, a psychiatrist whose evaluation the judge ordered on behalf of the prosecution for use in the competency hearing. After fifteen hours of interview spread over four sessions, Dr. Halleck diagnosed Susan as having a "dependant personality disorder," a person who "feels she can't do things on her own. She constantly needs affection and becomes terrified that she'll be left alone." Halleck felt the depression Susan suffered was not deep and only manifested when she was alone. It was at these times he felt she was suicidal. In addition, Halleck studied Smith's family history and surmised that she had a genetic predisposition for depression, which began to show itself in childhood. The high number of blood relatives who displayed symptoms of depression and alcoholism revealed this. Morgan's analysis was equally intriguing. After ten hours of interview spread over three sessions, he diagnosed Susan as "manifesting an adjustment disorder with mixed emotional features, including some depression." He had testified at Susan's previous competency hearing that, should Smith be allowed to take the stand, she might sabotage her own defense, inadvertently committing suicide.

Defense witness Arlene Andrews, a social worker at the University of South Carolina, testified that she had constructed the family tree of Susan Leigh Vaughan Smith and that it revealed a strong history of deep depression marked by several attempted and successful suicides by various family members. Dr. Halleck also testified on Friday, July 21, that in the months before the murders, Susan suffered from depression and entertained suicidal thoughts, which led to a "destructive cycle of sexual relationships in order to ease her loneliness." In the six weeks before she killed her sons, Smith had continued sexual activity with both Tom Findlay and J. Carey Findlay, as well as with Bev Russell, and David Smith. "Much of her sexual activity was not for her own satisfaction," he testified. "Susan was more concerned with pleasing others and making sure that they like her." He based this idea on her history of sexual abuse, as well as her statement to him during a session that having sex with Bev Russell "made her skin crawl." He dismissed the prosecutorial assertion that Susan's motive for murder was rekindling her relationship with Tom Findlay as an "absurd idea."

He believed their affair was only "passing," stating that she had "strong feelings for a lot of different men" and that "it was very unlikely that Tom Findlay was number one on her list." However, these claims starkly contrasted Susan's actions before and after the murder regarding Tom Findlay's relationship with her as well as the fact that his letter was found in her car and she was wearing his sweatshirt on the night of the murders. As for the reasoning behind Susan herself abandoning the suicide attempt in favor of a double homicide, Halleck chalked it up to a "survival instinct" that took over and "blocked out" her sons' presence at the instant she released the parking brake. "When she ran out of her car, her self-preservation instincts took over and although up to that moment she fully intended to kill herself, she got frightened." He also testified that, in his professional opinion, Susan would not have committed the murders if she had received proper treatment with Prozac®.

After several more witnesses testified to Susan's depression and suicidal tendencies throughout childhood, the defense ended its four days of testimony. It claimed that though Susan accepted responsibility for her actions, her depression drove them.

In closing arguments, the prosecution speculated passionately about the last moments of Michael and Alex's lives. "I submit to you," Solicitor Pope addressed the jury, " that they were in that car, screaming, crying, calling out for their father, while the woman who placed them in that car was running up the hill with her hands covering her ears. She used the emergency brake like a gun, and eliminated her toddlers so that she could have a chance at a life with Tom Findlay, the man she said she loved."

In a final plea to the bleeding hearts, defense attorney Judy Clarke declared that Susan Smith had never shown anything "except unconditional love for her children . . . There was no malice in what she did, so it was not murder . . . This is not a case about evil, but a case of sadness and despair. Susan had choices in her life, but her choices were irrational and her choices were tragic."

Before sending the jury out for deliberation, Judge Howard dismissed another juror on the grounds that he had a family tie to the case. The final alternate stepped in, and deliberation began. After only two and one-half hours, the jury returned a verdict of guilty on two counts of murder. As the words fell on the ears of the accused, she bowed her head in tears and trembled, her first real display of emotion, which came not for her sons but for herself.

The following Monday marked the beginning of the penalty phase. Defense attorney Bruck stated that "the greatest punishment for Susan Smith would be life in prison, not death," while Solicitor Pope reminded the jury of her "nine days of deceit and nine days of trickery." Three distinct revelations marked this phase of the prosecution's case. One was the testimony of SLED agent Eddie Harris, who had transported Susan during her numerous interrogations. He said he was "surprised by her calmness and disinterest in finding her children." He even told the court that at one point Susan had asked him how she looked on TV. Next came the heartbreaking testimony of devastated father David Smith. "All my hopes, all my dreams, everything that I had planned for the rest of my life, ended [with the death of Michael and Alex]." Smith broke down, and at least three jurors were in tears. Judge Howard was forced to call a recess to allow Smith time to regain his composure. As they led Susan away to her holding cell, she softly said, "I'm sorry David," but David Smith never looked at her. The next day jurors saw photographs of Michael's and Alex's discolored and decomposing arms and legs taken after their removal from their mother's car. The court banned images of the rest of the bodies. The prosecution rested with the horrific pictures burned into the minds of the jury.

In a last ditch effort to save Smith's life, the defense called three witnesses. The first was social worker Arlene Andrews who had testified at trial. She reiterated Susan's emotional turmoil and subsequent depression, stating that "Susan's suicidal despair set in and she began to think everything about her was bad." The second witness, Susan's brother, Scotty Vaughan, made an emotional plea for mercy.

"We've been devastated already with the loss of Michael and Alex, it seems sad and ironic that the tragedy of their loss is going to be used to sentence Susan to death. Susan's pain is in living, not in the fear of dying. I don't think the state could punish her any more than she's been punished."

The next day Bev Russell was the final witness to take the stand. He admitted to molesting Susan when she was a teenager and engaging in consensual sex with her as an adult. He testified that their encounters had occurred mostly at his home, once at Susan and David's home, and once at a motel in Spartanburg. He read from the letter he had written to Susan on Father's Day and pled for her life, contending that, "Susan was sick and even though she loved her children, what happened was from a sickness . . . It's horrible."

Solicitor Pope's closing argument centered around the theme of the choices Smith had made. "Susan Smith chose to drive to the lake. She chose to send Michael and Alex down that ramp. Then, as heinous as that act was, she carried it even further by choosing to lie." He contended that her claims of remorse were no more than an act intended to deceive, just like her claims of a carjacking during the nine-day search for her sons. He reminded the jury that Susan was "selfish" and "manipulative," characteristics which had hardened her heart toward her children in favor of her wealthy ex-lover, Tom Findlay.

In contrast, David Bruck again led the jury on a journey through Smith's "tragic" past in a last attempt at saving her life. He told jurors that the choices Solicitor Pope had talked about would "haunt her for the rest of her life" and quoted Scripture from the Gospel of John where Jesus intercedes in the stoning of an adulterous woman standing between the angry crowd and its target, Christ said, "He that is without sin among you, let him cast the first stone." When Bruck had finished, Judge Howard extended to Susan the opportunity to address the jury. She declined.

Within two and one-half hours, the same period it had taken to convict her of murder, the jury rejected the death penalty, sentencing her to life imprisonment. According to South Carolina law, she must serve thirty years of that sentence before becoming eligible for parole. This will occur in the year 2025, when Susan is fifty-three years old.

When questioned about their decision, jurors revealed they had considered the fact of Susan's confession. They did not believe there was enough evidence to tie her definitively to the crime without it. In their eyes, she had an opportunity to preserve her freedom but willfully choose not to do so, which they considered a reason to spare her life. In addition, they believed the defense's claim that she had intended to commit suicide and pitied her for her mental state during the commission of the crimes. The verdict, however, was not popular with everyone, especially David Smith. He felt that sparing Susan the death penalty was a great judicial miscarriage. He added that, should parole hearings ever occur, he plans to be present at each and every one to "make sure that her life sentence means life."

Aftermath

Today Susan Smith resides as inmate #221487 in the Women's Correctional Institute in Columbia, South Carolina, and is living proof that a severe psychopath's behavior never changes. In prison, her sexual improprieties in her "search for love" from authoritative men carried on just as they did before she murdered her sons. In September 2000, headlines broke the news about Smith's latest scandalous affair: having sex with South Carolina Department of Corrections prison guard Houston Cagle. An investigation of Smith's medical records revealed the affair. Investigators wanted to substantiate tabloid claims of an attack and beating in prison. While they discovered that the allegation of a beating was unfounded, they also learned that she had received treatment for a sexually transmitted disease "other than HIV." She admitted under questioning to having sexual relations with Cagle, who subsequently was jailed. At a bond hearing for the thirteen year correctional veteran, Cagle's wife Tammy, herself a former prison inmate, sobbed to the judge, saying, "I feel like I've been murdered too, just like those two little boys. She took my life from me."

In a pathetic effort to attract more perverted affection, and also more media attention, on June 1, 2003, Susan placed a personal ad at *WriteAPrisoner.com.* This organization matches incarcerated individuals to pen pals on the outside and operates under the slogan, "We'll see you at mail call!" Their disclaimer states that participating inmates are "reaching out for new beginnings and lasting friendships." Interestingly, her profile informs potential pen pals that she is a Methodist and a Libra. But although it states she is serving a life sentence, under the category, *Incarcerated For:* Susan has listed "N/A." In order to add to her angelic appearance, Susan also posted a picture of herself kneeling and smiling. However, the appearance of divine innocence is somewhat marred by the pants she wears that clearly read "South Carolina Dept. of Corrections." When word of the ad went public, society was outraged. Of course, attention, whether negative or positive, is exactly what Susan wanted. Naturally, she wants us to believe otherwise, and on July 17, 2003, she wrote the following letter to be posted on the Web site. It is enough to nauseate anyone.

> *To Whom It May Concern,*
>
> *Due to recent media coverage of my ad, I would like for you to remove my ad from the website.*
>
> *I apologize to you for all inconveniences this may have caused your company. Had I known it would have been sensationalized, I would not have placed the ad.*
>
> *I would like to commend you for creating this site as it is a great way for prisoners to meet new people and create friendships. There are many who have no family or friends supporting them. Your web site creates a way to ease the loneliness of so many.*
>
> *I thank you for your service during the time my ad was active and I thank you for your support. Your kindness has not gone unnoticed. I pray you will continue your good work.*
>
> *May God bless each one of you who has a hand in helping prisoners. Even prisoners need love and acceptance and we all need God and His grace, love, and mercy. You will be in my prayers.*
>
> *In Christ's love,*
> *Susan Smith*

UNIT V

Standing on the Shoulders of History

CHAPTER EIGHT

Standing on the Shoulders of History

In this the final chapter, we conclude our study of the *psychology of deception* with an illuminating mix of historical benchmarks. First, we disclose rumblings around a theoretical vortex that intermingled mind with sexuality and tangentially, with potential legal issues surrounding parental abuse; this landmark in our recap of history eventually became a revolutionary theory of mind and personality known as *psychoanalysis.* The setting for this foray into "human forensics" occurred in 1885 when Freud, a recent medical school graduate, visited the Paris morgue and observed evidence of graphic sexual abuse in children. Then, a few years later from the gas lit streets of Jack the Ripper's London (1888) autopsies again became pivotal in the deaths of East End prostitutes. Who was the perpetrator? Why did he hack his victims to death? Such historical events would forever merge psychology and law, eventually producing forensic psychology as we know it today.

Next, *behind the scenes* accounts of twelve selected founders of the new 19th century discipline of psychology will be presented. These early mavericks and those who preceded them allowed investigators to stand on their shoulders in the modern analysis of violent psychopathy and sexually psychopathic serial crime. Today, *psychologically-oriented forensics* would not be possible without *advanced brain imaging technologies* that characterize forensic neuropsychology—*the cause and effect of violent behavior at the tissue level.* As we have documented, the proof is in the high resolution neuroscans showing high or low blood flow as degrees of functionality.

As the story of the classical founders of psychology will show, they were afflicted with their own share of human inadequacies living their lives *through a glass darkly*—just the sort of thing that stimulated the modern understanding of violent psychopathy; we know today, for example, that *psychopaths see a different world* than those with more refined prefrontal cortices. The playground of creepy things—Things that Go Bump in the Night—might just be due to *high activity in the midbrain limbic system* versus *low activity in the prefrontal cortex* as we have documented in neuroscans.

Today, forensic psychology is an outstanding "new product" of behavioral analysis by applying psychology to criminal law; it has changed courtroom proceedings, crime scene investigation, the training of psychologists and criminal lawyers, and strategies used in criminal profiling. (The other new product of psychology is, of course, neuropsychology.)

Early Benchmarks of "Human Forensics"

In France (1885), twenty-nine year old Sigmund Freud joined Jean Charcot—the most celebrated neurologist of his day—to study hypnosis. Freud visited the Paris morgue on numerous occasions where he witnessed autopsies on young children appearing to be victims of physical and sexual abuse. This led directly to Freud's revolutionary Seduction Theory and soon thereafter, his surprising *abandonment of it;* this move ironically gave birth to psychoanalytic theory as a worldwide phenomenon. By abandoning his original theory, Freud speculated that *neurosis must be based upon illusion, not reality.* This reversal contradicted his original view that *real sexual abuse was the basis of neurosis*—a deduction that came from directly viewing bodies in the morgue. However, had Freud not abandoned his original theory, he and his perspective would have likely collapsed into obscurity due to the collective outrage it produced. How could a family allow abuse of its own children? From that time forward, autopsies and psychology, including the darker side of mind, would be forever fused together.

Specifically, for other pioneers with less controversial views—James Cattell, Alfred Binet, William Stern, Hugo Munsterberg, Lewis Terman, and William Marsten—applied psychology to the legal system and to crime scenes that helped to point psychology toward forensics. This trend continued after World War II with *criminal profiling* at the FBI's Behavioral Science Department in Quantico, Virginia. Agents John Douglas, Robert Ressler, and Roy Hazlewood were instrumental in launching yet another tool to catch society's elusive sexual predators.

While at Columbia University, James Cattell studied the *psychology of testimony* (1895) showing surprising degrees of inaccuracies. Meanwhile, future assessment tools used in forensic psychology came from the research of Alfred Binet (1911) and intelligence testing. Interestingly, Binet who possessed a law degree and was passionate about science, worked for a time with Charcot in Paris; he is credited with studying sexual perversity coining the term *erotic fetishism.*

Psychologist William Stern coined the term I.Q.—intelligence quotient—and launched studies in *witness recall* that led to a journal of applied psychology, while Hugo Munsterberg's book *On the Witness Stand* (1908) advocated the use of psychology to sharpen testimonial skills and appropriate courtroom demeanor.

Stanford professor Lewis Terman modified Binet's test that became known as the Stanford-Binet (1916), an assessment tool that provided the basis for the Army Alpha and Army Beta tests for classifying military recruits and alerting those in command to possible mental conditions. In 1917, William Marston discovered that systolic blood pressure had a strong correlation to lying that led to the development of *deception detection.*

The Brain & Neurochemistry

And so, supported by a cast of millions, the mix of law, psychology, and advances in clinical psychopathology and violent psychopathy as *originating within the brain and chemical networks* made more plausible by advances in medical technology became a symbol of the ever-present continua of normalcy versus abnormality. Today, headquarters for this ambitious study is firmly established in the human brain. The perfect metaphor for brain has always been mind; interestingly, to further the analogy, the perfect metaphor for mind might be captured in the myth of a modern Prometheus. (Prometheus, a mythical character in the writings of Hesiod and Aeschylus, claimed he gave humans "fire and the arts"—the passions of civilization—writing, mathematics, medicine, and science.) Have we learned enough through the centuries how to effectively "mind our own brains?" Who is prepared to ignore "WHY Things Go Bump in the Night?"

Now, to the early pioneers in the analysis of psychological causes and effects so vital to behavioral science. . .

Why Psychology Majors should rejoice that Neuropsychology is here to stay

The principle founders outlined here are presented in no particular order. They lived a combined and overlapping 934 years—almost a millennium—to provide psychology with places to be, things to say, and competition to join all those NEW schools of thought! The average lifespan of this special group was almost 79 years of age. All founders (except Skinner) were born in the nineteenth century and died in the twentieth century.

We begin with the oldest—Wilhelm Wundt eighty-eight years of age. Two other founders lived to be eighty-seven—Pavlov and Freud—Anna, not Sigmund. Anna Freud outlived her famous father by four years. Two other founders lived to be eighty-six—Jung and Skinner. Edward Titchener died the youngest at a youthful aged sixty. The longevity of this group is remarkable since the average lifespan in the twentieth century was most assuredly less. Might becoming a famous intellectual correspond to longevity?

It is more true than false that the Founders Club presented *speculative and philosophical tracks* for the engine of early psychology to run upon. Only Pavlov, Thorndike, Watson, Skinner—all from behavioral psychology (Pavlov would object)—could be defined as scientists in the modern definition; that is, of experimental design with measureable and observable behavior in controlled studies. Therefore, the rest may be famous, but *not as empirical gatherers of facts required of scientists,* but as courageous (and often arrogant) founders of a discipline that would change the world. In fact, Wundt would have loved to live among modern neuropsychologists as he so wanted to locate the "structures" of the brain that he felt ultimately explained behavior. The FORS BRAIN (an image celebrating the forensic science-oriented mind of early founders) will be attached to some of the historical luminaries in the early days of psychology as they directly or indirectly contributed to what would become forensic psychology.

Wilhelm Maximilian Wundt (1832–1920)

Lived Eighty-Eight Years

The German influence in psychology was none greater than its acknowledged founder, Wilhelm Wundt. He acted the part of a laboratory psychologist, not a philosopher, physiologist, or physicist (although he was clearly a physiologist). He established a laboratory dedicated to psychological research called *The Institute,* which eventually propelled him into lifetime membership of the Founders' Club and as principal founder of psychology. He established psychology as a separate discipline in 1879 and authored its first journal in 1881.

Universally, Wundt is credited with offering the first course ever in *scientific psychology.* The course emphasized experimental and physiological psychology. Wundt maintained the new science must be a *natural science* to study the relationship between the brain, behavior, and mind. His most important work may have occurred in 1874 with the publication of his textbook, *Principles of Physiological Psychology.*

Wundt sought to create paradigms to explain *immediate conscious experiences* (such as feelings and emotions). Could a person objectify, and therefore study, his own conscious experiences? If so, Wundt argued for the tool of *introspection*—self-examination of conscious experiences as perceived by the individual. Through introspection, Wundt sought to understand human behavior by identifying individual parts in the puzzle of mind. This was similar to breaking down chemical compounds into elemental parts (known as a *reductionistic perspective*).

However, it would be a student of Wundt's—Edward B. Titchener—who, upon leaving the University of Leipzig and travelling to America to assume his new post as professor of psychology at Cornell University, named the school he shared with his teacher the *School of Structuralism.* Structuralism represented the first Great School of the Masters delineating the "new science" of psychology. This perspective visualized *structures of mind*—thinking, emotions, and sensations—to be elemental like hydrogen and oxygen to water in the grand scheme of perception. By what method could this be proven? (Of course, it could never be proven empirically—observationally—without advances in medical technology with neuroimaging of live tissue of the brain in real time. This technology became available in the early 1970s.)

While structuralism was certainly meritorious and provided a focus to the new science, it could never become established as a serious attempt to understand behavior, due entirely upon its reliance on self-reporting *introspection.* More importantly, the living brain could not be viewed in real time with its cascading chemistry intact.

Today, graduate school psych majors view the living brain at work with the advent of non-invasive, high resolution neuroimaging such as PET Scans and fMRIs, soon to be addressed. Its cortical regions (structures) and functional (neurochemical) workings are imaged and studied in real time with a plethora of independent variables as stimuli.

No one could have imagined that shortly before his death, Wundt shifted his views from physiological psychology toward *social psychology* as he sought to explore social contingencies of nurture impacting one's nature or biological inheritance. This may have occurred due to his frustration with his own school of Structuralism driven by self-reporting introspection, clearly not a school of science.

As a founder of early experimental psychology, Wundt deserves a FORS BRAIN. Today, the laboratory of forensic psychology is NEUROSCANS requiring a dedication to scientific inquiry something very much in the brain of Wundt who indirectly influenced the later development of forensic psychology.

Edward B. Titchener (1867–1927)

Lived Sixty Years

A famous student of Wundt, Edward Titchener, migrated to America and Cornell University where he became professor of psychology. Upon the establishment of a physiological lab, he outlined the criteria and established the Structuralist School, erroneously attributed to Wundt. Among other outstanding contributions, he anglicized the word *empathy,* translated from the German, and wrote about the importance of empathy—the intellectual identification with, or vicarious experiencing of another's thoughts, feeling, or attitudes; the skills of which became absolutely necessary to becoming a successful clinician. He became an associate editor of the *American Journal of Psychology* in 1895.

William James (1842–1910)

Lived Sixty-Eight Years

Although Professor William James appeared cynical of the new science of psychology in the last decade of the 1800s—"This is no science, it is only the hope of a science"—he actually had great expectations for it, as evidenced by his introductory text *Principles of Psychology* (1890). As a Harvard professor of psychology (and brother to novelist Henry James), William began teaching *physiological psychology* in 1875.

Born into a wealthy and elite family headed by theologian Henry James Sr., James socialized with some of the most influential scholars and academicians of the day, including American essayist and philosopher Ralph Waldo Emerson (James' godfather), Justice of the Supreme Court of the United States, Oliver Wendell Holmes, Jr., American humorist and writer, Mark Twain, English science fiction writer, H. G. Wells, and fellow Founder Club alumnae Austrian physician, Sigmund Freud, and Swiss psychiatrist Carl Jung.

Interestingly, James suffered though a childhood filled with medical problems and mood disorders, especially depression, with one documented suicide attempt. As a young man, he traveled widely before contemplating what to do with his medical degree awarded in 1869. He never practiced medicine. He remained fascinated with philosophy and this new science called psychology. Soon, he became famous for saying: "The first lecture I ever heard in this new science of psychology was the one I gave."

He introduced physiological psychology to American students after procuring the services of Professor *Hugo Munsterberg* as his lab director from Professor Wundt.

Possessing witticism, and a natural charisma in classroom lectures, some historians argued James would have been a logical choice for the principle founder of psychology, rather than the more austere Wundt. Appearing in his classes were soon-to-be-famous colleagues such as G. Stanley Hall and Founder Club alum Mary Calkins.

James was one of the strongest proponents of what became known as the *Functionalist School* of psychology (the school competed for a time in the early days for new followers with the Structuralist School of Titchener and Wundt).

Functionalism focused upon behavior for its *adaptive qualities* above all in addressing social milieu. In this view, psychology must solve *practical problems and issues in real life* otherwise it was nothing more than an academic discipline, closely related to philosophy (James loved philosophy!).

Functionalism proved to be an early precursor to modern *evolutionary psychology* and *applied psychology*—methodologies not readily testable by controlled experiments. *Evolutionary psychology* is a theoretical approach that seeks to document how evolution has shaped the mind and behavior of humans through *natural selection* by successful adaptation to the natural environment.

Application of principles and methods of psychology to solve *practical problems* and to *launch careers* became known by a variety of names such as Industrial & Organizational (I/O) psychology, Educational psychology, and modern Sports Psychology. In addition child rearing and parenting, advertising and marketing soon became predominant fields of applied psychology.

James adhered to the *Pragmatist School* in Philosophy. Becoming a strong force in modern philosophy in the twentieth century the Pragmatist School considered *practical consequences verifiable for real life adaptation and ultimately to lie behind meaning and truth.* Francis Bacon's famous quote "Knowledge is Power" reflects philosophical pragmatism.

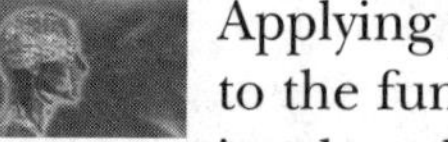

Applying psychology to solve the practical puzzles of life was central to the functionalism of Jamesian psychology. As forensic psychology involves the application of psychology to criminal law and legal procedure, James directly influenced the philosophical development of forensic psychology and richly deserves a FORS BRAIN.

Ivan Petrovich Pavlov (1849–1936)

Lived Eighty-Seven Years

The celebrated experimental physiologist, Ivan Pavlov, summarily dismissed anyone in his lab delusional enough to think they were following anything resembling psychology—the so-called psychic pseudoscience (in his opinion). He detested psychology. At the *Institute of Experimental Medicine* at St. Petersburg University, his pioneering work on digestion in dogs earned him the Nobel Prize in 1904. During his canine studies, Pavlov noted that the dogs commenced the *salivary response* PRIOR to the actual presentation of the bowls of food. What caused anticipatory salivation? Was it stimulated by *chaining*—an array of connective behaviors such as the sound of the footfalls of attendants approaching the lab or opening the door for morning shifts. This apparent *psychic secretion* troubled the sleep of the natural scientist; so he set out to discover the all-important WHY.

In a series of experiments, a bell was paired with the delivery of food (in addition to whistles, metronomes, tuning forks, and various visual stimuli). With continued *parings,* the bell came to stand for food—just like the sound of approaching footfalls of the attendants. The pairings soon became known as *conditioning.*

Pavlov's work became known a world away in North America at Johns Hopkins University in the lab of Professor John B. Watson. Conditioning became the catchword for the *objectification of behavior* within the Stimulus-Response (S→R) Psychology paradigm.

Interestingly, Pavlov's *conditional reflex* was mistranslated from the Russian as *conditioned reflex* so Western theorists believed conditioning to be an automatic form of learning due to *pairing two stimuli together.* In a paradigm for conditioned responses, the following formula became one of the most recognized paradigms in the early days of behavioral psychology. This type of learning became universally known as *classical conditioning* (also Pavlovian or respondent conditioning). It was a form of *associative learning* and resembled a mathematical formula. In relation to salivating dogs, it will forever ring a bell with psych majors.

The *unconditioned stimulus* (natural stimuli for eliciting salivation) is food as the UC. It naturally elicits salivation (UR) preparatory to eating as the *unconditioned response.* So . . .

US (food)→UR (salivation)
CS (bell)→CR (salivation)

By pairing the 'bell' (heretofore a neutral stimulus) with food, the conditioned stimulus (CS) produces the *conditioned response*—salivation—the CR.

Therefore, to both Pavlov and Watson's paradigms *stimuli* meant everything to behavior as it SET UP responses. Interestingly, Thorndike and B. F. Skinner held the opposite to be true, that is, responses meant everything due to REINFORCEMENT, that is, *what follows what we do.* We act (or behave) and what shapes and maintain responses ultimately becomes more important in determining behavior that continues. (What a friendly way to start a science with all the bickering over which is more important—stimuli or responses.)

Psych Sidebar

Pavlov's work not only stimulated scientists in North America to further along the development of *behavioral psychology,* it also became part of pop culture. For example, a *Pavlovian Dog* was someone who immediately reacted to a situation rather than using critical thinking. In literary circles, S→R psychology was the foundation for Aldous Huxley's novel *Brave New World,* just as *operant conditioning* was the foundation for Skinner's *Walden Two* utopian novel.

Edward Lee Thorndike (1849–1936)

Lived Seventy-Five Years

E. L. Thorndike, the graduate student, not the professor of psychology at Teachers College, Columbia University, was the person who became famous in the emerging school that focused upon the *objectification of behavior*—eventually known as Behaviorism. Thorndike studied under both William James at Harvard and James Cattell at Columbia.

Thorndike became famous for his Puzzle Box, featuring *trial and error learning* in cats. Observing cats escaping from the boxes led Thorndike to a theory of *connectionism* assembled around *neural networks in the brain.* Connectionism also contributed to modern educational psychology and to learning theory.

Development of medical technology that emerged in the latter half of the twentieth century was the only way to verify the great pathways and systems of neurochemistry within specific cortical regions of cascading chemical connectivity to behavior.

While Thorndike's work with animals depended upon observation considerable guesswork was required to propose how it might be connected to the brain. Nonetheless, observing the animals making right and wrong choices within puzzle box *payoffs* (with food) led to Thorndike's two well-known theories—*The Law of Effect* and *The Law of Exercise.*

The Law of Effect

Stimulus-response (S→R) connections between escape from the box to obtain food was easily observed to stamp in those responses as so-called *satisfiers,* while mistakes became known as *annoyers* and were not reinforced. Annoyers led to discarding wasted behavior. Hence, Thorndike observed trial-and-error learning slowly becoming *goal-directed* behavior. Hence his Law of Effect showed that cats stayed with the behavior that worked in affecting escape and procuring food so that such payoffs *shaped and maintained goal-directed behavior.*

Later, the *radical behaviorist,* B. F. Skinner would expand and deepen the Law of Effect in his theory known as *Operant Conditioning.* Clearly, Skinner stood on the shoulders of Thorndike, as both learning theorists championed the importance of 'what follows what we do' as payoffs—*reinforcers* that shape and maintain behavior.

The Law of Exercise

The related learning principal of the Law of Exercise states: a response will be more strongly connected to a stimulus in proportion to the number or times it has been connected with the stimulus and to the vigor and duration of that connection. Practice makes perfect is a shorthand version of this law and is observed everyday in almost everything we do to perform better (such as academic drills to athletic workouts).

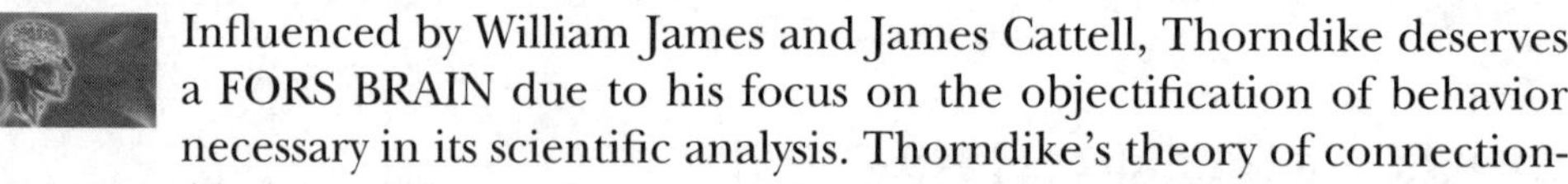

Influenced by William James and James Cattell, Thorndike deserves a FORS BRAIN due to his focus on the objectification of behavior necessary in its scientific analysis. Thorndike's theory of connectionism assembled around neural networks in the brain reads like a modern textbook.

John Broadus Watson (1878–1958)

Lived Eighty Years

Professor John B. Watson of Johns-Hopkins University is the recognized founder of an important school of psychology in North America known as

Behaviorism, an influential school in the early decades of the twentieth century (1920 to about 1970). In the strictest sense of the word, behaviorism is the North American perspective in the *objectification of behavior* that challenged the psychoanalytic perspective of Freud and his reliance upon introspection and unconscious experiences. A more global definition of behaviorism is a philosophy of psychology that demands objectivity and adherence to scientific methodology. In this view, all affect and mental behavior can be regarded as learned (conditioned) behaviors that can be re-learned or extinguished (discontinued). *Hypothetical constructs* (such as mind) are unnecessary in the objective analysis of behavior. The so-called Radical Behaviorism of B. F. Skinner differs from so-called traditional behaviorism of Watson, as an absolute reliance upon *schedules of reinforcement* independent of states of mind.

Watson's perspective in the early days of Behaviorism still showed periods of intense speculation, even with laboratory design. This period was known as the Age of Great Schools in Psychology. By the 1920s, the major schools had expanded to five:

- William James & Functionalism
- Wilhelm Wundt and Edward Titchener & Structuralism
- John B. Watson & Behaviorism
- Sigmund Freud's Psychoanalytic School
- the German school of *Gestalt psychology*

Psych Sidebar

A twentieth century investigation into *perception,* the so-called Gestalt theory of psychology evolved as a theory of perception where the brain must be analyzed as *holistically designed with self-organizing tendencies toward simplicity.* Rejecting reductionism, Gestaltists argued *perceptual holism* cannot be reduced to simple parts and always will be greater than their sum. The existence of a soap bubble is a classical example. The spherical shape is holistic and not defined by rigid templates or formulas but emerges spontaneously and disappears just as spontaneously.

S→R Psychology

A precocious child, Watson entered North Carolina's Furman University at aged sixteen, leaving with a Master's at twenty-one. The young scholar was influenced to study *behavioral contingencies* leading to his focus on *S→R (stimulus-response) psychology* or the objectification of behavior (behaviorism). Stimulus-Response Psychology (or Watsonian Behaviorism) focuses upon the importance of *environmental stimuli as predicators of behavioral responses.* This is best captured in a famous quote from Watson:

> "Give me a dozen healthy infants, well-formed, and my own specified world to bring them up in and I'll guarantee to take any one at random and train him to become any type of specialist I might select—doctor,

> lawyer, artist, merchant-chief and, yes, even beggar-man and thief, regardless of his talents, penchants, tendencies, abilities, vocations, and race of his ancestors" (Psychology: *As the Behaviorist Views It,* 1930).

Later at the University of Chicago, Watson wrote his doctoral dissertation on *rat intelligence,* the first scientific book on rat psychology. In 1908 at age thirty, Watson became chair of the psychology department at Johns-Hopkins. In a meteoric rise to an academic chair of one of the most prestigious universities in the world, Watson paved the way with his engaging personality and brilliance in animal learning and laboratory acumen.

In 1913, Watson wrote an article entitled: *Psychology: As a Behaviorist Views It*—a paper that remains a benchmark in his philosophy of behavioral psychology. The article left no doubt that a behaviorist views *mind as mostly behavior* resulting in the often snide remarks from competitors that "behaviorists' caused psychology 'to lose its mind!'"

The first two sentences of his article said it all: "Psychology as the behaviorist views it is a purely objective experimental branch of natural science. Its theoretical goal is the prediction and control of behavior."

Watson's behaviorism did not become popular until woven into textbooks in the 1950s. Also, the article is notable for its strong defense of the scientific status of *applied psychology,* considered inferior to the established experimental psychology (structuralism) of the day. Philosophically, behaviorism is clearly:

- Deterministic (hence tied to physiology)
- Reductionistic (scientific and skeptical)
- Empirical (based upon observation, not personal introspection)
- Evolutionary (consistent with human and animal behavior being studied together for shared implication)

Incredibly, forced out of academia at the pinnacle of his fame and influence, due to an inappropriate relationship with a graduate student—one Rosalie Rayner—whom he subsequently married, Watson packed his academic bags and left to become (eventually) a vice president of *J. Walter Thompson* advertising agency in New York. In this capacity, Watson became the first successful *advertising psychologist* blending psychological expertise with marketing. Single-handedly, Watson made the 'coffee break' a national pastime while handling the account for Maxwell House Coffee. He was the first advertising executive to use *celebrity endorsements* as he featured the Queens of Romania and Spain showcasing their flawless beauty as principle account executive for Pond's Cold Crème.

Watson focused his energetic marketing plans upon sizzle (sex appeal) not substance (facts and figures) a practice still followed today. In the process, he made himself a small fortune.

Watson had experienced success as a scholarly laboratory scientist, a University department chair, and finally, as an advertising executive. Yet, all of his career accomplishments may not have outweighed his blunders, distractions, and shortcoming, especially as a father and grandfather. Students can read reports of this from Mariette Hartley's book *Breaking the Silence;* Mariette was Watson's granddaughter.

In the end his passion and innovation for psychology as a behaviorist viewed it shaped his world and changed ours. While Freud disturbed our dreams, Watson showed how human behavior was quite unremarkable as simple chains of conditioning. Before his death, he personally destroyed all of his personal papers and diaries.

Psych Sidebar

How did Watson's own children fare?

No physical touching or affection (other than shaking hands before bedtime), along with a harsh and disciplined regime characterized Watson's parental skills. The union with his first wife Mary produced daughter Mary (Polly) and son John (Little John). Polly made multiple suicide attempts later in life. Little John was plagued by bleeding ulcers and headaches throughout his life; he died in his early fifties largely from *psychosomatic* (stressed-induced) illnesses.

As mentioned earlier, a scandalous affair with graduate student Rosalie Rayner provoked Mary to divorce him, and Johns Hopkins to fire him. Soon after, Watson married Rosalie, producing two sons William (Billy) and Jimmy.

In adulthood, Billy openly rebelled against behaviorism and established a successful career as a Freudian psychiatrist for a time. Nevertheless, he attempted suicide. His first attempt was thwarted by his younger brother; he killed himself on the second try at age forty. As with Little John, Jimmy suffered chronic stomach ailments for years, but managed not to attempt suicide after intensive analysis.

What about Mary, the daughter from his first marriage? She and her husband produced a daughter, Mary (or Mariette), Watson's granddaughter, who became actress Mariette Hartley. In a recent book she recounted her life with her parents and her famous grandfather.

She portrayed her mother as a full-of-rage, secret drinker. Reports show she tried to commit suicide on more than one occasion. Her father, a retired advertising executive, took his own life at the age of sixty-seven after bouts with chronic depression.

Disillusioned, Mariette descended into alcoholism and with suicide ideation; then as if nothing else could go wrong, her career hit bottom. Somehow, she persisted in rebuilding her life; she's not exactly sure how it happened. Later on, she wrote a memoir of her experiences, *Breaking the Silence* (1990). On her grandfather's childrearing, she said: "Grandfather's theories infected my mother's life, my life, and the lives of millions. How do you break a legacy? How do you keep from passing a debilitating inheritance down, generation to generation, like a genetic flaw?"

Watson deserves a FORS BRAIN for the power he assigned to environmental influences upon learning and behavior. Giving Watson "12 healthy infants" might not be a good idea considering his own track record. Ironically, he stands as one of the best examples of how "toxic parenting" can produce emotionally crippled children and grandchildren—his own!

Sigmund Freud (1856–1939)

Lived Eighty-Three Years

From the first lecture Freud delivered to Clark University in 1909, the informed general public, as well as intellectuals and academicians in North America, embraced him as something of an enigma, albeit an appealing one, and a pop icon. All this adulation came, despite the fact that Freud was not a psychologist, or for that matter a psychiatrist—he was trained as a neurologist. His *Psychoanalytic Theory* was, nonetheless, seductive and strangely hypnotic.

Freud's theory focused upon the central importance of *unconscious experiences* (and dreams) and the corrupting influences they imposed upon personality functioning, especially when defense mechanisms (such as denial, repression, and rationalization) were overused or weak.

Viewed as energy systems—id, ego, and superego—developed from biology (id) and parent-taught systems of reward and punishments (superego). Highly conflicted, ego (as conscious mind) was a balancing act between the irrational demands of id with the restrictions of superego. From this all confusing mix so often arose conflict so that anxiety often consumed our best laid plans. Freud concluded that we were destined to live *neurotic lives*—existing in emotional pain. A seldom used term today in psychiatry or psychology, *neurosis* (also psychoneurosis) referred to mental disorders that cause distress, but not psychosis (or withdrawal from reality). A related term is *ego-dystonic* describing a person who feels conscious distress, anxiety, or depression due to emotional issues or unresolved conflicts.

Did North Americans anticipate that he would disturb our sleep with his controversial theories? Had we experienced enough conflicts in our lives to be drawn to his theories? Throughout his long life, colleagues were often bewildered, shocked, and offended by his theories as well as his own behavior. For example, Freud wrote a series of papers extolling praise for cocaine which, while it was not illegal at the time, was nonetheless a view that nearly destroyed his career.

Critics have been particularly harsh in relation to his theories of sexuality, repression, unconscious influences, and especially, his reversal of the seduction theory, soon to be addressed.

A magnanimous departure from the prevailing theories of his day—structuralism, functionalism, behaviorism, and Gestalt psychology—psychoanalysis focused upon *the darker more conflicted side of personality* rich in *archetypal metaphors* such as Oedipal Complex, Electra Complex, Freudian slips, penis envy, and transference. Critics, balked at his irreverence and the unreliability of unconscious experiences—how could this theory be proven? This strange man and his theories prompting some to suggest that psychoanalysis was more of a pseudo-science or an esoteric religion—maybe as some have suggested: The Study of the ID by the ODD.

First interested in studying law, and then medicine, Freud undertook a puzzling problem that if solved would have made him instantly famous among the medical scholars of his day: he spent a month in dissection studies trying to find the testicles of eels! Disgusted by the blood and stench of the autopsies and his frustration in failure, Freud decided to embrace a more cerebral problem.

In 1874, German physiologist von Brücke and physicist von Helmholtz, proposed a 'psychodynamic' theory suggesting that all living organisms are 'energy systems' governed by energy conservation. At the University of Vienna, von Brücke supervised Freud and instilled in him the radical view at the time that *living organisms have dynamic (changeable energy systems).* Obviously, this began the ball rolling for Freud's psychodynamic views of the psychology of unconscious mind with energy systems that would come to be known as id, ego, and superego with elaborate ego defense mechanisms (such as repression and denial).

Interestingly, French theoreticians Pierre Janet and Alfred Binet had used the terms 'unconscious' and 'subconscious' in prior writings. Seeming, Freud was, in fact, NOT the first theoretician to theorize its importance; he was just the first to *present ways of systematically studying and defining them, ways of analyzing them through dreams.*

Freud soon opened his own medical practice to support his new wife and growing family. He experimented briefly with, and then abandoned, hypnosis used with neurotic patients suffering from *hysteria.* Like neuroses, hysteria has fallen out of favor with psychology and psychiatry owing to advances in diagnostic criteria within the DSM (2002). Originally alluding to a *somatization disorder,* hysterical individuals focused upon a diseased body part and fear associated with the imagined disease. The afflicted person would often lose self-control and become 'hysterical' due to overwhelming fear.

Rejecting hypnosis, Freud favored copying colleague Josef Breur's success with a 'talking cure' that produced *catharsis*—a healing of the neurosis—by simple talk. Talking about whatever entered the patient's mind.

As a component of his insistence upon the central importance of sexuality in life, Freud created the theory of *sublimation* of the sexual drive by *denying expression of it.* Freud himself denied sexual expression with his wife after age forty; he wanted to show how this could be done by channeling libido (sex drive) into more creative endeavors—as in creating the elements of his own theory of psychoanalysis. Sublimation was theorized to occur when the sex drive (libido) was redirected by ego to more socially accepted modes of expression such as becoming a writer, a painter, or a volunteer for the betterment of society, instead of ruminating over sexual fantasies.

Yet, did Freud have an illicit affair with his sister-in-law Minna Bernays? Was archrival Carl Jung to blame for stoking the fires of this rumor? More oddities abound in the mystical life of Sigmund Freud.

After age fifty, Freud suffered many psychosomatic (stressed-induced) disorders, phobias, fear of dying, and a cancerous mass that required many surgeries over the remainder of his life. He turned introspectively to his own dreams for consolation as he meticulously recorded every dream in his memory. As Nazi occupation threatened all Jews in Vienna, Freud left Austria and moved his family to London in 1938.

The Seduction Theory

Did Freud know of the rampant sexual abuse of children in sexually-repressed Victorian Vienna and Austria? Many critics, including the former director of the Freudian archives, Jeffrey Masson, agrees that he did know and suppressed it to save his own neck and theory.

In 1896, Freud presented 'The Seduction Theory'—that the origin of neurosis (known in his day as *hysteria*) lay in early sexual traumas—"infantile sexual scenes" or "sexual intercourse in childhood" in Freud's terms. It was his belief that these early experiences were real, not fantasies, and had damaging and long-lasting effect upon the lives of the children who had experienced them. There was no doubt that Freud believed the sexual acts were FORCED upon the children, who in no way, encouraged or desired it.

However, nine years later, he *reversed himself on the seduction theory and said he was mistaken*—the memories of childhood seduction admitted by analysands (term for psychoanalytic patients) were *fantasies that never happened.*

Many consider the Seduction Theory to be the cornerstone of psychoanalysis and connection to repression (automatic suppression) of the sexual experience.

Were the sexual experiences real or were they imagined?

Freud saw no way out of the semantic pretzel except to conclude the experiences were imagined, BUT the imagination by itself caused hysteria. He had to admit he had been wrong! Admitting error and attributing the experience to fantasies of sexual seduction gave birth to his psychoanalytic school. This singular event occurred in 1908. Abandoning this theory made psychoanalysis a major school of thought, and helped to shape the world we live in today. Neuroses could be based upon an illusion. Interestingly, Freud viewed religion in the same vein detailed in his short book *Future of an Illusion* (1927), where he characterized illusions as derived from human wishes alone.

As Freud neared death, he had prearranged with his physician to administer an overdose of morphine—an instance of *euthanasia* (easy death)—upon consultation with his beloved daughter, Anna Freud. The old stoic remained in control of his life until the end. Upon his death, Freud's body was cremated and his ashes deposed in a Grecian urn, a present given him by Marie Bonaparte.

Freud deserves a FORS BRAIN due to his Seduction Theory and its reversal. In the days of visiting the Paris morgue, Freud indirectly influenced the development of forensic psychology by observing the inescapable connection between sexuality and potential violence often requiring legal intervention.

Carl Jung (1875–1961)

Lived Eighty-Six Years

Carl Gustav Jung was a solitary and painfully shy child convinced from childhood that he felt comfortable being *two people*—a modern Swiss citizen and a person who would have been more comfortable living in the eighteenth century (Jung was never diagnosed *schizophrenic* or with multiple personality disorder if it has existed in his zeitgeist as far as historical documents denote, so his feeling was an interesting peculiarity, not a psychotic disorder!). His explanation later in life provided a clue to the formation of his theory of archetypes. From the Jungian perspective (known as Analytic psychiatry) *archetypes* refer to innate and universal stereotypes, or ideas as models for personality integration (and ways of interpreting perceptions). Personal, self, shadow, anima, and animus are specific to Jungian Analytic Psychiatry. In relation to his own personality and his assertion of being two people Jung stated, "personality Number 1 referred to my feeling like a typical schoolboy living in my era, while personality Number 2 was a feeling of a dignified, authoritative, and influential person more comfortable living in the past."

He interpreted this connection to the past as 'unconscious memories' from an *esoteric collective unconscious* he attributed to all persons appearing spontaneously. This universal unconscious inheritance functions as a "reservoir for the experiences of our species." Jung theorized the collective unconscious helped to direct Self (an archetype) toward self-actualization. According to Jung "archetypes are innate, universal prototypes (or ideas) that help guide our comprehension in everyday life; they are inherited from our ancestral past." Psychological and mental in origin, archetypes require no prior learning as they are "psychological organs." Jung discussed four central archetypes:

Self: regulating center of the psyche (mind and/or personality) and facilitator of *individuation,* we become what we are supposed to become
Shadow: images evoking fear response such as snakes or spiders that are avoided because they threaten life
Anima: the feminine, nurturing side of males
Animus: the masculine, courageous side of females

Additional Jungian concepts important today include *introversion* (psychological type characterized by less sociability while being more reserved) and *extraversion* (psychological type of high sociability and gregariousness) that influenced the modern Myers-Briggs Personality test.

As a schoolboy, Jung suffered fainting spells. This oddity seemed to coincide with conditioned stimuli tied to an incidence of unconsciousness when push to the ground by a bully at school. Upon being revived, he unconsciously (apparently) tied this experience with 'not having to go to school.' Fainting spells, therefore, always preceded his departure to school. Hearing from his father that he might suffer from epilepsy, he immediately feared he would be separated from his family, so he 'recovered' from his fainting spells and became a brilliant scholar, so the story goes. Interestingly, years later when both Freud and Jung were world famous, Freud fainted during a speech given by Jung. Jung immediately stopped and collected his colleague and carried him to a couch (of course, in reference to the Freudian couch of therapy).

As university studies neared, his first choice was archeology, but he settled upon medicine as the cost was prohibitive. Toward the end of his medical studies, he became interested in psychiatric medicine by reading a book by Krafft-Ebing, a noted sex researcher of his day.

In 1906, six years after Freud published his seminal psychoanalytic book on dreams and their relation to the unconscious, *The Interpretation of Dreams,* the future Founders Club Alumnae met and would forever be connected, first as collegial, then as competitors, and finally as archrivals.

The famous split with Freud was attributed to Freud's dogged insistency upon the strength of libido (and fanciful seduction themes), and overemphasis upon dark childhood memories and negativism of unconscious determents of behavior. Jung countered psychoanalysis and his rival with his theory of synchronicity—the importance of balance, harmony, and spirituality in life. He rejected the scientific method, reductionism, and skepticism in favor of appreciation and celebration of more esoteric realms, no doubt enhanced by his travel to India and subsequent integration of Eastern mysticism into his theory.

Psych Sidebar

Jung & Wolff

Although Jung seemed to flourish in his marriage, a union that produced five children, there was widespread rumor that he had affairs with other women, most notably Toni (Antonia Anna) Wolff (1888–1953), who began as his patient and then his lover. Wolff later became a Jungian psychoanalyst. The extramarital affair between Jung and Wolff was openly enacted over ten years.

Burrhus Fredrick (B.F.) Skinner (1904–1990)

Lived Eighty-Six Years

B. F. Skinner is universally known as the founder of Operant Conditioning, a radical extension of behaviorism. Harvard psychologist B. F. Skinner proposed *Radical Behaviorism* as his philosophy of psychology as the *experimental analysis of behavior.* This approach relied upon Operant Conditioning, schedules of reinforcement as independent variables and rates of responding as dependent variables.

Over his long life he excelled as an author—especially *Verbal Behavior, Beyond Freedom and Dignity,* and the utopian novel *Walden Two,* but also as an inventor, poet, and advocate for social reform. He achieved world fame along the lines of Freud.

'The Skinner Box,' an experimental space encased in a plastic box and a cumulative recorder to measure rates of responding during schedules of reinforcement, was invented by Skinner. Experimentally, he favored pigeons over rats and monkeys; nonetheless, his brilliant schedules of reinforcement showed how habits could be pulled out of pigeons. Expanding upon Thorndike's *Law of Effect,* Skinner became the world's most recognized psychologist for three decades. The novel *Walden Two* has gradually found its way into many college Freshmen and Sophomore literature and psychology classes. The book champions pacifism and gratifying social relationships, where happiness is attained through passionate work and creative leisure.

After graduating from college with a B.A. in English, he tried (unsuccessfully) to write fiction for over a year. Disillusioned, he chanced across a copy of Bertrand Russell's *Outline of Western Philosophy* in a passage devoted to the behaviorist John B. Watson. He became a psychological devotee almost overnight; he graduated from Harvard University with a Ph.D. in 1931.

Radical behaviorism is data-driven, empirical, and seeks to expand human learning through inductive logic. Prediction and control of behavior was

Skinner's passion as he carefully crafted animal studies with implication to human populations. He applied the same experimental approach to education as he did in the lab, stating there were five obstacles to learning:

- Fear of failure
- Lack of directions
- Lack of clarity and meaning
- Lack of positive reinforcement
- Meaningful assignments must be divided into small steps

It could be argued forcefully that Skinner spent fifty years enlarging and refining Thorndike's Law of Effect. Regarding personality, Skinner believed it evolved due to rewards and punishments of external events. 'We become the way we are' because of rewards and punishments along the way. It is this belief that discounts emotions, thoughts, and even *human freedom of choice* as contingencies subject to conditioning presented in his book *Beyond Freedom and Dignity*.

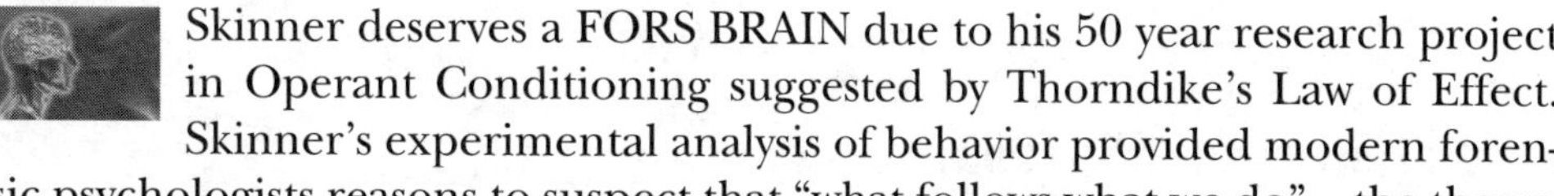

Skinner deserves a FORS BRAIN due to his 50 year research project in Operant Conditioning suggested by Thorndike's Law of Effect. Skinner's experimental analysis of behavior provided modern forensic psychologists reasons to suspect that "what follows what we do"—the theory of reinforcement—might be overpowering in some forms of behavior. Why not sexually psychopathic serial crime?

Anna Freud (1895–1982)

Lived Eighty-Seven Years

Anna Freud was the baby of the Freudian family and her father's (Sigmund's) favorite. At the tender age of fifteen she began to read her father's works. Later in life she confessed that she had learned more from him and his frequent visitors (imagine the long list of intellectual luminaries) than from formal schooling. However, Sophie, the best looking of the children, gained most of the attention from Freud's visitors, resulting in a keen competition developing between the two siblings. For Anna, this situation may have contributed to depression, low self-esteem, and uncertainty as what direction to follow in life. She would use this experience later as she focused upon *childhood psychoanalysis*. Interestingly, Anna underwent psychoanalysis herself from her father that lasted four years. In fact, Freud insisted that all future 'lay analysts'—practitioners without medical degrees—undergo psychoanalysis before they began to practice.

When the family was forced to move to London due to Nazi occupation in Austria, Anna was eager to start her own practice, but had a brief tiff with Melanie Klein, a well-known British psychoanalyst. The women differed on several points of therapeutic techniques to be taught in child psychoanalysis; however, the disagreement was settled by the British Psychoanalytic Society creating (1) Kleinian, (2) Anna Freudian, and (3) independent divisions for training, a tradition still practiced today.

Ego functions, the constant struggle and conflict in deciding how to navigate through desires and values, were central tenants in Anna's work. She focused on research, observation and treatment of children as they developed through stages. An emotionally healthy child would expect to keep pace with most of his or her peers—such as eating habits, sleep, personal hygiene, play, communication, and relationships with other children. When one aspect of development seriously lagged behind the rest, clinicians could assume (perhaps) that there was a problem, and could communicate the problem by describing the particular glitch.

While in London, Anna established a group of prominent child developmental analysts (including the developmentalist Erick Erikson). She observed that childhood symptoms were ultimately analogous to adult personality disorders and thus related to *developmental stages.*

Anna Freud deserves a FORS BRAIN for her work in exposing how problems can arise in developmental stages of children and the importance of parental guidance to eclipse future development of adult personality disorders that can lead to violence.

Karen Horney (1885–1952)

Lived Sixty-Seven Years

Freudian psychoanalyst of Norwegian and Dutch descent, Karen Horney (orr-na) was the first women to focus almost entirely upon *feminine* psychiatry. Rebellious and ambitious, Horney would later question traditional psychoanalysis determining many of the perspectives were too narrow, deterministic, and confining, especially relating to women and children.

As a child, she felt rejected by her father who was a strict disciplinarian and emotionally vacant. She was emotionally closer to her mother, yet Horney battled depression throughout her life. Because she felt unattractive, even though others considered her attractive, she decided to direct her energy into intellectual pursuits.

She entered medical school in 1909, although at the time it was unknown for women to seek medical degrees. During her education, she met Oskcar Horney whom she married; the union produced three daughters and a lot of unhappiness.

As Horney became well-regarded in psychoanalytic circles, her practice and teaching assignments blossomed; however, this was not so on the home front. As Oskar's business failed, so did the marriage. Soon thereafter, she and her three daughters moved to Brooklyn, New York—a sanctuary state for many Jewish intellectuals at the time who migrated there from Nazi-occupied Europe.

Battling depression yet again, she slowly widened her circle of acquaintances, including Erick Erikson and Harry Stack Sullivan. She enjoyed a brief intimate relationship with Erikson that, nonetheless, ended bitterly. In 1937, Horney published a book, *The Neurotic Personality of our Time.* She established the *Association for the Advancement of Psychoanalysis* as traditional psychoanalysis seemed discordant with the times in her view.

Horney continued to be a pioneer in the new discipline of feminine psychiatry. As a woman, she felt most cultures and societies encouraged woman to be wholly *dependent upon men* for love, prestige, care, protection, and wealth. As a result of the imbalance, *women constantly over-strived to please men*. To men, women were regarded as objects—toys—due to their charm and beauty, which is completely at variance with every human being's *ultimate purpose to self-actualize, achieve success, and find happiness.*

According to Horney, women achieve credibility through their own children, emotional bonds with other women, and extended family members. An essay, T*he Distrust Between the Sexes,* compares the husband-to-wife relationship to the parent-to-child relationship. Both propagated misunderstanding and bred *detrimental neuroses* (unhappiness). Another essay, *Maternal Conflicts,* broke new ground on the problems women experience raising adolescents.

Finally, Horney believed that both sexes have strong motives to be ingenious and productive. Women satisfy this need internally by becoming pregnant and giving birth, yet are capable of accomplishment in so many other areas of life. Men seek external ways—outside of home and family—to accomplish these needs. Horney believed the striking accomplishments of men at work can be viewed as compensation for their inability to give birth to children. Perhaps tongue-in-cheek, did Horney counter Freudian 'penis envy' with 'womb envy'?

In 1946, Horney was the first analyst to write a 'self-help' book, *Are You Considering Psychoanalysis?* Self-awareness continued to be the banner of her personal battles against depression as well as the education of young women to be more than society narrowly expects them to be.

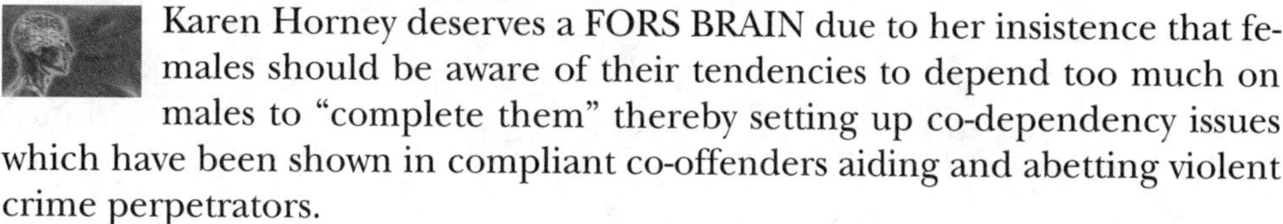

Karen Horney deserves a FORS BRAIN due to her insistence that females should be aware of their tendencies to depend too much on males to "complete them" thereby setting up co-dependency issues which have been shown in compliant co-offenders aiding and abetting violent crime perpetrators.

Mary Whiton Calkins (1863–1930)

Lived Sixty-Seven Years

Starting her career as a Greek professor at Wellesley College, Mary Calkins developed a keen interest in the new science of psychology. Calkins established the first psychology lab at Wellesley College, a small women's liberal arts college in Massachusetts. Interestingly, many luminaries are alumnae of Wellesley, such as the first female United States Secretary of State, Madeleine Albright, Democratic Presidential Candidate and New York Senator, Hillary Rodham Clinton, actresses Ali McGraw and Elisabeth Shue, as well as journalists Cokie Roberts and Diane Sawyer.

Calkins would later study with Professor William James at Harvard. She is credited with developing the *paired-associate technique* for studying memory. Calkins is best remembered not so much for what she did, but what Harvard refused to grant her—her hard earned Ph.D. degree although she recorded a near-perfect examination score.

Why did Harvard refuse to grant her degree that she so richly deserved? She was a woman! Calkins was offered a Ph.D. from Radcliff College but she refused it. Eventually, she rose to the top of her new profession and her first love, philosophy, by becoming the first female president of BOTH the *American Psychological Association* (1905) and the *American Philosophical Association* (1918), all the while never to receive her Ph.D. (as a humane gesture, one can only wonder why Harvard did not award the earned degree posthumously).

Students are urged to Google® the next 10 famous founders not presented as members of the Founders' Club due to space limitation. Wikipedia will do for starters. They are:

- John Money, Ph.D. (1921–2006) lived seventy-nine years
- Hervey Cleckley, M.D. (1903–1984) lived eighty-one years
- Alfred Adler, M.D. (1870–1937) lived sixty-seven years
- Albert Bandura, Ph.D. (1925-) currently eighty-three year old
- Eric Erikson, Ph.D. (1902–1994) lived ninety-two years
- Abraham Maslow, Ph.D. (1908–1970) lived sixty-two years
- Carl Rogers, Ph.D. (1902–1987) lived eighty-five years
- Edward Tolman (1886–1959), lived seventy-three years
- Piaget Jean, (1896–1980) lived eighty-four years
- Philippe Pinel, (1745–1826) lived eighty-one years

Glossary

A

Ach (acetylcholine) (uh-see-tull-co-leen) Ach is a ubiquitous neurotransmitter and parasympathetic agent responsible for conservation of energy, attention and memory, thirst, sex, mood, and REM sleep. It is present at muscle receptors for muscle contraction.

ADD (attention-deficit disorder) A common learning disorder of inattention and lack of focus affecting mostly males.

ADHD (attention-deficit hyperactivity disorder) A common learning disorder of inattention with a hyperactive component affecting mostly males.

aberrant (ab-er-unt) Literally, "to go astray;" atypical behavior or thinking.

actus reus (act-us ray-us) Literally, "criminal act;" the physical part of a crime.

addiction Pertains to a compulsive need for a habit-forming substance to produce a feeling state that "compels" the user to continue. Addiction is more about the feeling it produces and the desire to replicate the feeling.

addictionology The science of addiction related to substance abuse, tolerance, psychophysiological habituation, and neurochemistry of addiction.

adrenaline Powerful hormone secreted by the adrenal glands related to fight or flight.

affective (AF-ekk-tive) Pertains to feelings; contrast affect ("feeling") with cognitive ("thinking").

agonist (AG-un-st) A chemical substance capable of combining with a receptor, liberating the targeted substance.

amygdala (uh-mig-duh-luh) Limbic system structure of gray matter in the anterior temporal lobe, center of emotional memory and aggression.

analgesic (anal-jee-sik) Literally, "insensitivity to pain."

androgenic (and-dr-gen-ik) Pertains to the action of the powerful male sex hormone testosterone.

anecdotal (ann-ik-dough-tul) Literally "unpublished." Refers to a short narrative or story of real life experiences.

anhedonia (ann-ha-doan-ee-uh) Experiencing displeasure.

anima (ann-uh-muh) Refers to Carl G. Jung's analytic theory of the archetypes of masculinity and femininity. Literally "soul"—pertains to the feminine part of a male's personality.

animism (ann-uh-miz-um) The belief that spirits occupy all things and influence outcomes.

animus (ann-uh-muss) Refers to Carl G. Jung's analytic theory of archetypes of masculinity and femininity. Literally, "courage"—pertains to the masculine part of a female's personality.

antecedent (ann-tuh-see-dunt) Pertains to former events as shaping current behavior or conditions.

antisocial developmental programming Refers to the slow, stage-by-stage effect of detrimental influences contributing to psychopathic (or antisocial) personality. May include a combination of physical, sexual, or verbal abuse, pornography, drug abuse, or other factors leading to full-blown antisocial behavior with a pronounced feeling of isolation, emotional disconnectedness, and inability to bond, or to feel love.

antisocial parenting (also: predatory, toxic parenting) Pertains to the most damaging and destructive type of loveless and hateful parenting most often observed in the development of sexual psychopathy. Antisocial parenting includes a combination of physical, sexual, and/or verbal abuse, alcoholism, poly-drug abuse, where compulsive viewing of pornography, prostitution, and spousal abuse occur routinely.

antisocial personality disorder In the DSM, a pervasive pattern of disregard for, and violation of, the rights of others that begins in childhood or early adolescence and continues into adulthood.

anxiety Characterized as a fearful concern or uneasiness of mind due to the imbalance of GABA, a neurotransmitter.

aphorism (AF-o-riz-um) Literally, "to define." A concise statement of a principle or sentiment.

aphrodisiac (aff-ro-dee-zee-ak) A sexual stimulant; a substance that arouses sexual desire.

apocalyptical (uh-pok-uh-lip-tee-kul) Prophetic; foreboding imminent disaster.

archetype (ARK-type) Refers to a perfect example or a specimen, an inherited idea according to the analytic psychology of Carl G. Jung derived from the experiences of the race and present in the unconscious of the individual, such as anima and animus.

assessment Literally "to sit beside" another; refers to making a psychological appraisal of another who is "up close and personal."

association cortices (cor-tuh-seez) Loosely referred to as cognition; comprising most of the cerebral surface of the human brain responsible for complex processing between the arrival of input and the generation of behavior.

autoerotic (auh-toe-e-rot-ik) sexual self-stimulation; masturbation.

autoerotic asphyxiation (azz-fix-ee-a-shun) Refers to the practice of reducing oxygen to the brain through self-strangulation (often by hanging) accompanied by masturbation. Intensifying orgasm is the goal behind this dangerous practice that can lead to accidental death.

B

behaviorism (also known as behavioral psychology) The study of observable behavior without regard to cognition or affective (feeling) states, which are considered introspective tools and not scientifically valid or reliable.

blocking reuptake Pertains to a chemical antagonist preventing neurochemical reclamation of molecules of neurotransmitters back into axonal vesicles.

Blunt affect (also flat affect) Emotionless.

body language Refers to non-verbal body "language," such as tone of voice, facial expression, or posturing.

borderline personality disorder (or borderline features) In the DSM, a disorder characterized by severe distortions of self-image, mood, and interpersonal relations. "Borderlines" experience marked mood shifts, impulsivity, difficulty tolerating loneliness, and are emotionally needy, typified by pronounced fear of abandonment.

brainstem One of the savage architects of mind responsible for autonomic functions, obsessive-compulsive behavior, and territoriality. It lies below the midbrain limbic system (MLS), the second savage architect. Also known as the reptilian brain from the triune brain paradigm.

British empiricism Chiefly due to John Locke's influence upon philosophy, the intellectual climate (zeitgeist) of North America moved more rapidly toward association of experiences as causes of behavior influencing the study of psychology. Psychology "turned the corner" away from philosophy toward behavioral explanations because of British empiricism.

C

CNS (central nervous system) Pertains to the brain and spinal cord.

CPS (child protective service) A state child advocacy agency aimed at preventing child abuse or interceding on a child's behalf when abuse becomes proven.

catecholamines (cat-uh-col-uh-means) Neurotransmitter amines (a chemical compound containing one or more halogen atoms attached to nitrogen) related to catachol, namely norepinephrine and dopamine.

catharsis (or cathartic) (kah-thor-sis) Purging of emotions (such as anger, guilt) through discourse with a therapist (or trusted friend); literally, "talk therapy."

cerebellum (sara-bell-um) Prominent hindbrain structure characterized by motor coordination, posture, and balance.

chaining Behavioral term relating to sequencing of behavior where small muscular movements become learned and "chained" together to produce a desired action, such as behavior required to hit a baseball.

classical conditioning Behavioral paradigm characterized by the law of association where one stimulus (the unconditioned stimulus—UCS) is paired with another, a formerly neutral stimulus, to produce a new behavioral contingency, the conditioned stimulus—CS. Through association, the bell (CS) came to stand for the original stimulus (the food—the UCS) in Pavlov's famous experiment.

co-dependency A psychological condition where a person becomes controlled or manipulated by another person who has a pathological condition, such as addiction, obsessive-compulsive personality disorder, or psychopathy.

cognitive Refers to "thinking."

cognitive-behavioral perspective Specialty within general psychology focusing on how cognition (thinking) influences behavior, especially in forming powerful "cognitive maps" of learning and thinking. Also referred to as "soft" behaviorism.

cognitive mapping Powerful thinking "maps" of experience that affect behavior in far-reaching and profound ways.

cognitive neuroscience Neuropsychological specialty that deciphers the structural and functional organization of specific brain regions connected to cognition by case studies of neurological patients, the advent of noninvasive brain imaging, and primate studies.

cognitive psychology Psychological specialty that seeks to understand the psychophysiological process of thinking and how thought processes impact behavior.

collective unconscious Jungian concept related to a genetically shared unconscious by all members of a race or ethnicity.

colloquial (kuh-lo-qwe-ull) Used in conversation; non-technical words and/or expressions.

concussion A potentially serious head injury where the soft tissue of the brain gets violently jarred against the bony cranium.

contusion A serious head injury that results in bruising, bleeding, and trauma to the brain.

compliant co-offender Refers to a compliant person "recruited" by an offender.

compulsion An irresistible impulse to perform an irrational act.

conduct disorder In the DSM, refers to a repetitive and persistent pattern of behavior in which the basic violation of the rights of others, or major age-appropriate societal norms or rules, occurs.

conflict theory Freudian theory relative to cross-purposes of id, ego, and superego with resulting anxiety.

consumerism Marketing of consumer goods to increase consumption.

continuum A line denoting possible causes or answers to given questions in behavioral science.

"cool-coded" Refers to blue colored ("cool-coded") brain scans (according to Jacobs) of neurologically damaged individuals in brain areas such as the prefrontal lobes. In undamaged brains, the scans glow pink or red, indicating normal blood flow.

corpus delicti (cor-pus di-lek-ti) Refers to the body of evidence required to prove the commitment of a crime such as ***mens rea***—the mental part, ***actus reus***—the physical part, and the dead body of the victim.

cri de coeur (kree-duh-kur) Literally, passionate "cry from the heart."

criminologist One who studies crime as a social phenomenon, criminal activity and behavior, and incarceration variables.

cynicism Contemptuously distrustful of human nature; a sneering disbelief in sincerity.

D

DID (Dissociative Identity Disorder) In the DSM, a serious mental disorder related to multiple identities residing within the same person, formerly referred to as Multiple Personality Disorder (MPD).

DSM (Diagnostic and Statistical Manual of Mental Disorders) Diagnostic and descriptive "textbook" of mental disorders used by mental health professionals such as clinical psychologists and psychiatrists.

decadence (deck-uh-dunce) Marked by decline and deterioration.

deductive Pertains to speculative logic or "hunches" (termed a priori reasoning in philosophy), where no systematized knowledge exists as a yardstick to measure present or future behavior. Contrast inductive, the logic of science, the opposite of deductive.

deny (denial) Freudian defense against anxiety where the person consciously disclaims the importance of an issue. Contrast to repression, which operates as "unconscious denial."

dependent personality disorder Characterized by a passive, "clingy," submission to others; a "doormat" personality with separation anxiety and fear. A compliant co-offender often has a full-blown dependent personality disorder.

developmental psychology Psychological specialty that studies stages or periods of physical and psychological development and corresponding expectations across the lifespan.

dimorphic nuclei (die-mor-fick) An identifiable part of the anterior hypothalamus show male/female differentiation.

dissonant (diss-uh-nunce) Pertains to insufficiency of agreement resulting in lack of consonance.

dopamine (DA) (do-pa-meen) A powerful neurotransmitter known to lie behind pleasure across a wide continuum.

dogma Presented as authoritative or established opinion without adequate grounds.

dualism A view that human beings consist of two irreducible elements—matter and spirit.

dual diagnosis A diagnosis made when a chemical dependency exists alongside a pure psychopathology such as depression or anxiety.

Dull Hypothesis Theory hypothesized by Don Jacobs (2004) that weak signals (or other abnormalities) within the brain of violent psychopaths drive them to seek over-the-top stimulation often evidenced at crime scenes.

E

efficacy (ef-uh-kuh-see) Power to produce an effect such as truth.

egocentric Limited to outlook or interest to one's own activities; self-centered and egotistical.

emitted behavior Pertains to everyday behavior occurring due to no identifiable stimulus, a subject of operant conditioning.

empiricism (em-peer-uh-ciz-um) Refers to laboratory analysis and reporting, objectivity, observation, and replication of studies in experimental psychology.

enabler One who helps another to persist in self-destructive behavior.

enuresis (en-yur-ree-sis) Pertains to bed-wetting.

environmental psychology Concerned with influences of the environment—neighborhoods, noise and congestion, and employment-and how these influences affect behavior.

epiphany (uh-piff-uh-nee) Pertains to an illuminating discovery or insight.

eroticism Pictorial, literary, or graphic portrayal of sexual desire or sexual themes.

esoteric (es-o-ter-ick) Literally "within." Refers to understanding by a specially initiated person.

ethnocentrism (Eth-no-sin-tris-um) Pertains to race and characterized by the belief that one's own racial group is superior to all others.

etiology (ee-tee-oll-o-jee) Refers to causation.

evisceration (ee-viss-err-ra-shun) To disembowel; to remove the entrails of another.

evolutionary neuroanatomy Neuropsychological analysis of the evolution of the brain into three layers—reptilian, "old mammalian," and neo-cortex—for purposes of delineation of neuroanatomy and concomitant neurochemical influences.

exacerbate (ig-saz-sir-bate) To make worse.

excitement phase In Masters and Johnson's human sexual response, the first stage of sexual excitement prompted by erotic thoughts, sights, or contact.

existential (eggs-uh-sten-shul) Grounded in the experience of being or existing.

exorcism In Catholicism, to expel Satan or an evil spirit.

extrapolate To project or extend known data into an unknown area so the use of conjecture guides the way.

extravert Refers to a "people person"—a gregarious and unreserved individual who seeks people and people-orientated activities to feel jazzed.

F

fait accompli (fait-au-kom-plee) Literally, "the accomplished fact."

fellatio (fuh-LAY-she-oh) Stimulation of the penis by the mouth.

fetish Sexual arousal caused by objects (ex: high-heeled shoe) or body parts.

forensic neuropsychologist A specialist (and often an expert witness) within neuropsychology who presents courtroom testimony (often in the form of brain scans) in criminal cases related to potential causes of criminality from medical sources such as neurology, biology, and brain scanning technology.

forensic pathologist An MD-trained physician who analyzes pathology such as cause of death and related issues from autopsy protocols headed to the courtroom.

forensic psychologist A Ph. D.-trained psychologist who testifies in court as an expert witness regarding sanity or insanity, or pathological issues of a given participant.

forensic science Pertains to the science of criminal evidence gathering by forensic pathologists, who often testify in a courtroom.

frontal lobes The largest of four lobes of the brain with a wide repertoire of functions, including cognition.

frottage (fraw-taazh) Secretly rubbing against another for sexual pleasure or while fantasizing a caring relationship.

futurist A journalist or intellectual who writes or speculates about the future.

G

GABA (Gamma-aminobutyric acid) Found primarily in the hippocampus, hypothalamus, and amygdala, a powerful inhibitory neurotransmitter that reduces arousal, aggression, and anxiety.

garrote (guh-rott) Pertains to a device or a noose used in strangulation.

glib Literally, "slippery"; marked by ease and informality; nonchalant.

glutamate The major excitatory neurotransmitter in the CNS believed to underlie all learning thought synaptic cleft sequencing.

H

hard core porn Refers to pornographic pictures, material, or any visual manifestation of explicit and often deviant sexual behavior usually rated as XXX.

hawthorne effect Refers to Elton Mayo's famous field study showing that the mere fact of being observed stimulates output or achievement.

hedonistic (He-dun-is-tik) **hedonism** Pleasure-seeker.

hemispherectomy (hemmah-sfeer-eck-tuh-me) Removal of one or both cerebral hemispheres.

hippocampus (hippo-camp-us) A cortical structure in the medial portion of the temporal lobe; in humans the center for learning and short term declarative memory.

histopathy (hist-op-uh-thee) Branch of physiology concerned with tissue changes characteristic of pathology.

histrionic (histrionic personality features) (Usually a female) who acts dramatic and theatrical for affect.

holistic (holism) Concerned with complete systems rather than dissection into parts.

homicidal triad Refers to bed wetting at an inappropriate age, setting fires, and violence against peers or pets as precursors to severe conduct disorder showing lack of control.

hubris (hew-bris) Exaggerated pride or self-confidence.

humanism (humanistic psychology) Focus on human interests and values.

hypocrisy Pertains to presenting a false appearance of virtue

hypothalamus A collection of nuclei in the diencephalon governing reproduction, homeostasis, and circadian rhythm.

I

I/O Psychology Refers to Industrial/Organizational Psychology, a specialty of applying psychological principles to business and industry.

id Pertains to the biological inheritance of "mind" or "personality" as instincts of a sexual and aggressive nature according to Freud's psychoanalytic school.

idiosyncratic (id-e-oh-sin-krad-ik) Pertains to the uniqueness of an individual.

inductive (IN-duc-tive) Pertains to the methodology of science. Philosophy calls inductive logic a posteriori reasoning, where systematized knowledge is useful (such as known offender characteristics in criminal profiling) as a yardstick in analyzing present and future behavior. The opposite of inductive is deductive, or speculative logic used by Sherlock Holmes as "elementary observation."

ignominious (igg-no-minnee-uss) Dishonorable, despicable.

in loco parentis (en-lo-co-puh-ren-tus) A legal term meaning "in the place of the parents."

incompetent parenting Pertains to ambivalence in parenting—withholding love, inconsistent punishment, inattentiveness, and lack of adequate nurturing skills that produce children who may seek therapy later in their lives.

intellectualize Freudian defense against anxiety where the person verbalizes elaborate excuses without addressing feelings.

internal dialogue Refers to "talking" or thinking to oneself.

intrapsychic (en-truh-sigh-kick) Literally "inside one's head" as might occur in internal dialogue.

internal locus of control (ILOC) Refers to social learning theorist Julian Rotter's theory of expectations. Internal LOC refers to a responsible person who trusts his plan or strategy for success accomplished by hard work. (Contrast External Locus of Control where luck or timing controls one's expectations of the future.)

introvert Pertains to a type of personality marked by robust intrapsychic activity so that territoriality and time alone become valued over the "people person" who seeks additional social interaction.

inventory test Refers to any of a number of standard personality tests.

L

LSD (lysergic acid diethylamide) Pertains to a chemical substance that produces vivid hallucinatory (psychedelic) experiences.

law of association In behavioral psychology, the law that states when two stimuli are paired together in time, one (the conditioned stimulus—the CS) comes to stand for the other, or the natural stimulus—the UCS. UCS is natural while CS is learned.

law of effect In behavioral psychology, the famous law proposed by Edward Thorndike stating that behavior increases when followed by "satisfiers" and decreases when followed by "annoyers." Precursor law to operant condition.

law of frequency In behavioral psychology, the law that states the more frequently we perform a certain behavior the more likely we will continue it in the future; literally, practice makes perfect.

law of recency In behavioral psychology, the law that states the more recently we performed a certain behavior the more likely we will continue to perform it in the future.

Lesch-Nyhan Syndrome (LNS) A rare and usually fatal genetic disorder transmitted as a recessive trait on the X chromosome. LNS shows hyperuricemia (excessive uric acid), spasticity, rigidity, and compulsive biting of the lips and fingers caused by an absence of the HPRT enzyme and the resulting damage to the midbrain.

libido (luh-bee-doe) The sex drive.

learned helplessness Pertains to a condition when a person feels helpless when he encounters conditions over which he has no prior experience or control.

litigious (lie-tee-jus) Pertains to law suits.

lothario A seducer of women.

lycanthropy (LIE-can-throw-pee) A psychiatric condition in mind only where a patient imagines himself to be a wolf.

M

MFB (medial forebrain bundle) Pertains to the medial forebrain bundle of the hypothalamus, dubbed the "pleasure pathway" due to prevalence of dopamine receptors.

macabre (muh-cob) Morbid preoccupation and fascination with death or thoughts of death.

magical thinking Pertains to psychotic episodes where patients feel their thinking produces precognitive or paranormal events that defy the physical world.

manie sans delire Literally, "obsession without insanity," which later becomes the basis of psychopathic personality.

malum in se Literally "evil in itself." Pertains in criminal law to serious crimes against personal safety, such as murder or rape.

masochism (maz-uh-kism) masochist A sexual deviation characterized by the enjoyment of having pain inflicted.

Munchausen syndrome by proxy Deliberately making another person sick, evidently for attention and pity. Also known as Factitious Disorder by proxy.

mass murderer Pertains (usually) to a male killer who kills four or more people in one incident, in one location, and in one emotional experience.

Megan's law Law intended to protect citizens against sexual offenders that came into effect on October 31, 1994. The New Jersey State Legislature enacted the law requiring certain convicted sexual offenders to register with law enforcement, thus providing community notification of sex offenders.

melatonin (mel-uh-tone-un) Powerful hormone secreted by the pineal gland involved in the sleep-wake cycle. Implicated in the depressive disorder SAD (seasonal affective disorder).

metencephalon (met-in-sef-uh-lin) One of the six divisions of the brain; pertains to the pons and the cerebellum.

mens rea (menz ray-uh) Literally "criminal mind."

milieu (mi-lyoo) Refers to a social context where learning and experiences occur, such as family, school, or peers.

modus vivendi (MO-dus vuh-vin-dee) According to Jacobs (2003), the five components of sexualized serial crime, including mens rea, actus rea, modus operandi, signature, and aftermath.

"morally fuzzy" Pertains to popular semantics where meaning is defined from the perspective of the speaker. How "spin" controls meaning.

morphology Pertains to the body or body type.

motive Literally, "to move." Emotion or desire acting on will; motivation.

medulla oblongata (muh-doo-luh-Ob-lun-got-uh) The most caudal of the brain stem serving sensory and motor systems and including involuntary "automatic" actions such as heart rate and blood pressure.

mesencephalon (mez-in-sef-uh-lun) Pertains to midbrain structures of the red nucleus and ***substantia nigra*** involving movement.

mullerian ducts Tiny microscopic ducts that determine internal and external female genitalia.

multiaxial system Refers to domains of patient information reviewed by clinicians in clinical psychology

myelencephalon (my-e-len-sef-uh-lun) Refers to the medulla or medulla oblongata.

myelinized (my-uh-lun-eye-zd) The process of axonal insulation by a white fatty substance known as myelin.

N

narcissistic personality disorder In the DSM, a pervasive personality disorder characterized by egocentrism. Often observed in conjunction with psychopathic personality characteristics.

narcissism Literally "self-love."

nature Refers to biological inheritance. Contrast to nurture, or social learning.

necrophilia (neck-row-feel-ee-yah) A sexual perversion marked by an obsession with having sex with a corpse.

nefarious (nu-fair-ee-us) Flagrantly wicked.

neocortex Literally "new bark." Refers to the micro-thin, most recently evolved part on the upper-most part of the brain.

neoplasm Literally "tumor."

netherworld Literally "underworld." Pertains to a counter-culture of antisocial parenting that produces sexual psychopaths.

neurology The medical specialty and scientific study of the nervous systems of the body in regard to structure, function, and abnormalities.

neuropeptide An endogenous chain of proteins (peptides) that influences neural activity.

neurophysics Proposed by Jacobs (2003) as the science responsible for merging quantum theory with computer chip technology thereby developing a "bionic brain" transplant to correct diseased or pathological cortices in the brain due to disease, trauma, or neglect.

neuropsych paradigm Pertains to the central importance of neurotransmitters and neurohormones that give rise to thinking, behavior and personality—ultimately normal versus abnormal, and criminal behavior.

neuropsycholinguists (new-row-sigh-ko-lin-gwis-tiks) The neuropsych specialty that studies how word knowledge and usage changes brain chemistry and behavior.

neuropsychology (new-row-sigh-kol-o-gee) **neuroscientist** Psychological specialization that merges psychology, neurology, and neurochemistry to study underpinnings of behavior at the tissue level of interneurons.

neuropsych-bio-social perspective Pertains to an eclectic perspective of systematized knowledge and influences from neurology, psychology, biology, and sociology in a comprehensive analysis of behavior.

neurotheology (new-row-thee-ol-o-gee) Pertains to biological basis observable in brain scans that produces euphoria for spiritual or mystical experiences.

neuroscan Pertains to brain imaging via tomogram technology

nihilist (nigh'l-ist) Belief that society is so corrupt that traditional values, moral truths, or beliefs are unfounded and that existence is senseless and useless

norepinephrine (NE) Neurotransmitter acting as a "focuser" of "jazzed" activities.

nosology (noz-oll-o-gee) Refers to a classification of diseases.

nurture Refers to care of others in social learning, as contrasted to nature or biological inheritance.

nymphomania Literally "inner lips of the vulva." Excessive sexual desire by a female.

O

ODD (oppositional defiant disorder) In the DSM, a pervasive behavioral disorder of opposition to authority displayed by adolescents.

obsession A persistent disturbing preoccupation with an often unreasonable idea or feeling; compelling motivation.

obsessive-compulsive Pertains to persistence in thinking (obsession) and action (compulsion). Rigidity of thinking and doing exemplified as "my way or the highway."

obsessive-compulsive disorder In the DSM, a psychological disorder related to rigid behavior characterized by routine and protocol, termed "anal retentive" by Freud.

operant (OP-er-unt) In behaviorism, any behavior capable of being reinforced.

operant conditioning In behaviorism, a form of learning characterized by consequences or "payoffs" where behavior is shaped and maintained by positive and negative consequences.

operational definition Pertains to defining terms used in intellectual discourse so participants are on the same page.

orgasmic phase The summit of Master and Johnson's model of human sexual response characterized by orgasm—ejaculation and release of muscular tension.

overkill Pertains to excessive brutality in ***actus reus*** phase of ***modus vivendi,*** such as fifty stab wounds perpetrated on the victim, when one wound could have brought death.

oxytocin (OXY) Powerful hormone known as the "cuddle chemical" related to social bonding.

P

PEA (phenylethylamine) (fee-nul-eth-ul-ah-meen) An endogenous amine similar to methamphetamine responsible for the "romantic" rush of physical attraction.

PMS (premenstrual syndrome) Denotes mood changes and other emotional and physical discomforts just prior to the menstrual period.

PNS (peripheral nervous system) Refers to the portion of the nervous system outside the brain and spinal cord such as the somatic and autonomic systems.

PTSD (post-traumatic stress disorder) Denotes an enduring, distressful emotional disorder that follows exposure to a severe fear inducing threat.

pan-hedonism Literally "all sexual."

paranoia Literally "demented mind." Denotes a psychosis characterized by systematic delusions (false beliefs) of grandeur or persecution.

paranoid Overly suspicious or fearful.

Parkinson's disease A chronic movement disorder characterized by jerky movements and muscle rigidity.

paradigm (par-a-dime) A framework for presentation of systematic or scientific ideas or concepts.

paraphilia Literally "abnormal love." Refers to unusual sexual practices or special erotic activities that may victimize non-consenting persons, such as voyeurism or exhibitionism.

pedophilia (ped-uh-fee-lee-uh) **pedophile** (ped-uh-file) Sexual perversion in which children are preferred sexual objects.

pejorative (puh-jor-uh-tiv) Having a negative connotation.

perception How a person organizes sensation. Pertains to one's view of reality or any subset through physical sensation.

personality disorder In the DSM, characteristic behavior listed taxonomically with the most salient presenting symptoms of a variety of personality disorders, such as antisocial personality disorder, the most frequent personality disorder observed in psychopathy.

personality disorder, not otherwise specified (PD NOS) In the DSM, a personality disorder category with features of more than one personality disorder so that a person diagnosed PD NOS may display narcissism along with antisocial personality.

persona Literally, "social mask;" the many social demeanors expressed by a single individual.

personation Pertains to a killer's emotional "call card" or signature.

phenomenology (fuh-nom-uh-noll-uh-gee) Refers to phenomena that make up conscious experiences and self-awareness.

physiological psychology Refers to experimental psychology and the interaction of body-brain.

picquerist (pick-er-ist) Refers to a killer who becomes sexually stimulated by penetrating the skin with knife cuts or wounds.

plateau phase In Masters and Johnson's human sexual response model, the psychophysiological experience and feeling that sex is imminent.

pleasure principle Denotes Freudian concept of the role of fantasy within libido (or sex drive).

pluripotent (plur-ip-uh-tent) Refers to developmental plasticity.

poly-addiction Addiction to more than one chemical substance.

pontine tegmentum (pon-teen teg-men-tum) The extension of midbrain covering at the level of the pons.

pop culture USA Refers to whatever is "hot"—such as fashion, movies, music, or lingo in North American marketing; refers to what is highly influential as *au courant* (in the "current") of culture.

pop semantics Pertains to "spin" or whatever meaning the sender attaches to it.

pornography Literally, "to write about prostitutes." Depicting erotic, sexual behavior, pictures, or text intended to cause sexual excitement.

pornography (hardcore) Pornography rated XXX for violence, anal sex, bondage or sadomasochism. More explicit, varied, and perverted forms of pornography.

postmortem Literally "after death."

postpartum Literally, "after birth."

predator parenting Refers to antisocial or toxic parenting.

prefrontal (cortex) lobes Cortical regions in the frontal lobes thought to be involved in planning complex cognitive behavior, expression of personality, and appropriate social behavior.

prescient (pre-zee-unt) Forward thinking.

proclivities (pro-kliv-uh-tecz) Inclination toward something objectionable

profiling Pertains to gathering information at a crime scene in order to construct personality and other personal proclivities (habits and behavior) of an UNSUB.

profundity Refers to gravity or depth of a condition.

projective test Refers to psychometrics (testing) aimed at disclosing "deeper" dynamics of unconscious conflict.

protocol A detailed plan of a scientific procedure

pseudo-intellectual Literally, "false" intellectual. A person who pretends to be intellectual by using scholarly words.

pseudo-sapien Literally, "false" sapient. Refers to a brain damaged sexual predator, who is not psychotic, but acts irrationally and criminally by preying upon others.

psyche (Seye-kee) Greek word meaning "soul."

psychiatric social worker Refers to a professionally trained person who is often part of a psychiatric treatment team who specializes in family dynamics and pathological influences that threaten to disrupt normal interactions.

psycho-behavioral profile Older term for criminal profiling denoting the importance of psychological principles and behavioral analysis in construction of the profiles.

psychobabble Refers to using terms and concepts related to psychology in everyday conversation.

psychological perspectives Psychological "schools" of thought—such as behaviorism or cognitive neuroscience—used in analyzing personality, behavior, or mind.

psychopathic personality (psychopath); psychopathic parenting Severe emotional and behavioral state characterized by clear perception of reality except for social and moral obligations, which get overridden by the pursuit of immediate personal gratification in criminal acts, drug addiction, or sexual perversion. Psychopathic parenting pertains to dysfunctional parents who prey on their children through abuse and neglect.

psychopathology (sigh-ko-pa-tho-lo-gee) Refers to dysfunctional or deviant influences upon behavior, such as disordered family dynamics, drug dependency, and aberrant thinking and how they influence the observation and documentation of abnormal behavior.

Psychopathy (sigh-kop-uh-thee) Pertains to psychopathic personality, psychopathic features, or in the extreme, sexual psychopathy.

psychosocial history A clinical document prepared usually by a social worker and detailing the family dynamics of a client seeking therapy.

psychotic Pertains to insanity.

psychotropic Psychoactive medication that influences mood and/or behavior.

purposive behaviorism Edward Tolman's "school" of behaviorism related to goal-directed behavior.

Pyromania An irresistible impulse to start fires.

Q

QM (quantum mechanics) Quantum mechanics pertains to the structure, motion, and interaction of particles (atoms and molecules) where the discrete nature of the physical world is unimportant in contrast to classical physics.

R

RAS (reticular activation system) A midbrain collection of medial nuclei important in regulation of sleep, motor activity, and diffuse integrative functions.

radical behaviorism Refers to Skinnerian principles of operant conditioning without regard to mentalistic concepts.

rapacious (ruh-PAY-shus) **(or rapacity)** (ruh-pa-suh-tee) Preying upon a victim by a perpetrator with intent to do physical bodily harm (such as rape or murder in sexual psychopathy) or to inflict psychological harm (as observed in brainwashing).

rape kit Pertains to accessories brought to the crime scene by the perpetrator (such as duct tape and/or rope) to incapacitate the victim.

rationalism The use of reason over emotion in perceiving and handling conflict.

rationalization Freudian defense against anxiety by making "convincing" reasons to explain behavior.

reptilian brain (or reptilian brain theory) Refers to the brain stem or R-complex (reptilian complex) as one of three overlapping brains, proposed by neurologist Paul MacLean, former Chief of brain evolution and behavior at the National Institute of Health. The R-complex brain relates closely to physical survival, ritualistic behavior, and dominance.

resolution phase In Masters and Johnson's human sexual response, the return to pre-arousal psychophysiological state.

rootedness Pertains to feeling emotionally grounded and safe.

ruse To deceive.

S

S→R psychology In behaviorism, the reliance on cause and effect as determining behavior, presented as stimulus response.

SSRIs (selective serotonin reuptake inhibitors) Refers to the chemical action of blocking reuptake (liberating) serotonin in brain chemistry.

sadist (SAY-dist) Pertains to the sexual perversion of feeling stimulation by inflicting pain or causing another to suffer.

sadomasochism (SAY-doe-mass-uh-kiz-um) Being sexually stimulated by both giving and receiving pain.

schizophrenia (skit-zoe-free-nee-uh) Refers to a serious mental disorder characterized by a thought disorder and disintegration of personality.

schizotypal personality disorder (skezz-o-tip-ul) In the DSM, a pervasive personality disorder characterized by acute discomfort with, and a reduced capacity for, close relationships, as well as by cognitive or perceptual distortions and eccentricities of behavior.

second messenger peptides (or second messengers) Type of neurotransmitter that prolongs and modulates mood and emotion in contrast to fast acting ion-channeled linked receptors.

secondary follower role Pertains to a person working in tandem as an accomplice with a serial killer

serial Sequential, or one act following another act with a time lapse in between.

serial killer Refers (usually) to a male killer in his twenties to mid-thirties characterized by sexual psychopathy, a violent subcategory comprising the larger classification of antisocial personality disorder and psychopathic personality. He kills at least three victims in sequence with a "cooling off" period—time spent away from killing.

serotonin (5-HT) (SARA-toe-none) Powerful inhibitory neurotransmitter producing a calm, cool, and collected mood.

sex addict Pertains to a person addicted to sexual stimulation and the accompanying brain chemistry lying behind the feeling.

sexual dysfunction NOS Pertains to sexual dysfunctions that do not meet the criteria for any specific sexual dysfunction, such as diminished erotic feelings or whether or not a sexual dysfunction is due to a general medical condition or substance induced.

sexual psychopathy (sigh-kop-uh-thee) **or sexually psychopathic serial crime** Pertains to a cold-blooded sexually psychopathic serial killer characterized by perverted sexuality, egocentrism, and entitlement. Viewed as the personality type of society's most elusive predators.

sexual sadism Erotic desire caused by inflicting pain upon another.

shaken baby syndrome Neurological damage to delicate brain tissue due to vigorously shaking the body of an infant.

signature Refers to the emotional reason evidenced at the crime scene as the "calling card" that drives serial crime. It's what "jazzes" him every time he commits the crime.

skepticism A philosophical appeal of doubt and to gather more information before a conclusion is drawn. Educated doubt.

social modeling Pertains to observational learning—watching and copying the behavior and mannerisms of others.

sodomized (sodomy) Pertains to various forms of sexual intercourse considered unnatural or abnormal, especially anal intercourse or bestiality.

spree killer Pertains (usually) to a male killer who kills in two or more locations with no "cooling off" period between kills.

staging Following a crime, the placement of the body in a certain position by the perpetrator or family members in anticipation of the body's discovery.

sublimation Freudian term relating to the redirection of libido away from pure sexuality into alternative social pursuits and achievements such as art and science.

substantia nigra (sub-stan-shuh nee-gruh) Literally "black substance," part of the basal ganglia and a rich source of dopamine.

subterfuge Dishonesty.

successive approximations The "small steps," or increments of behavior, required to shape (or perfect) behavior, such as the "small steps" of behavior developed in hitting a baseball.

sui generis Literally "of its own kind." Constituting a unique class alone.

surrogate A substitute.

T

tableaux mortido Literally "death pictures,"

tactile stimulation Refers to touch.

taxonomy General principles observed in systematic classification (such as the Periodic Tables).

testosterone (tess-toss-tuh-rone) Powerful anabolic steroid; hormone of aggression and libido.

thalamus Brain structure responsible for integration of sensory information from lower centers to higher centers of the brain.

toxic parenting Pertains to loveless, hateful, parenting.

trephination (treph-uh-nay-shun) Ancient medical practice of perforating the skull (such as drilling holes)

U

UNSUB Refers to an unidentified subject in FBI lingo.

V

vasopressin (Va-zo-pres-n) Hormone secreted by the posterior pituitary gland that increases blood pressure, urine flow, and social bonding in males.

vignette (vin-yet) Pertains to a short, descriptive literary sketch

voyeurism (voy-yur-iz-um) Literally "one who sees." Obtaining sexual gratification from viewing nudity or sexual acts; one who habitually seeks sexual gratification from visual means.

W

wolffian ducts (wolf-un ducks) Pertains to tiny embryonic ducts responsible for male sexual differentiation under the influence of testosterone.

Z

zeitgeist (zit-gist) Literally "spirit of the times." The moral and intellectual climate in a given time in history as it impacts society and popular culture.

References

Adams, R. D., and Victor, M. *Principles of Neurology*. 5th ed. New York: McGraw-Hill, 1993.

American Psychiatric Association Diagnostic and Statistical Manual of Mental Disorders. 4th ed. DSM-IV-TR. Washington, D.C.: American Psychiatric Association Press, 2002.

Barlow, David and Durand, V. Mark. *Abnormal Psychology*. 2nd ed. Belmont: Wadsworth, 2001.

Beatty, Jackson. *Principles of Behavioral Neuroscience*. Dubuque: Brown & Benchmark, 1995.

Bugliosi, Vincent. *Outrage*. New York: Island Books, 1995.

Burgess, A., Douglas, J. and Ressler, R., Eds. *Crime Classification Manuel*. San Francisco: Jossey-Bass Publishers, 1997.

Burgess, A., Douglas, J. and Ressler, R., Eds. *Sexual Homicide: Patterns and Motives*, New York: Lexington Books, 1988.

Charles Manson: Journey into Evil. A & E Biography Videocassette. Arts & Entertainment, 1995.

Cohen, Sidney. *The Chemical Brain*. Irvine: CareInstitute, 1988.

Cooper, Jack R., Bloom, Floyd E., and Roth, Robert H. *The Biochemical Basis of Neuropharmacology*. 6th Ed. Oxford: Oxford University Press, 1991.

Douglas, John E., and Olshaker, M. *Mind Hunter* New York: Pocket Books, 1996.

Douglas, John E., and Olshaker, Mark. *Obsession*. New York: Scribner, 1998.

Douglas, John E., and Olshaker, M. *The Anatomy of Motive*. New York: Scribner, 1999.

Douglas, Ressler, Burgess, and Hartman. 1986. "Criminal Profiling from Crime Scene Analysis." *Behavioral Science and the Law*, 1986, 4–401–406.

Erikson, E. H. *Identity: Youth in Crisis*. New York: W. W. Norton, 1968.

Erikson, E. H. *The Life Cycle Completed: A Review*. New York: W. W. Norton, 1982.

Everitt, David. *Human Monsters*. Chicago: Contemporary Books, 1993.

Emmons, Nuel. *Manson: In His Own Words*. New York: Grove Press, 1986.

Films for the Humanities and Sciences. *Mind of a Murderer 2. Damaged: When Trauma Leads to Violence*. Princeton, N.J., 2002.

Films for the Humanities & Sciences. *Mind of Murderer 2. Men: The Killer Sex*. Princeton, N.J., 2002.

Films for the Humanities & Sciences. *Psychopaths: Natural Born Killers?* Princeton, N.J., 2002.

Films for the Humanities & Sciences. *The Mind of a Serial Killer.* Princeton, N.J., 1992.

Freeman, Lucy. *Before I Kill More.* New York: Crown Publishers, 1955.

Freud, Sigmund. *An Outline of Psychoanalysis.* In Standard Edition of the Complete Works of Sigmund Freud (Vol. 23). London: Hogarth Press, 1938.

Freud, Sigmund. *Civilization and its discontents.* College Edition. New York: W. W. Norton, 1961.

Gay, Peter. *Freud: A Life for Our Time.* New York: W. W. Norton, 1990.

Haas, Kurt and Haas, Adelaide. *Understanding Sexuality.* 3rd ed. St. Louis: Mosby, 1993.

Hare, Robert D. *Without Conscience.* New York: Guilford Press, 1999.

Hare, R. D. (2003). "The Psychopathy Checklist—Revised, 2nd Edition." Toronto: Multi-Health Systems.

Harlow, H. F., and Zimmerman, R. R. 1959. Affection Responses in the Infant Monkey. *Science,* 1959. 130, 421–432.

Hock, Roger, R. *Forty Studies That Changed Psychology.* 3rd ed. New Jersey: Prentice Hall, 1999.

Holmes, Ronald M., and Holmes, Stephen T. *Profiling Violent Crimes: An Investigative Tool.* London: Sage Publications, 1999.

Horgan, John. *Rational Mysticism.* New York: Houghton Mifflin, 2003.

Hunt, Morton M. *The Story of Psychology.* New York: Doubleday, 1993.

Isenberg, Sheila. *Women Who Love Men Who Kill.* New York: Simon & Schuster, 1991.

Jack the Ripper: Phantom of Death. A & E Biography Videocassette. Arts & Entertainment, 1995.

Jacobs, Don. *Inside the Clinical Picture.* 3rd ed. Dubuque: Kendall Hunt Publishing, 2002.

Jacobs, Don. *Personality: Compositions of Mind.* Boston: McGraw-Hill, 2002.

Jacobs, Don. *Psychology: Brain, Behavior, and Popular Culture.* 4th ed. Dubuque: Kendall Hunt Publishing, 2002.

Jacobs, Don. *Psychology of Adjustment.* 3rd ed. Dubuque: Kendall Hunt Publishing, 2002.

James, William. *The Varieties of Religious Experience.* New York: Collier, 1961.

Kennedy, Dolores. *Bill Heirens: His Day in Court.* Chicago: Bonus Books, 1991.

Keppel, Robert, D. *Signature Killers.* New York: Pocket Books, 1997.

Masters, W. H., Johnson, V. E. and Kolodny, R. C. *Human Sexuality.* Boston: Little, Brown, 1982.

Merriam-Webster. *Webster's Medical Desk Dictionary.* Springfield: Merriam-Webster, 1986.

Millenson, J. R. and Leslie, Julian C. *Principles of Behavioral Analysis.* 2nd ed. New York: McMillan, 1979.

Neitzel, M. *Crime and Its Modification: A Social Learning Perspective.* Elmsford, New York: Pergamon, 1979.

Norris, Joel. *Serial Killers.* New York: Anchor Books, 1988.

Ornstein, Robert. *The Roots of Self.* San Francisco: HarperCollins, 1995.

Peck, M. Scott. *The Road Less Traveled.* New York: Simon & Schuster, 1978.

Ressler, Robert K., and Shachtman, Tom. *Whoever Fights Monsters.* New York: St. Martin's, 1992.

Restak, Richard M. *Receptors.* New York: Bantam Books, 1995.

Russell, Bertrand. *A History of Western Philosophy.* New York: Simon & Schuster, 1945.

Siegel, Larry J. *Criminology,* 8th ed. Belmont: Wadsworth, 2003.

Skinner, B. F. *Science and Human Behavior.* New York: Macmillan, 1953.

Skinner, B. F. *Beyond Freedom and Dignity.* New York: Bantam Books, 1971.

Sundberg, Norman D. and Winebarger, Allen A. and Taplin, Julian R. *Clinical Psychology: Evolving Theory, Practice, and Research.* Upper Saddle River, N.J.: Prentice Hall, 2002.

The Wrong Man? PrimeTime Video, ABC Television, aired August 7, 1996.

Turvey, Brent. *Criminal Profiling: An Introduction to Behavioral Evidence Analysis.* London: Academic Press, 1999.

Watson, J. B. *Psychology from the Standpoint of a Behaviorist.* Philadelphia: Lippincott, 1919.

Watson, J. B. *Behaviorism.* New York: W. W. Norton, 1930.

Wormser, Rene A. *The Story of Law.* New York: Simon & Schuster, 1962.

Wrightsman, Nietzel, and Fortune. *Psychology and the Law.* 4th ed. New York: Brooks/Cole Publishers, 1997.

References for POD Readings:

II. Scott Peterson: A Diagnostic Evaluation

Ablow, K. *Inside the mind of Scott Peterson.* New York: St. Martin's Press, 2005.

American Psychiatric Association. *Diagnostic and Statistical Manual of Mental Disorders.* 4th ed. Text revision. Washington, D.C.: American Psychiatric Association, 2002.

Associated Press. San Quentin doors clang shut behind Peterson. *NBC News.* 2005, March 18. Retrieved March 31, 2008, from http://www.msnbc.msn.com/id/7217582/

Durand, V. M., & Barlow, D. H. *Essentials of Abnormal Psychology.* 4th ed. Belmont, CA: Thomson Wadsworth, 2006.

KNBC News. People magazine: Scott Peterson's life in prison. *KNBC News,* 2005, August 25. Retrieved March 31, 2008, from http://www.knbc.com/news/4895108/detail.html

Peterson, S. Perhaps a commendation. Message posted to Canadian Coalition Against the Death Penalty, Scott Peterson page, 2005, December 26. archived at http://www.ccadp.org/scottpeterson.htm

Peterson, S. Thank you. Message posted to Canadian Coalition Against the Death Penalty, Scott Peterson page, 2005, July 21. archived at http://www.

ccadp.org/scottpeterson.htm

III. Minding the Twenty-First Century Brain

American Psychiatric Association Diagnostic and Statistical Manual of Mental Disorders. 4th ed. DSM-IV-TR. Washington, D.C.: American Psychiatric Association Press, 2002.

George Bush. Filed with the Office of the Federal Register, 12:11 P.M., July 18, 1990

Mary Kilbourne Matossian. *Poisons of the Past.* 1989.

Myers, D. G. *Exploring Psychology in Modules.* 7th ed. 2008.

VIII. The Critical Difference: Between Psychopathic Personality & Antisocial Personality Disorder

American Psychiatric Association Diagnostic and Statistical Manual of Mental Disorders. 4th ed. DSM-IV-TR. Washington, D.C.: American Psychiatric Association Press, 2002.

IX. Neuro School Behavioral Sink

Brown, K., and Grunberg, N. 1996. Effects of environmental conditions on food consumption in female and male rats. *Physiology and Behavior, 60*(1). 293–297.

Evans, G. W. 1979. Behavioral and psychological consequences of crowding in humans. *Journal of Applied Psychology, 9,* 27–46.

Freedman, J. L., Heshka, S., and Levy, A. 1975. Population density and social pathology: Is there a relationship? *Journal of Experimental Social Psychology, 11,* 539–552.

Kitamura,T., Shima, S., Sugawara, M., and Toda, M. 1996. Clinical and social correlates of Antenatal depression: a review. *Psychotherapy and Psychosomatics, 65* (3), 117–123.

Marsden, H. M. 1972. Crowding and animal behavior. In J. F. Wohlhill & D. H. Carson (Eds.), E*nvironment and the social sciences.* Washington, D.C.: American Psychological Association.

McCain, G., Cox, V. C., and Paulus, P. B. 1976. The relationship between illness, complains, and Degree of crowding in a prison environment. *Environment and Behavior, 8,* 283–290.

Paydarfar, A. 1996. Effects of multifamily housing on marital fertility in Iran: Population-policy Implications. *Social Biology, 42*(3/4), 214–225.

X. Behind the Monster's Eyes: The Role of the Orbitofrontal Cortex in Sexually Psychopathic Serial Crime

Arana, F. S., Parkinson, J. A., Hinton, E., Holland, A. J., Owen, A. M., and Roberts, A. C. 2003. Dissociable Contributions of the Human Amygdala and Orbitofrontal Cortex to Incentive Motivation and Goal Selection. *The Journal of Neuroscience,* 23 (29), 9632–9638.

Berlin, H. A., Rolls, E. T., and Iversen, S. D. 2005. Borderline Personality Disorder, Impulsivity, and the Orbitofrontal Cortex. *The American Journal of Psychiatry,* 162, 2360–2373.

Burns, J. M. and Swerdlow, R. H. 2003. Right Orbitofrontal Tumor With Pedophilia Symptom and Constructional Apraxia Sign. *Arch Neurology,* 60, 437–440.

Camille, N., Coricelli, G., Sallet, J., Pradat-Diehl, P., Duhamel, J., and Sirigu, A. 2004. The Involvement of the Orbitofrontal Cortex in the Experience of Regret. *Science,* 304, 1167–1169.

Gur, R. C., Gunning-Dixon, F., Bilker, W. B., and Gur, R. E. 2002. Sex Differences in Temporo-limbic and Frontal Brain Volumes of Healthy Adults. *Cerebral Cortex,* 12 (9), 998–1003.

Jacobs, D. and Mackenzie, E. *The Savage Architects of Mind: What Lies Beneath Violent Predatory Criminality.* In virtual format by Vicon Publishing, 2006.

Jacobs, D. *Sexual Predators: Serial Killers in the Age of Neuroscience.* Dubuque: Kendall Hunt Publishing Company, 2003.

Joseph, R. *Neuropsychiatry, Neuropsychology, Clinical Neuroscience.* New York: Academic Press, 2000.

Rule, R. R., Shimamura, A. P., and Knight, R. T. (2002). Orbitofrontal cortex and dynamic filtering of emotional stimuli. *Cognitive, Affective, and Behavioral Neuroscience,* 2(3), 264–270.

Winstanley, C. A., Theobald, D. E., Cardinal, R. N., and Robbins, T. W. 2004. Contrasting Roles of Basolateral Amygdala and Orbitofrontal Cortex in Impulsive Choice. *Journal of Neuroscience,* 24(20), 4718–4122.

www.dangoldstein.com/dsn/archives/2004/08/the_orbitofrontal.html. 2004. The Involvement of the Orbitofrontal Cortex in the Experience of Regret. *Decision Science News.*

medinfo.ufl.edu/year2/neuro/review/cc.html www.medicalnewstoday.com/medicalnews.php?newsid=36593&nfid=rssfeeds

Griffiths, D. 2006. Overactive Bladder Related to Orbitofrontal Cortex Activity. *Medical News Today.*

Related Websites

www.apt213.com

www.aristotle.net

www.crimelibrary.com

http://www.galenpress.com/demon_doctors

http://www.crimelibrary.com/serial_killers/predators/heirens/heirens_1.html

http://www.interpol.int

www.paloaltodailynews.com

http://www.4degreez.com/misc/personality_disorder_test.mv

http://mentalhelp.net/poc/center_index

www.serialhomicide.com

www.serialkillers

Other Related Websites

http://www.corbis.com (Type in Name of Serial Killer)

Click here: Serial Murder

Click here: The Neuropathology of (Violent) Aggression

Click here: Types of Disorders: Sexual Disorders, including sexual dysfunctions, paraphilias, and gender identity disorders

Click here: Special: Violence as a Biomedical Problem

http://www.crimelibrary.com/

http://www.zodiackiller.com

http://www.courttv.com

http://www.carpenoctem.tv

http://serialkillers.coolbegin.com

http://www.fortunecity.com/roswell/streiber/273/index. html

http://www.criminalprofiling.ch

http://flash.lakeheadu.ca/~pals/forensics

http://faculty.ncwc.edu/toconnor/401/401lect11.htm

http://www.crimelibrary.com/serial_killers/predators/index.html

Suggested Reading List

Book	Author(s)
A to Z Serial Killers	Harold Schechter and David Everitt
The Cases That Haunt Us	John Douglas and Mark Olshaker
Mindhunter	John Douglas and Mark Olshaker
Signature Killers	Robert D. Keppel and William J Birnes
I Have Lived in the Monster	Robert K. Ressler and Tom Shachtman
Jeffrey Dahmer	Joel Norris
The Evil That Men Do	Stephen G. Michaud with Roy Hazelwood
Dark Dreams	Roy Hazelwood and Stephen G. Michaud
Serial Killers—The Insatiable Passion	Joel Norris
Inside the Criminal Mind	Stanton E. Samenow
Step Into My Parlor—Serial Killer Jeffrey Dahmer	Ed Baumann
Serial Killers	David Lester
Alone with the Devil	Ronald Markman, M.D. and Dominick Bosco

Casebook of a Crime Psychiatrist	James A. Brussel, M.D.
Why They Kill	Richard Rhodes
Tears of Rage	John Walsh
Texas Crime Chronicles	Editors of Texas Monthly
William Heirens: His Day in Court	Dolores Kennedy
I Know You Really Love Me—A Psychiatrist's of Obsessive Love	Doreen Orion, M.D.
Helter Shelter: The Manson Murders	Vincent Bugliosi
Before I Kill More	Lucy Freeman
Basic Instincts—What Makes Killers Kill	Jonathan H. Pincus M.D.
And Deliver Us From Evil	Mike Cochran
Cracking Cases	Dr. Henry C. Lee and Thomas W. O'Neal
Evil—Inside Human Violence & Cruelty	Roy F. Baumeister, Ph.D.
Will You Die For Me?	Charles "Tex" Watson
Psychology of Crime	S. Giora Shoham and Mark Seis
Dead Man Walking	Sister Helen Prejean
Portrait of a Killer: Jack the Ripper	Patricia Cornwell
The Boy Next Door	Gretchen Brinck
A Father's Story	Lionel Dahmer
In Cold Blood	Truman Capote
The Killers Among Us	Colin Wilson and Damon Wilson
Defending the Devil—My Story as Ted Bundy's Lawyer	Polly Nelson
The Phantom Prince—My Life with Ted Bundy	Elizabeth Kendall
A Love to Die For	Patricia Springer
Blood Rush	Patricia Springer
Kids Who Kill	Charles Patrick Ewing
Guilt By Reason of Insanity	Dorothy O. Lewis, M.D.
Profiles in Murder	Russell Vorpagel
Famous Crimes Revisited	Dr. Henry C. Lee and Dr. Jerry Lab

Index